Residual Stresses 2018
ECRS-10

10th European Conference on Residual Stresses (ECRS10), Leuven, Belgium, 11-14th September, 2018.

Marc Seefeldt

Chairman ECRS10
KU Leuven, Department of Materials Engineering

Peer review statement

All papers published in this volume of "Materials Research Proceedings" have been peer reviewed. The process of peer review was initiated and overseen by the above proceedings editors. All reviews were conducted by expert referees in accordance to Materials Research Forum LLC high standards.

Cover pictures

The SEM pictures were taken by Maxim Deprez, KU Leuven

Published under License by **Materials Research Forum LLC**
Millersville, PA 17551, USA

Published as part of the proceedings series
Materials Research Proceedings
Volume 6 (2018)

ISSN 2474-3941 (Print)
ISSN 2474-395X (Online)

ISBN 978-1-94529188-3 (Print)
ISBN 978-1-94529189-0 (eBook)

This book contains information obtained from authentic and highly regarded sources. Reasonable efforts have been made to publish reliable data and information, but the author and publisher cannot assume responsibility for the validity of all materials or the consequences of their use. The authors and publishers have attempted to trace the copyright holders of all material reproduced in this publication and apologize to copyright holders if permission to publish in this form has not been obtained. If any copyright material has not been acknowledged please write and let us know so we may rectify in any future reprint.

Distributed worldwide by

Materials Research Forum LLC
105 Springdale Lane
Millersville, PA 17551
USA
http://www.mrforum.com

Manufactured in the United State of America
10 9 8 7 6 5 4 3 2 1

Table of Contents

Diffraction Methods

Mechanical Relaxation Methods

Acoustic and Electromagnetic Methods

Composites, Nano and Microstructures

Films, Coatings and Oxides

Cold Working and Machining

Heat Treatments and Phase Transformations

Welding, Fatigue and Fracture

Stresses in Additive Manufacturing

Preface

The European Conference on Residual Stresses (ECRS) series is the leading European forum for scientific exchange on internal and residual stresses in materials. It addresses both academic and industrial experts and covers a broad gamut of stress-related topics from instrumentation via experimental and modelling methodology up to stress problems in specific processes such as welding or shot-peening, and their impact on materials properties. After ECRS-8 2010 in Riva del Garda (Italy) and ECRS-9 in Troyes (France), ECRS-10 takes place in Leuven (Belgium). Because of the past and present research activities of KU Leuven and imec on stresses in textured materials, in Microelectronics and Additive Manufacturing, ECRS-10 features the two latter as focus topics, including tutorials for junior researchers. In total, about 170 invited, oral and poster contributions are presented.

Proceedings remain a very valuable source for in-depth reading on the work presented at the conference, and an important reference for the community. After reviewing by two referees each, from the Scientific Comittee and/or the Local Organizing Committee, almost 50 contributed papers are included. On behalf of the Organizing Committee, I would like to thank Materials Research Forum and Dr. Thomas Wohlbier for the nice collaboration.

Enjoy reading!
Marc Seefeldt, Chairman ECRS-10

Committees

Scientific Committee

- Andrzej Baczmanski, AGH University of Science and Technology, Poland
- Sabine Denis, Institut Jean Lamour, Université de Lorraine, France
- Michael Fitzpatrick, The Open University, United Kingdom
- Manuel François, University of Technology of Troyes, France
- Christoph Genzel, Helmholtz-Zentrum Berlin für Materialien und Energie GmbH, Germany
- Josef Keckes, MU Leoben and Erich Schmid Institute for Materials Science, Austria
- Alexander Korsunsky, University of Oxford, United Kingdom
- Petr Lukáš, Nuclear Physics Institute, Czech Republic
- Eric J. Mittemeijer, MPI for Intelligent Systems and University of Stuttgart, Germany
- João P.S.G. Nobre, University of Coimbra, Portugal and University of Witwatersrand, South Africa
- Ru Lin Peng, Linköping University, Sweden
- Walter Reimers, TU Berlin, Germany
- Paolo Scardi, University of Trento, Italy
- Berthold Scholtes, University of Kassel, Germany
- Marc Seefeldt, KU Leuven, Belgium
- Olivier Thomas, IM2NP, University of Aix-Marseille, France
- Philip Withers, University of Manchester, United Kingdom

Local Organizing Committee

- Marc Seefeldt, KU Leuven (Chairman)
- Kim Vanmeensel, KU Leuven
- Dimitri Debruyne, KU Leuven
- Barbara Rossi, KU Leuven
- Laurent Delannay, UC Louvain
- Pascal Jacques, UC Louvain
- Thaneshan Sapanathan, UC Louvain
- Ingrid De Wolf, imec and KU Leuven
- Mario Gonzalez, imec
- Maria Strantza, as guest from Los Alamos National Laboratory
- Bey Vrancken, as guest from Lawrence Livermore National Laboratory

Diffraction Methods

Residual Stresses 2018 – ECRS-10   Materials Research Forum LLC
Materials Research Proceedings 6 (2018) 3-8   doi: http://dx.doi.org/10.21741/9781945291890-1

Single Tilt Method for Residual Stress Evaluation with 2D Detectors

Bob B. He

Bruker AXS Inc, 5465 East Cheryl Parkway, Madison, WI 53711, USA

bob.he@bruker.com

Keywords: 2D detector, Residual stress, Coating, Thin films, Polymer.

Abstract. When X-ray diffraction is used for residual stress measurement, high 2θ peaks are typically used for enhanced 2θ shift and better tolerance to the sample height error. But for thin films, coatings, or polymer materials, high 2θ peaks may not be available or appropriate for stress measurement. As a result of large angular coverage with a 2D detector, residual stress can be measured with a single tilt angle. The diffraction vector coverage from low 2θ angle diffraction ring can satisfy the stress or stress tensor measurement at a single tilt angle. The single tilt method can avoid the sample height error caused by changing the tilt angle, which is especially critical when measuring stress with a low 2θ peak. Another advantage is the consistent depth of penetration due to a constant incident angle, which is especially suitable for residual stress measurement on coatings, thin films or samples with steep stress gradient. This paper introduces the single tilt method for stress evaluation with two-dimensional detectors, including experimental examples on coatings and polymers.

Introduction

Measurement of residual stresses in thin films or coatings by X-ray diffraction is always a challenge due to weak diffraction signals from the thin layer, sharp stress or strain gradients, preferred orientation, anisotropic grain shape and inhomogeneous phase and microstructure distribution [1]. When residual stresses are measured on a coating or thin film sample, it is preferable to keep a small incident angle or control the incident angle to get the most X-ray scattering from the thin film layer. Generally, high 2θ peaks are preferred for stress measurement due to the more significant 2θ shift and less sensitive to the sample height error. But for thin films, coatings, or polymer materials, high 2θ peaks may not be available or appropriate for stress measurement. With low 2θ peaks, it is more difficult or even impossible to measure stress with the conventional $\sin^2\psi$ method. With iso-inclination method, the incident angle varies during data collection so the incident angle cannot be kept low during data collection, while with side-inclination method, the actual incident angle to the sample surface is further reduced, and the measurement results become extremely sensitive to the sample height error.

With two-dimensional X-ray diffraction (XRD^2), stress measurement is based on a direct relationship between stress tensor and diffraction cone distortion [2]. For a diffraction ring with low 2θ, the diffraction vectors cover more directions at each measurement so that sufficient angular coverage can be achieved with a single tilt. Therefore, the data collection can be done at a fixed ψ angle with only ϕ rotation. Since the incident angle is constant, the depth of penetration as a function of γ is consistent at all ϕ angles. For most goniometer with Eulerian geometry, the ϕ axis is typically built on precision bearing with very small spherical error, while ψ rotation is achieved by a circular track which tends to have much more significant spherical error. Avoiding ψ rotation can significantly reduce the sample height variation during the data collection,

Residual Stresses 2018 – ECRS-10   Materials Research Forum LLC
Materials Research Proceedings **6** (2018) 3-8   doi: http://dx.doi.org/10.21741/9781945291890-1

therefore improve the measurement accuracy. As long as the sample height is consistent at various ϕ angles, deviation of the sample surface from the instrument center may affect the pseudo-hydrostatic stress, but has minimal effect on the stress results.

Geometry and $\Delta\psi$ Coverage

Figure 1 illustrates the diffraction vector distribution for the diffraction pattern collected with a point (0D) detector and a 2D detector. The hemisphere represents all the possible orientations from the origin O of the sample coordinate $S_1S_2S_3$. With a point detector, at $\psi=0°$, the diffraction vector points to the sample normal direction S_3. In order to measure stress, the sample has to be tilted at several ψ angles, for instance 0°, 15°, 30° and 45° as indicated by the purple mark $\otimes$. With a 2D detector, the trace of the diffraction vector covers a range as shown by the red curve. The diffraction vectors H_1 and H_2 correspond to the two extreme values of γ_1 and γ_2 on the diffraction ring covered by the 2D detector, $\Delta\psi$ is the total angular range of the diffraction vector distribution, and $\Delta\gamma$ is the γ range. At a given tilt angle ψ, for example 22.5°, the diffraction vector covers a range as shown by the green curve. For low 2θ diffraction rings at proper detector distance, it is possible to cover sufficient angular range for stress evaluation with a single tilt. The complete data set for stress tensor can be collected at several ϕ angles, for instance 360° scan with 45° steps. Therefore the complete data set are collected with ϕ scan only.

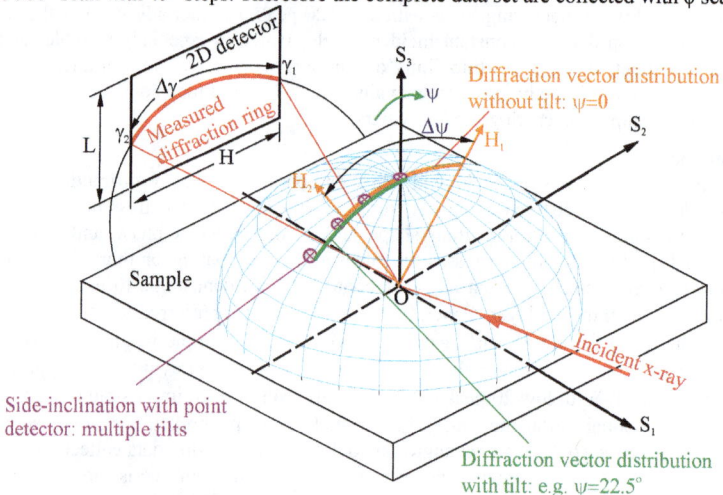

Figure 1. Diffraction vector distribution for 0D and 2D detectors.

The diffraction vector distribution range ($\Delta\psi$) is determined by the detector distance D, detector height H, detector width L, 2θ angle and detector swing angle α. Calculating the $\Delta\psi$ coverage for a flat detector involves solving some implicit equations [2]. For simplicity, an equation derived from cylindrical detector can be used for $\Delta\psi$ calculation with negligible error:

$$\Delta\psi = 2\arcsin\left(\frac{H}{2\sin\theta\sqrt{4D^2 + H^2}}\right) \tag{1}$$

Residual Stresses 2018 – ECRS-10   Materials Research Forum LLC
Materials Research Proceedings **6** (2018) 3-8   doi: http://dx.doi.org/10.21741/9781945291890-1

Here we assume the γ range is limited by the detector height H. The measured γ range may also be limited by the detector width L when L is too small or the detector swing angle is not properly set. Figure 2 shows the diffraction vector distribution range $\Delta\psi$ as a function of 2θ calculated for EIGER 2R 500k™ detector at various detector distance and in γ-optimized orientation (H=77.2mm and L=38.6mm). For stress measurement with single tilt, $\Delta\psi$ of more than $30°$ is acceptable, but the desired $\Delta\psi$ coverage is $45°$ or above. A shorter detector distance can be used to increase $\Delta\psi$. In general, the angular coverage $\Delta\psi$ is significantly higher with low 2θ angles. Therefore, the single tilt method is more suitable for middle or low 2θ angles.

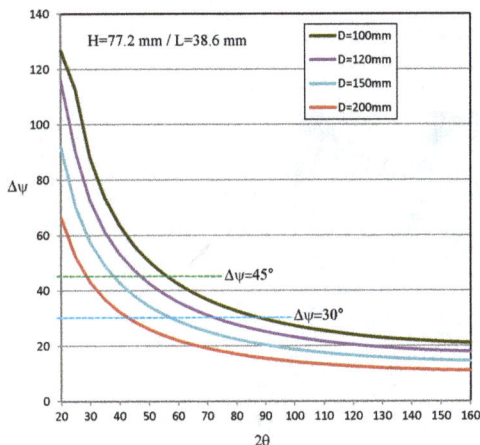

Figure 2. $\Delta\psi$ range as a function of 2θ for EIGER 2R 500k™ detector.

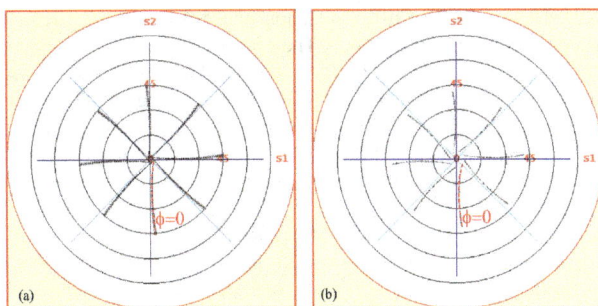

Figure 3. Data collection strategy schemes with single tilt at $\psi = 22.5°$ and complete ϕ rotation of 45° steps. (a): PE polymer (020), $2\theta=36.3°$ and D=20 cm; (b): $Al_2O_3(116)$, $2\theta=57.5°$ and D=15 cm.

Figure 3 illustrates the single tilt scheme generated with GADDS software for VÅNTEC-500 2D detector. The left (a) is for PE polymer (020) with $2\theta=36.3°$, $\psi=22.5°$ and detector to sample distance D=20 cm and the right (b) is for $Al_2O_3(116)$ with $2\theta=57.5°$, $\psi=22.5°$ and D=15 cm. The arcs represent the trace of the diffraction vector corresponding to the data set. S_1 and S_2 are two sample orientations. The red broken curve marks the diffraction vector distribution covered by the frame collected at $\phi=0$. With eight frames collected with ϕ scan at 45° steps, the scheme produces comprehensive orientation coverage in a symmetric distribution. The data set collected with this strategy can be used to calculate the complete biaxial stress tensor components.

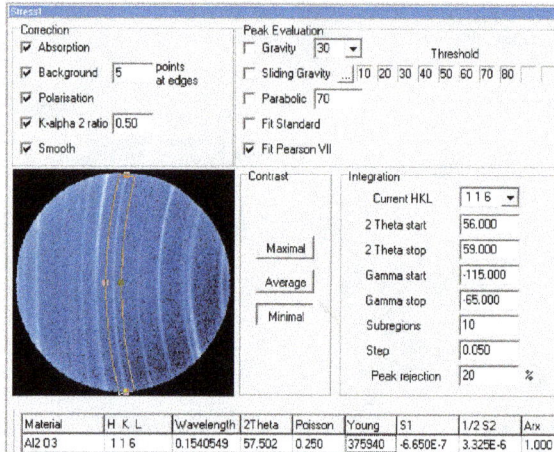

Figure 4. Data evaluation setting for 1μm thick Al_2O_3 coating on cutting insert.

The residual stress in the Al_2O_3 coating of less than 1μm thick on a proprietary cutting insert is measured with a Bruker D8-DISCOVER system containing centric Eulerian cradle and VÅNTEC-500 2D detector. With Cu-Kα radiation, the diffraction ring from (116) planes at $2\theta=57.5°$ is used for stress evaluation. The stress calculation is done with Bruker DIFFRAC.LEPTOS software version 7.9. Figure 4 shows the data evaluation setting. The data integration region is defined by 2θ range of 56° to 59° and γ range of -65° to -115°. The 50° γ range is divided into 10 subregions, 5° for each subregion. The counts within each subregion are integrated into a diffraction profile and the 2θ peak position is determined by one of the five peak evaluation algorithms. In this experiment, Pearson VII function is used to fit the profile and evaluate the 2θ peak position.

Figure 5 shows the stress evaluation results from one of the data set. The charts above "A" are the fitted data points on 2D frames. The charts above "B" are fitted data points in γ-2θ rectangular coordinates with magnified 2θ scale, in which, black line indicates $2\theta_0$, blue cross and line indicates the data points from the profile fitting of each subregion, and red line represents the calculated diffraction rings from the stress results. The scattering of the crosses about the red line represents the quality of the data, affecting the standard deviation of the stress results. Any roll error of the detector will change the trend of the fitted data points and the red line, thus the stress results. By click on any data point, the integrated profile displays above "C". With 60 seconds per frames, the total data collection time is 8 minutes. The measured stress values are given in the region "D" as σ_{11}=954.7MPa, σ_{22}=957.9MPa and standard deviation 26.5MPa (<3%).

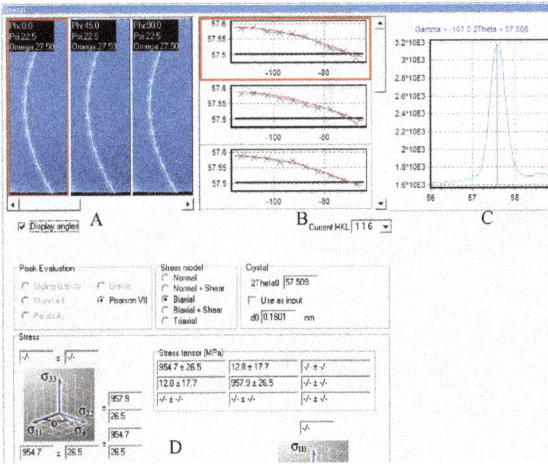

Figure 5. Data evaluation results for 1μm thick Al₂O₃ coating on cutting insert.

The single tilt stress measurement method is especially suitable for measuring residual or loading stresses in polymers. For stress measurement by X-ray diffraction, the polymer sample must contain sufficient crystalline phase. The crystalline peaks from polymers are typically with low 2θ angles. For instance, polyethylene (PE) polymer contains mainly three diffraction rings at 2θ about 21.4°, 23.6° and 36.3° with Cu-Kα radiation, corresponding to crystalline planes of (110), (200) and (020) respectively. Even with the 36.3° peak, a large error in stress result can be introduced by the sphere of confusion with multiple sample tilt angles. The single tilt method can overcome this problem. Other challenge for stress measurement on polymers is the low stress value. However, due to the extremely low Young's modulus, such as 1070 MPa for high density polyethylene, the 2θ shift (strain) is also more significant for the same stress if compared with metals. Due to large depth of penetration, the zero normal stress assumption (σ_{33}=0 on sample surface) for polymers may not be as accurate as for metals.

Residual stresses on a high-density polyethylene (HDPE) pipe are measured with the single tilt method [3]. An XRD[2] system in vertical θ-θ configuration (Bruker D8-DISCOVER™) with IμS™ Cu microsource and VÅNTEC-500™ 2D detector is used for the measurement. The HDPE pipe has a diameter of 32 mm and wall thickness of 3 mm. The length of the sample is 50 mm, cut from a pipe. Residual stresses of a total 7 points in the outer surface of the pipe and along the axial direction were measured. The depth of penetration corresponding to 50% of the total diffracted intensity is 0.3 mm estimated from the diffraction condition in diffractometer plane (γ=-90°). To avoid relaxation effect near the cutting edges, the measurement starts at 10 mm from one end with 5 mm steps and completes at the last point 10 mm from the other end. A total of 8 frames are collected for each measurement point with the data collection strategy given in Figure 3(b). Figure 6 shows the data evaluation setting and fitting results with LEPTOS software for one of the measurement point. The measured stresses are tensile in the pipe extrusion direction with variation of 1.3~2.2 MPa over the 7 points, and compressive in the hoop direction with variation of 6.5~8.1 MPa.

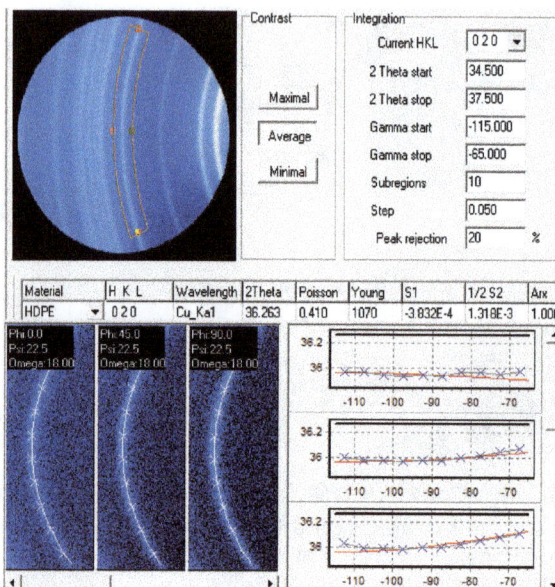

Figure 6. Data evaluation setting and fitting results for the HDPE pipe.

Summary

For thin films, coatings, or polymers, when the diffraction peaks at high 2θ angles are not available or appropriate, a low 2θ peak may be used for stress evaluation. With diffraction rings at low 2θ angle, the diffraction vector distribution can satisfy the angular coverage for stress or stress tensor measurement at a fixed tilt angle (ψ). Without ψ change during data collection and with φ rotation only, the sample height is accurately maintained and data collection time is reduced. The single tilt method is a unique method only achievable with an XRD^2 system, which can measure residual stress with high accuracy and high speed for thin films, coatings and polymers with low to medium 2θ peaks.

References

[1] B. B. He, Measurement of residual stresses in thin films by two-dimensional XRD, Materials Science Forum, Vols. 524-525, (2006) 613-618.
https://doi.org/10.4028/www.scientific.net/MSF.524-525.613

[2] B. B. He, Two-dimensional X-ray Diffraction, 2nd ed., John Wiley & Sons, 2018.
https://doi.org/10.1002/9781119356080

[3] M. Ren, C. Zheng, Y. Shi, Y. Tang, B. He, Residual stress measurement of high-density polyethylene pipe with two-dimensional X-ray diffraction, Adv. X-ray Anal. 61 (2018).

Residual Stresses 2018 – ECRS-10
Materials Research Proceedings 6 (2018) 9-14

Materials Research Forum LLC
doi: http://dx.doi.org/10.21741/9781945291890-2

Use of Symmetry for Residual Stress Determination

Robert Charles Wimpory[1,a] * and Michael Hofmann [2,b]

[1] Helmholtz Centre Berlin for Materials and Energy, Hahn-Meitner-Platz 1, 14109 Berlin, Germany

[2] Heinz Maier-Leibnitz Zentrum (MLZ), Technische Universität München, Lichtenbergstr. 1, 85748 Garching, Germany

[a]robert.wimpory@helmholtz-berlin.de, [b]Michael.Hofmann@frm2.tum.de

Keywords: Surface Effect, Absorption Effect, Neutron Diffraction, Aberrations, Round Robin, Standardization, Grain Size Effects, Plane Stress Calculations, Nickel Alloy

Abstract. Instrumental and certain sample characteristics can affect the detected Bragg peak shifts which are not related to the strain being measured. Three major effects can influence the measurement: the surface effect, where the instrumental gauge volume (IGV) is not fully immersed at a surface or interface, the grain size effect where there is random positioning of large grains in the sample within the gauge volume and the relative shift in position of the centre of gravity of measurement due to absorption of neutrons. All of these effects can be reduced/eliminated by making pairs of neutron diffraction measurements 180 degrees to each other at the same location. Results are presented from a round robin benchmark weldment, denoted TG6, from the European Network on Neutron Techniques Standardization for Structural Integrity (NeT). This is made from a nickel alloy which has large grains and strains and has a high neutron attenuation coefficient.

Introduction

The surface effect, where the instrumental gauge volume (IGV) is not fully immersed at a surface or interface, normally results in a detected shift of the Bragg-peak and is not related to strain. Webster et al [1] noticed that the instrumental surface effect is essentially symmetrical. This aberration is influenced by a number of parameters, such as the bending radius when using a perfectly bent monochromator [2, 3] and is therefore instrument specific. Figure 1 shows the surface effect when measuring with a $2\times2\times10$ mm^3 gauge volume on E3 at the HZB [4] (where 10 mm is the gauge volume height) on a nominally strain-free 10mm thick ferritic steel sample. Making pairs of measurements 180° to each other show the symmetrical nature of the surface effect. The bending radius of the Si [400] monochromator is set fortuitously so the surface effect is almost canceled out for the out-of-plane normal direction for this particular 2theta scattering angle. Averaging the pairs of measurements for each corresponding position results in values that are consistent from the surface to the inside of the specimen (see table 1). Holden et al [5] found that for large-grained samples making pairs of neutron diffraction measurements 180° to each other is a way of minimizing the scatter coming from the random positioning of grains within the gauge volume. Also the relative shift in position of the centre of gravity of measurement due to absorption of neutrons [6] can be cancelled out using this 180° technique. The following results presented are from a round robin benchmark weldment, denoted TG6, from the European Network on Neutron Techniques Standardization for Structural Integrity (NeT). This is made from a nickel alloy which has large grains, large strains and a high neutron attenuation coefficient, making it ideal for testing out the '180°' technique.

Residual Stresses 2018 – ECRS-10   Materials Research Forum LLC
Materials Research Proceedings **6** (2018) 9-14   doi: http://dx.doi.org/10.21741/9781945291890-2

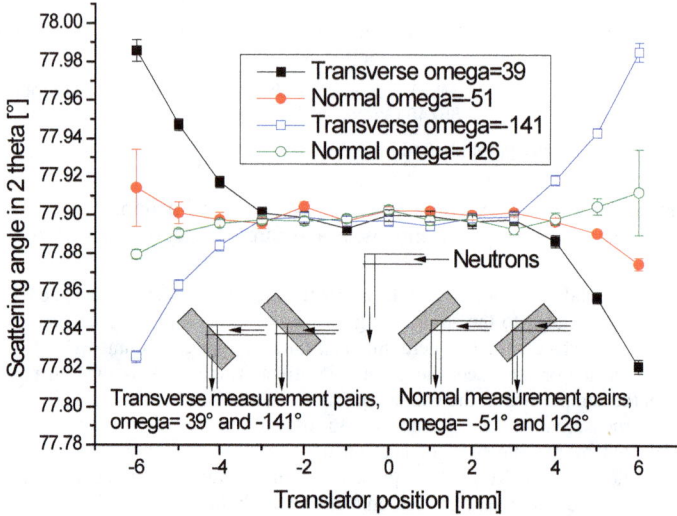

Figure 1. The surface effect of in-plane (transverse) and out of plane (normal) direction on E3.

Table 1. Measurement on a nominally strain-free 10mm thick ferritic steel sample.

Translator position [mm]	Actual position [mm]	Transverse ω=39 [°]	Normal ω =-51 [°]	Transverse ω=- 141 [°]	Normal ω =126 [°]	*Transverse Average* [°]	*Normal Average* [°]
-6.00	-4.86	77.9858	77.9140	77.8259	77.8794	77.9059	77.8967
-5.00	-4.51	77.9472	77.9010	77.8632	77.8906	77.9052	77.8958
-4.00	-3.88	77.9172	77.8974	77.8840	77.8956	77.9006	77.8965
-3.00	-3.00	77.9011	77.8958	77.8978	77.8974	77.8995	77.8966
-2.00	-2.00	77.8988	77.9045	77.8990	77.8970	77.8989	77.9008
-1.00	-1.00	77.8930	77.8969	77.8967	77.8981	77.8949	77.8975
0.00	0.00	77.9000	77.9025	77.8971	77.9028	77.8986	77.9027
1.00	1.00	77.8994	77.9021	77.8944	77.8973	77.8969	77.8997
2.00	2.00	77.8967	77.8999	77.8986	77.8978	77.8977	77.8989
3.00	3.00	77.8977	77.9014	77.8991	77.8928	77.8984	77.8971
4.00	3.88	77.8866	77.8968	77.9183	77.8982	77.9025	77.8975
5.00	4.51	77.8570	77.8906	77.9430	77.9047	77.9000	77.8977
6.00	4.86	77.8212	77.8747	77.9852	77.9120	77.9032	77.8934
					Average of averages	77.9002	77.8977
					Standard deviation	0.0032	0.0023

Residual Stresses 2018 – ECRS-10   Materials Research Forum LLC
Materials Research Proceedings 6 (2018) 9-14   doi: http://dx.doi.org/10.21741/9781945291890-2

Results

The TG6 test component comprises a 200mm × 150mm ×12mm rectangular base plate made from Inconel 600 with three passes of Alloy 82 weld metal deposited in a slot of length 76mm. This not only has large grain issues in the weld region but a large interplanar spacing variation due to high strain gradients and a change in material composition. Each position was measured on E3 in steps of 1 degree, -3,-2-1,0,2,3° about the scattering vector and the value was then averaged. This was to reduce grain size effects [7]. Figure 2 shows the measurement of longitudinal direction (which is made in transmission geometry) along the central line of the TG6 specimen (denoted the BD line) using a 2×2×2 mm^3 gauge volume. This is the third measurement of the BD line on E3 (July 2017). The bulge at the back of the weld was set to y=12mm. The back plate surfaces immediately to the sides of the bulge were y=11.5mm. The total length of the BD line is 14.59 mm. The reference pin (denoted pin 3) is actually 14.36 mm, one surface however should correspond to the back face of the weld, i.e. the surface of the bulge at y=12mm. The bottom of the weld is at about y=5mm, where the parent material is from y=5 to 12mm.

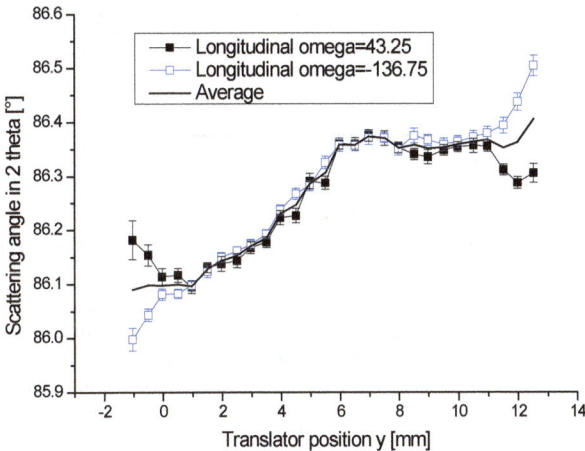

Figure 2. Measurement of the longitudinal direction along the BD line in the TG6 specimen.

The figure shows that there is indeed the surface effect as the gauge volume exits the surface. The coincidence of the lines where the gauge volume is fully submersed proves the good alignment of the primary slit over the centre of rotation of the diffractometer's omega table. The measurement was made in transmission geometry. Conversely the normal direction was measured in reflection geometry. This is susceptible to an absorption effect [6] which can be clearly seen in figure 3. This is because the centre of gravity of the gauge volume in terms of scattered neutrons in refection geometry is at slightly different position to the geometrical centre of the gauge volume. Measurement however is difficult on E3 over the complete BD line from one side as the neutron absorption of the nickel alloy is very high. Taking an average of the

11

values where there are two values gives the correct value for the position [6]. It is advised that the region where one can do overlapping measurements is measured first so one can find the absorption shift, so one can work out the translator positions that correspond to the intended measurement positions.

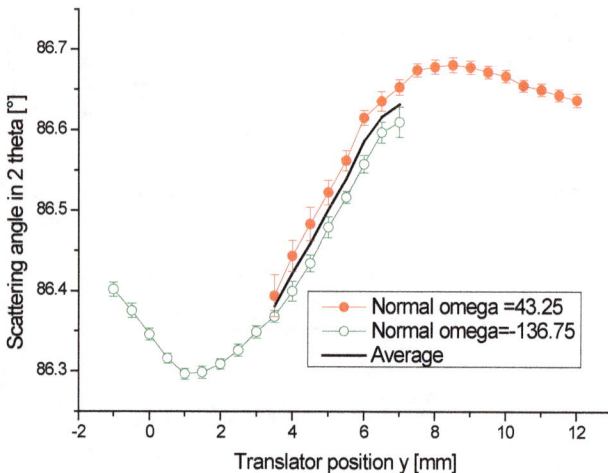

Figure 3. Measurement of normal direction along the BD line in the TG6 specimen on E3.

A 5mm slice of the weld was also available which turned out to be better for relative positioning of corresponding reference values for the welded plate compared to the pin (see figure 4). The stress is much smaller in the slice compared to the plate. Two plane stress calculations were made in the slice (assuming the longitudinal stress and normal stress directions were zero respectively) and these agree with each other well. For a comparison, the reference value needed to set the normal direction =0MPa in the welded plate is also depicted and this agrees with the slice positioning. One can fit the distribution with a Sigmoidal function and the centre positon of the three values obtain agrees to within 4.87± 0.06mm. There is a slight offset with respect with the reference pin which was 4.56 ±0.04 mm. This corresponds to the difference in length of the pin (14.36 mm) and the actual length of the BD line (14.59 mm) to within experimental uncertainty.

Final stress calculations agree mostly very well with measurements made on Stress-Spec [8] at the FRMII (see figure 5). The normal strain was also made using the '180°' technique but one was able to measure along the whole BD line due to the higher flux at FRMII and so the average values were made across the whole of the BD line. Agreement of the normal stress in the weld near the surface is not so good; this may be due to grain size issues that were not canceled out due to normal strain not being completely measured from both sides on E3. The best agreement is in the parent region where the grain size is smaller. The normal stress at the surface is 0 MPa

Residual Stresses 2018 – ECRS-10   Materials Research Forum LLC
Materials Research Proceedings **6** (2018) 9-14   doi: http://dx.doi.org/10.21741/9781945291890-2

as expected. A value of near 0 MPa was also obtained on the Stress-Spec measurement in the weld region.

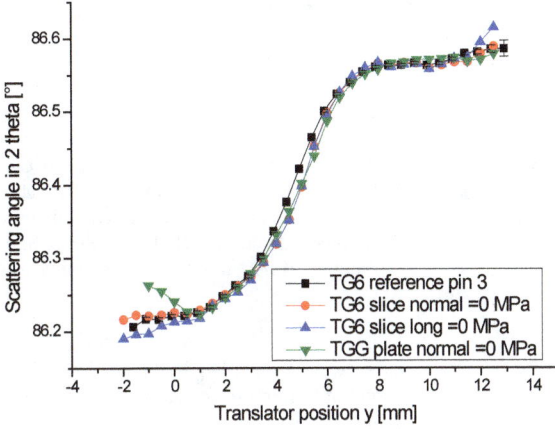

Figure 4. Comparison of reference values obtained from the TG6 pin, slice and plate.

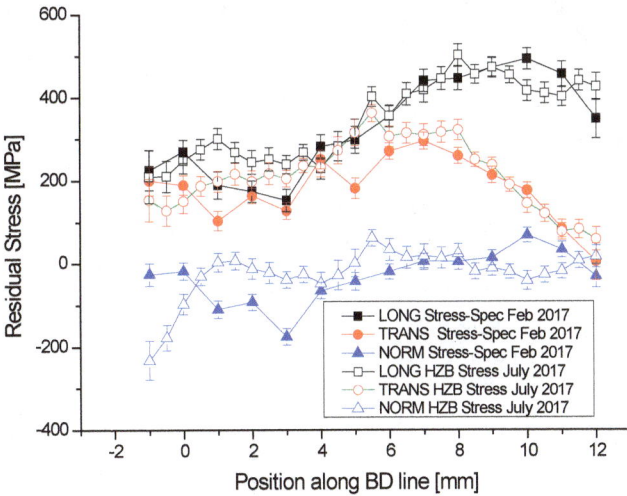

Figure 5. Comparison of stress from Stress-Spec and E3 at the HZB along the BD plate line.

Summary

The Net TG6 round robin specimen presented itself to be a challenge when determining the strain and stress using neutrons. Measuring the normal strain can be difficult when there is a large neutron path length and getting a complete overlap of measurements is challenging when using the 180° technique. Agreement of the normal stress in the two measurements presented in the weld near the surface is not so good; this may be due to grain size issues that were not canceled out due to normal strain not being completely measured from both sides on E3. The best agreement is in the parent region where the grain size is smaller. The normal stress at the surface is 0 MPa as expected. A value of near 0 MPa was also obtained on the Stress-Spec measurement in the weld region. In general using the 180° technique is useful for measuring near surfaces to get a more accurate value of strain. This technique also cancels out the absorption effect [6] and also can be used for grain size effect mitigation [5]. The sample has a large interplanar spacing variation due to high strain gradients and a change in material composition. The gradients due to only the material composition can be seen in figure 4. A Sigmoidal function can be used to fit these reference distributions to obtain a better positioning to obtain accurate strain values in the welded plate.

References

[1] Webster, P.J., Mills, G. , Wang, X.D. , Kang, W.P. , Holden, T.M. 'Impediments to Efficient Through-Surface Strain Scanning', Journal of Neutron Research, vol. 3, no. 4, pp. 223-240, 1996. https://doi.org/10.1080/10238169608200197

[2] M. Vrána and P. Mikula, "Suppression of Surface Effect by Using Bent-Perfect-Crystal Monochromator in Residual Strain Scanning", Materials Science Forum, Vols. 490-491, pp. 234-238, 2005. https://doi.org/10.4028/www.scientific.net/MSF.490-491.234

[3] Saroun, J., Kornmeier, J. R., Hofmann, M., Mikula, P. & Vrána, M. (2013). J. Appl. Cryst. 46, 628-638. https://doi.org/10.1107/S0021889813008194

[4] Wimpory, R. C., Mikula, P., Šaroun, J., Poeste, T., Li, J., Hofmann, M. and Schneider, R. Neutron News, 19, (2008) , 16 - 19. https://doi.org/10.1080/10448630701831995

[5] Holden, T. M., Traore, Y., James, J., Kelleher, J. & Bouchard, P. J. (2015). J. Appl. Cryst. 48, 582-584. https://doi.org/10.1107/S1600576715002757

[6] "Precise measurement of steep residual strain gradients using neutron diffraction in strongly absorbing materials with chemical compositional gradients", Robert C. Wimpory, Michael Hofmann, Vasileios Akrivos, Mike C Smith, Thilo Pirling and Carsten Ohms., Accepted for publication in Materials Performance and Characterization 2018

[7] Robert Charles Wimpory, René V. Martins, Michael Hofmann, Joana Rebelo Kornmeier, Shanmukha Moturu, Carsten Ohms, International Journal of Pressure Vessels and Piping, 2017, ISSN 0308-0161. https://doi.org/10.1016/j.ijpvp.2017.09.002

[8] Hofmann, M., Schneider, R., Seidl, G.A, Kornmeier, J, Wimpory, R., Garbe, U. und Brokmeier, H.G., Physica B, 2006, 385 – 368, 1035 - 1037

Residual Stresses 2018 – ECRS-10
Materials Research Proceedings 6 (2018) 15-20

Materials Research Forum LLC
doi: http://dx.doi.org/10.21741/9781945291890-3

Neutron Strain Scanning of Duplex Steel Subjected to 4-Point-Bending with Particular Regard to the Strain Free Lattice Parameter D_0

S. Pulvermacher[1,a], J. Gibmeier[1,b], J. Saroun[2,c], J. Rebelo Kornmeier[3,d], F. Vollert[1,e] and T. Pirling[4,f]

[1]Karlsruher Institut für Technologie, IAM-WK, Kaiserstraße 12, 76131 Karlsruhe, Germany
[2]Nuclear Physics Institute of the ASCR, 250 68 Řež, Czech Republic
[3] Heinz Maier-Leibnitz Zentrum (MLZ), TU München, D-85748 Garching, Germany
[4]Institut Laue-Langevin, 38042 Grenoble, France

[a]samuel.pulvermacher@kit.edu, [b]jens.gibmeier@kit.edu, [c]saroun@ujf.cas.cz, [d]joana.kornmeier@frm2.tum.de, [e]florian.vollert@kit.edu, [f]pirling@ill.eu

Keywords: Neutron Strain Scanning, Phase Specific Strain, 4-Point Bending, Stress Free Lattice Parameter, Duplex Steel

Abstract. Neutronographic residual stress analysis on multiphase materials is challenging with regard to phase-specific micro residual stresses and to the consideration of an appropriate stress free lattice parameter for meaningful lattice strain calculation. Even in case of randomly textured materials stress analysis becomes more elaborate due to plastic anisotropy effects. According to literature for stress analysis using neutron diffraction lattice planes should be chosen that are less prone to plastic anisotropy. These are the {311} austenite and the {220} ferrite planes in case of duplex steels. Here, we report about phase-specific in-situ neutron strain scanning at SALSA@ILL, Grenoble during defined 4-point-bending of duplex steel X2CrNiMoN22-5-3 using exactly these two recommended diffraction lines. It is shown that due to the local texture of the bending bars, which was cut from a hot rolled cylindrical rod, strong plastic anisotropy was determined. This effect must be taken into account for diffraction based residual stress analysis to prevent from erroneous stress determination.

Introduction

For non-destructive residual stress analysis for the inside of technical components beside high energy synchrotron X-ray diffraction, neutron diffraction is often the method of choice. In this regard a meaningful measure of the stress / strain free or stress / strain independent lattice parameter D_0 is required for calculating lattice strains from the determined interplanar lattice spacings. Different procedures are explained how to provide appropriate data for D_0 as e.g. determining the interplanar lattice spacings for the ´stress free state´ in an area with negligible (residual) stress or to cut cubes or comb structures out of the material, which in turn leads to a partial and adequate stress release [1]. In case of multi-phase materials, phase specific micro residual stresses can impede these well established procedures since it is questionable if the micro residual stresses will be released during cutting the samples. Assuming that the macro residual stresses are sufficiently released significant phase specific residual stresses might remain and will in consequence affect the local stress free lattice parameter.

For diffraction stress analysis in multiphase materials it is recommended to determine phase specific lattice strains in all phases if the phase content exceeds a volume share of about 10%. By this means macro (residual) stresses can be determined using a rule of mixture and the volume share of the contributing phases as weighting factor. Duplex steels represent typical applicants in this field. Here, large amount of austenite phase coexists with ferrite, hence for this gross two-phase material neutronographic stress/strain analysis strongly requires consideration of both phases and thus also the determination of appropriate phase specific stress free lattice parameters $D_{0,\alpha}$ and $D_{0,\gamma}$. For the assessment of manufacturing processes by means of process induced residual stresses often the separation of macro and micro residual stresses and in this regard the load partitioning on the two phases is of special interest. In this context in [2] the phase specific lattice strains for various {hkl} lattice planes for duplex steel subjected to defined macroscopic elastic and elasto-plastic uniaxial loading were studied using neutron diffraction. It has been shown that apart of the elastic anisotropy plastic anisotropy effects occur due to the fact that crystallites do not deform homogeneously since the deformation depends on the slip systems. Regarding the load partitioning for different lattice planes this plastic anisotropy in part strongly deviates from the elastic behavior. For practical applications the authors recommended to consider only lattice planes for (residual) stress analysis which have no or only a weak anisotropic effect. In accordance with [3,4] for fcc materials the {111}, {311} and {422} planes and for bcc materials the {110} or {211} planes are less prone to plastic anisotropy effects. In preliminary studies on load partitioning in duplex steels we noticed that crystallographic texture might have a strong impact on the plastic anisotropy. However, crystallographic texture was not discussed in the above mentioned works. The knowledge about the impact of crystallographic texture on plastic anisotropy is essential for the planning, the evaluation and assessment of neutronographic stress analysis, not least since semi-finished parts or technical components made of duplex steels often exhibit characteristic textures due to the processing route. To study the effect of texture on the load partitioning behavior for this first approach rectangular shapes bars were extracted from cylindrical rods of duplex steel X2CrNiMoN22-5-3 exhibiting phase fractions of about 50% ferrite (α) and 50% austenite (γ). For samples subjected to defined 4-point-bending using purely elastic and elasto-plastic loading neutronographic strain scanning in both phases was carried out with respect to the coordinate of the bending height on instrument SALSA@ILL, Grenoble. 4-point-bending experiments have the charm that within a single experiment different uniaxial loading states can be studied for different location at the bar. Considering purely elastic and elasto-plastic loading allows for the separation between elastic and plastic anisotropy effects. To provide meaningful data for the phase specific stress free lattice parameters $D_{0,\{hkl\}}$ the locally exiting lattice parameters of the sample prior to the loading experiment is determined and applied for local lattice strain calculation.

Material and experimental Procedures:
The used material is a hot-rolled duplex steel X2CrNiMoN22-5-3 (mat.no. 1.4462) with a diameter of 30 mm exhibiting a ferrite to austenite ratio of approximately 50:50. The microstructure is shown in Fig. 1 for a cut in longitudinal direction. Metallographic analysis revealed an average grain size in rolling direction (later longitudinal direction) of approx. 74 µm, 6 µm in transverse direction and 5 µm in normal direction, respectively. The sample geometry is 160 x 15 x 10 mm. Figure 2 shows a schematic view of the sample with the assigned coordinate system together with indication of the applied 4-point-bending loading.

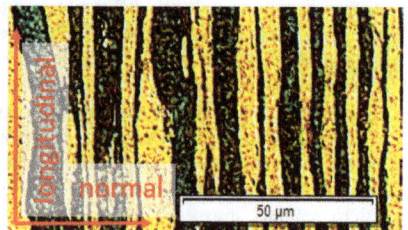

Figure 1: Microstructure of the duplex steel for a section from a cut in longitudinal direction of the used bars, etchant: Behara II, bright regions: austenite, dark regions: ferrite

Residual Stresses 2018 – ECRS-10   Materials Research Forum LLC
Materials Research Proceedings 6 (2018) 15-20   doi: http://dx.doi.org/10.21741/9781945291890-3

In the as received state the duplex steel was de facto stress free as verified by means of hole drilling and X-ray stress analysis. As indicated in the introduction part the bars exhibit a rather strong crystallographic texture in both phases with a gradient over the bending height. Since the bending bars were not exactly cut from the centre of the as delivered steel rod the texture distribution is not symmetric to the mid layer of the bars. Fig. 3 shows as an example the texture distribution for the later compressive loaded layer, the mid layer and for the later tensile loaded layer for both phases for the $\phi2 = 45°$ sections of the ODF (orientation distribution function) determined by means of X-ray diffraction prior to the neutron diffraction experiment. In total the ODF was determined for 10 locations distributed over the bending height. The ODF cuts exemplary shown in Fig. 3 clearly indicate that (a) a gradient in the crystallographic texture exist in both phases and (b) that the sharpness of the texture is much higher in case of the ferrite phase (in particular on the later compressive loaded side).

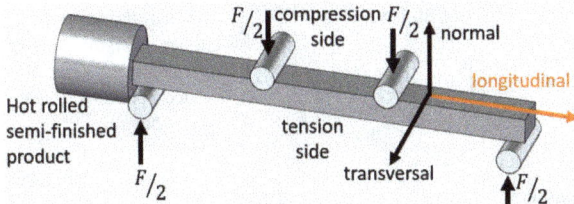

Figure 2: Schematic sketch of the bending bar with indication of the applied coordinate system

Neutron strain scanning with respect to the bending height was carried out in all three principal directions, i.e. in longitudinal, transverse and normal direction (see Fig 2) in both phases for various {hkl} planes at the SALSA experiment at ILL [5], Grenoble using a wavelength of 1.6 Å. A nominal gauge volume of $0.6 \times 0.6 \times 10$ mm was defined by radial collimators at the primary and secondary beam paths; the 10 mm axis was always parallel to the longitudinal direction of the bar. In total two ferrite (α-Fe) lattice planes ({211}, {220}) and three austenite (γ-Fe) lattice planes ({220}, {222}, {311}) were measured separately. Due to the rather strong local texture not for all directions interference lines could be recorded with intensities being sufficient for meaningful evaluation of the line positions. That means that the local texture impeded stress determination at all. Here, we only focus on the {220} α-ferrite and on the {311} austenite lattice planes, which are (according to literature) expected to be less prone to plastic anisotropy. Furthermore, we only focus on the strain scanning of the longitudinal component, which corresponds to the loading direction, to study the load partitioning in both phases. The peak fitting was realized using a Gaussian function subsequent to background subtraction. Only measurement locations, where the entire nominal gauge volume is immersed into material are considered. The phase specific values of the stress free lattice parameters $D_{0,\{hkl\}}$ were determined on the bending bar prior to the bending loading at the same positions, which were used for later in-situ lattice strain scanning during loading. Since, the D_0 scanning was carried out for less measuring points as in case of the in-situ experiments linear interpolation was applied. For in-situ neutron strain scanning during defined bending loading two different load stages were considered: (i) purely elastically deformed (0.22% total strain) and (ii) elasto-plastically deformed (approx. 1.5 % total strain). The loading was controlled via the total strain measured by strain gauges at both outer fibres of the bar. The loading stress were assigned using a reference bending stress-strain curve determined on an instrumented universal testing machine prior to the neutron beamtime using the same support distances as given by the 4-point bending device applied at SALSA (see Fig. 2).

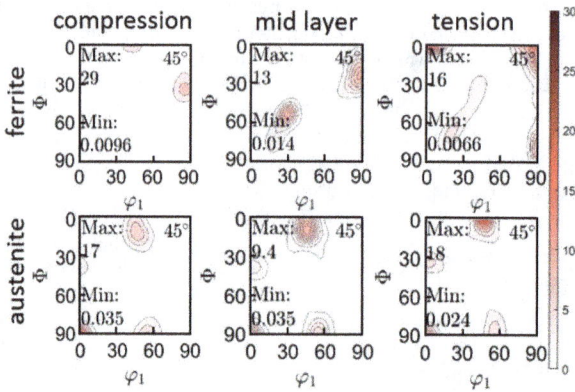

Figure 3: $\phi2 = 45°$ sections of the ODF determined by means of X-ray diffraction for three selected positions of the bending bar.

Experimental results and discussion

Figures 4 (A) and (B) show the interplanar lattice spacings for the initial state of the bending bar determined for the {311} γ-Fe and the {220} α-Fe lattice planes, respectively. The data indicate that the stress free lattice parameters $D_{0,\{hkl\}}$ shows a characteristic distribution presumable due to the processing route (hot rolling) of the rods, which leads to local texture and the generation of phase specific micro residual stresses. On the side with negative coordinates (later compressive loaded side) large errors bars result for the ferrite phase, which is due to local texture distribution and the connected weak intensities of the {220} interference lines. Regarding lattice strains that can be assigned to the changes in lattice parameters D_0, these changes in lattice parameters correspond to rather high strain (up to about 0.0016). Hence, neglecting this initial state might result in erroneous stress data. Consequently, the lattice strains presented in Fig. 4 (C) for the purely elastic case ($\varepsilon_t = 0.22\%$) and Fig. 5 (A) for the elasto-plastic case ($\varepsilon_t = 1.5\%$) are lattice strain differences $\Delta\varepsilon$ calculated with reference to this initial state.

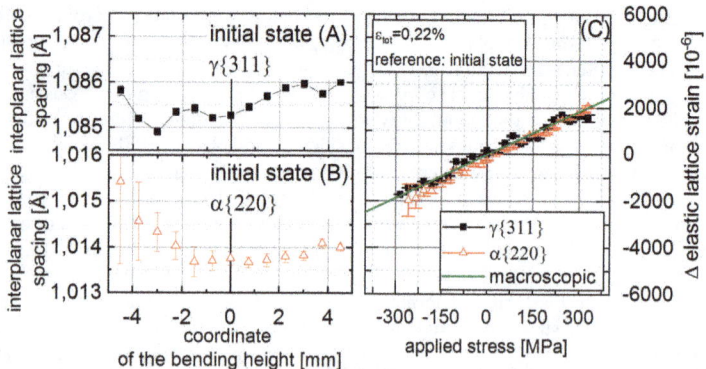

Figure 4: Phase-specific interplanar lattice spacings determined for the in the initial state of the bending bar for (A) the {311} austenite lines and (B) for the {220} ferrite lines. In (C) the change in lattice strains $\Delta\varepsilon$ with respect to the initial state for the macroscopically purely elastic loading state is shown. For better orientation the macroscopic ε vs. σ distribution is plotted.

Residual Stresses 2018 – ECRS-10   Materials Research Forum LLC
Materials Research Proceedings 6 (2018) 15-20   doi: http://dx.doi.org/10.21741/9781945291890-3

In Fig. 4 (C) the lattice strain difference $\Delta\varepsilon$ for the purely elastic loading vs. the applied stress is shown. For the {311} γ-Fe and the {220} α-Fe lattice planes almost linear courses are determined. Furthermore, both distributions roughly follow the macroscopic distribution, which can be expected for the orientation factors of $3\Gamma = 0.471$ for the {311} austenite and $3\Gamma = 0.75$ for the {220} ferrite planes. Both values are close to $3\Gamma = 0.6$, which generally reflects the macroscopic behavior.

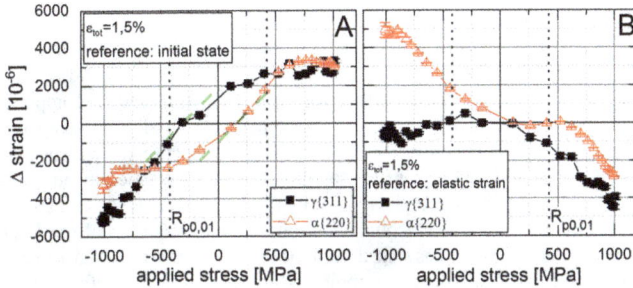

Figure 5: (A) change in phase-specific lattice strains $\Delta\varepsilon$ with respect to the initial state for elasto-plastic loading to $\varepsilon_t = 1.5\%$. For better orientation the macroscopic 0.01% proof stress $R_{p0.01}$ is added to the plot; (B) change in phase-specific lattice strains $\Delta\varepsilon$ after subtraction of the fictitious elastic materials response for the same loading state.

In Fig. 5 (B) the phase-specific load partitioning for the duplex steel in case of elasto-plastic loading is displayed for the loading with total strains in the outer layers of the bending bar of approx. 1.5%. Again the change in lattice strain with reference to the initial state is presented. In Fig. 5 (B) the same data is plotted as used in Fig. 5 (A), but here the fictitious elastic strain is subtracted for both phases using the trend determined for the purely elastic loading as shown in Fig. 4 (C). Here, the zero passage determined in Fig. 5 (A) is maintained for both phases. By this means the change in phase specific lattice strain plotted in Fig. 5 (B) indicate, when the individual phases start to plastify and by this means, differences in the phase specific materials response can be noticed more conveniently. To provide better orientation the 0.01% proof stress $R_{p0.01}$ (423 MPa) as determined by the macroscopic bending work hardening curve is added. The changes in phase specific lattice strains with respect to the initial state (Fig. 5(A)) indicate that in both phases large deviations to the behavior expected based on the work of [2] occur. In particular for the {220} α-Fe interference line it is expected that no significant plastic anisotropy exists. However, for both phases strong plastic anisotropy effects occur, which can be explained by the phase-specific texture exhibiting a strong gradient through the bending bar (see Fig. 3). According to [2] the strain response for the {220} ferrite lattice planes should follow the trend from the elastic regime. However, in the present case on both sides, i.e. the tensile and the compressive loaded side strong deviations occur, indicating that local phase-specific plastic deformation is strongly affected by the local crystallographic texture. That means that the texture is such that crystallites deform inhomogenously over the cross section of the bars. The reason is that the crystallites are orientated such that slip planes and slip directions are oriented in a way that dislocation gliding during bending loading is promoted in contrast to a more random orientation. In detail this will be studied by means of simulation based on crystal plasticity modelling taking into account the local gradient in crystallographic texture and is part of current research efforts. Here, the results of neutronographic strain scanning will be used for validation of the modelling approach.

A closer look to the data plotted in Fig. 5 (A) further reveals that the observed behavior is asymmetric when comparing the compressive with the tensile loaded side. The asymmetric phase-specific plastic deformation obviously causes a shift of the neutral fiber in both phases, i.e. in ferrite slightly towards the tensile and in austenite slightly towards the compressive loaded side. Obviously, the neutral fiber resulting for the macroscopic behavior is slightly shifted towards the compressive loaded side under consideration of the volume content, which is about 50:50. This shift of the neutral fiber (characterizing the macroscopic behavior) is nothing odd and can also be explained by means of the texture gradient.

The change in phase specific lattice strains when subtracting the fictitious elastic response supports the above mentioned statements. From this plot it can be clearly derived that for the $\{220\}$ ferrite line a very pronounced plastic anisotropy in tension and in compression can be noticed, while on the tensile loaded side the slope appears to be slightly steeper. In contrast, strong plastic anisotropy can be observed for the $\{311\}$ austenite interference line. On the compressive loaded side the local texture causes that almost no plastic anisotropy occurs. The local crystallographic texture induces a phase specific mechanical response that strongly deviates from literature. Hence, following the recommendations for consideration of the $\{311\}$ austenite and the $\{220\}$ ferrite interference lines will definitely result in erroneous phase specific residual stresses, when the effect of local crystallographic texture on the plastic deformation behavior is unattended during data evaluation and assessment. Hence, in case that texture is expected from the processing route, special care must be taken in regard to phase-specific plastic anisotropy.

Summary

In-situ neutronographic phase specific strain scanning during elastic and elasto-plastic 4-point bending was performed for duplex steel X2CrNiMoN22-5-3. The strain in load direction was determined for the $\gamma\{311\}$ and the $\alpha\{220\}$ interference lines under consideration of the local initial state of the bending bars prior to plastic deformation as D_0 reference. Inconsistent with literature strong plastic anisotropy occurs for both interference lines, while a neglectable effect was expected. However, in literature local texture was not particularly taken into account. In the present case, local crystallographic texture cause this strong plastic anisotropy for interference lines, which are often recommended and applied for local neutron stress analysis. Based on the findings special care on the texture induced plastic anisotropy must be taken for data evaluation and assessment for duplex steels, which often show pronounced texture induced by processing.

Acknowledgements

The work was partly supported through a joint project funded by the Czech Science Foundation (project No. 16-08803J) and the German Research Foundation (DFG project GI 376/11-1 and HO 3322/4-1) and by the Czech Ministry of Education, Youth and Sports (project no. LM2015050). The authors are grateful for granting beamtime at ILL (proposal No. 1-02-199).

References

[1] Withers, P. J., Preuss M., Steuwer A., Pang J. W. L., Appl. Crystallogr. 40 (2007), pp. 891–904. https://doi.org/10.1107/S0021889807030269

[2] Allen, A. J.; Bourke, M.; David, W. I. F.; Dawes, S.; Hutchings, M. T.; Krawitz, A. D.; Windsor, C. G., ICRS 1989, pp. 78-83

[3] Clausen B., Lorentzen T., Bourke M., Daymond M., Mat. Sci. Eng. A, 259 (1999), pp. 17-24. https://doi.org/10.1016/S0921-5093(98)00878-8

[4] Pang, J.W.L; Holden, T.M; Mason, T.E., Acta Mat., (1998), pp. 1503-1518. https://doi.org/10.1016/S1359-6454(97)00369-8

[5] Pirling, T. Bruno, G., Withers, P., Mat. Sci. Eng. A, 437 (2006), pp. 139-144. https://doi.org/10.1016/j.msea.2006.04.083

Residual Stresses 2018 – ECRS-10
Materials Research Proceedings 6 (2018) 21-26

Materials Research Forum LLC
doi: http://dx.doi.org/10.21741/9781945291890-4

Plastic Deformation of InSb Micro-Pillars: A Comparative Study Between Spatially Resolved Laue and Monochromatic X-Ray Micro-Diffraction Maps

Tarik Sadat[1,a,*], Mariana Verezhak[2,b], Pierre Godard[1,c], Pierre Olivier Renault[1,d], Steven Van Petegem[2,e], Vincent Jacques[3,f], Ana Diaz[2,g], Daniel Grolimund[2,h], Ludovic Thilly[1,i]

[1]Institut Pprime, CNRS, Université de Poitiers, SP2MI, Futuroscope Chasseneuil, France

[2]Paul Scherrer Institut, CH-5232 Villigen PSI, Switzerland

[3]Lab. Physique des Solides, CNRS, Université Paris-Sud, Orsay, France

[a]tarik.sadat@univ-poitiers.fr, [b]mariana.verezhak@psi.ch, [c]pierre.godard@univ-poitiers.fr, [d]pierre.olivier.renault@univ-poitiers.fr, [e]steven.vanpetegem@psi.ch, [f]vincent.jacques@u-psud.fr, [g]ana.diaz@psi.ch, [h]daniel.grolimund@psi.ch, [i]ludovic.thilly@univ-poitiers.fr

Keywords: Micro-Pillars, InSb, Laue Microdiffraction, Monochromatic Microdiffraction, Synchrotron Radiation, Plasticity

Abstract. Indium Antimonide (InSb) single-crystalline micro-pillars were mechanically deformed by uniaxial compression loading-unloading cycles up to the beginning of the plastic regime. After deformation, 2D spatially resolved maps were collected via two X-Ray Diffraction (XRD) techniques: polychromatic micro-Laue and monochromatic micro-diffraction. In both techniques, the integrated diffracted intensity shows strong variations inside the pillar. Moreover, the shift and streaking of one spot in polychromatic XRD as well as the lattice strain and tilt components derived from monochromatic XRD reveal that the plastically deformed area is localized on the top of the pillar, in agreement with scanning electron microscopy images. The two XRD techniques are thus providing correlated but yet complementary information.

Introduction

Indium Antimonide (InSb) and other semiconductors are known to be crystalline materials with brittle behavior at room temperature in the bulk state, but become ductile in the form of micro- and nano-objects [1-2]. At room temperature, InSb micro-pillars plastically deform by the nucleation of partial dislocations at surfaces: after gliding through the pillar, they escape at opposite surfaces (creating slip traces forming a deformation band) but leave in the crystal parallel Stacking Faults (SFs). This mechanism has been verified by post-mortem Scanning Electron Microscopy (SEM) observations and destructive Transmission Electron Microscopy (TEM) characterization [1-2] but also by non-destructive post-mortem study based on coherent X-Ray diffraction [3]. In the latter case, the profile of coherent diffraction patterns of 202 reflection evolves significantly between pristine and plastically-deformed regions: the splitting and streaking of patterns arise from interferences induced by the presence of SFs that are phase-defects creating phase-shifted regions within the pillar [3]. The maximum intensity of diffraction patterns could be related to the number of SFs present in the illuminated volume, and also to the volume of defected region (i.e. the size of the band containing the SFs). This diffraction technique thus proved to be very interesting as a non-destructive way to detect and characterize the presence of plastic defects in nano-and micro-objects.

Residual Stresses 2018 – ECRS-10   Materials Research Forum LLC
Materials Research Proceedings **6** (2018) 21-26   doi: http://dx.doi.org/10.21741/9781945291890-4

In the present work, the same system (InSb micro-pillar) is used to further assess the capacities of micro-diffraction techniques to characterize the impact of plastic deformation (lattice rotation, residual strain, defects storage, etc.) and in particular compare two techniques, Laue and monochromatic micro-diffraction, via spatially resolved 2D maps that are compared and discussed.

InSb micro-pillars – Mechanical tests

InSb micro-pillars were fabricated by Ga^+ Focused Ion Beam (FIB) milling from a bulk <123>-oriented single-crystalline wedge: at all milling steps, the ion beam current and acceleration voltage were kept as low as possible to avoid creation of lattice defects and/or surface amorphisation; for instance, the last milling step was performed at a beam current of 24 pA. The resulting geometry is a series of free-standing micro-pillars on top of pedestals. Fig. 1 (a) shows a SEM micrograph of the InSb micro-pillar of interest, with a dimension of $2.5 \times 2.5 \times 7.5$ μm^3. Fig. 1 (d) provides a 3D schematic view of the pillar on its pedestal as well as the relevant crystallographic directions (top view is also schematized, as discussed later). The <123> orientation was chosen to favor "single-slip" deformation during compression, i.e. partial dislocations will nucleate and glide only in a single glide plane, (1-11), resulting in the accumulation of SFs in parallel (1-11) planes [1-3].

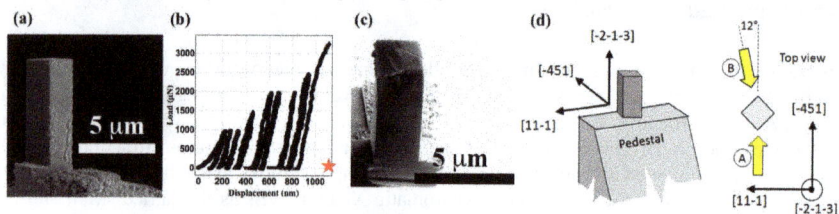

Figure 1: (a) SEM image of studied InSb pillar before deformation. (b) Load versus displacement compression curves applied to the pillar; the (red) star illustrates the condition at which 2D XRD maps were collected. (c) SEM image of the same pillar after deformation. (d) 3D schematic of pillar on pedestal with crystallographic directions and top view with the two diffraction geometries: "A" refers to the incident beam direction for polychromatic Laue micro-diffraction, "B" refers to the incident beam direction for monochromatic micro-diffraction

A custom-designed micro-compression device (MCD), equipped with a diamond flat tip and a standard 1D Triboscope transducer from Hysitron Inc. for force and displacement readout (see [4] for details), was used to in-situ perform uniaxial compression tests combined with a Laue polychromatic micro-diffraction as well as monochromatic micro-diffraction. Here, we focus on the post-mortem study of one specific pillar deformed by a series of loading-unloading cycles until the early plastic regime was attained, as shown by the mechanical curves presented in Fig. 1 (b). After the mechanical test, SEM observations were performed on the pillar, as presented in Fig. 1 (c): the top third of the pillar exhibits permanent bending as a sign of plastic deformation, as well as the presence of two cracks at the very top.

Spatially resolved 2D micro-diffraction maps: comparison between polychromatic and monochromatic cases

After plastic deformation, the pillar was spatially mapped with two different diffraction techniques: polychromatic Laue micro-diffraction and monochromatic micro-diffraction.

Post mortem Laue micro-diffraction mapping was performed at the MicroXAS beamline from the Swiss Light Source (SLS) in Switzerland, with a 1×1 μm^2 pink beam and energy range of [4-

23] keV. X-Ray Fluorescence (XRF) was used to localize each micro-pillar before and after deformation: XRF maps were obtained by summing the signals from In–Lα, Sb–Lα and Sb-Lβ emission yields. A 2D Eiger 4M detector with a pixel size of 75×75 μm^2 was used to collect Laue patterns in transmission mode. The sample-to-detector distance was fixed at 86.81 mm. 47×24 patterns were collected from top to bottom of the pillar, with a step size 300 nm. The polychromatic incident beam was parallel to the [-451] direction, corresponding to configuration "A" in top view of Fig. 1 (d). All Laue micro-diffraction data presented here are related to one of the many diffraction spots available in the patterns: the 1-1-1 reflection, as presented in Fig. 2 (a) and (b). A computer code was developed to analyze the data (XRF and XRD) using Python.

Post mortem monochromatic micro-diffraction mapping was performed at the ID01 beamline from the European Synchrotron Radiation Facility (ESRF) in France, with a 8 keV monochromatic beam. A 2D Maxipix detector with a pixel size of 55×55 μm^2 was used to collect diffraction patterns obtained in reflection mode, with a beam size of about 0.2x0.2 μm^2. The sample-to-detector distance was fixed to 140 cm. 60x60 patterns were collected from top to bottom of the pillar, with a step size of 250 nm. Only the (11-1) reflection was considered (associated to lattice planes that are parallel to pillar and compression axes, see Fig. 1 (d)), as presented in Fig. 2 (c) and (d): constraints from the goniometer and setup imposed that the incident monochromatic beam was at 12° in opposite direction to [-451], corresponding to configuration "B" in top view of Fig. 1 (d). *This will result in mirrored maps obtained from polychromatic and monochromatic micro-diffraction experiments.* A rocking curve made of 20 ϕ-scans was performed to probe the 11-1 Bragg peak in reciprocal space: the ϕ-scans were done between 18.2° to 19.15° with a step size of 0.05°. Homemade programs written in Python and Mathematica were used to analyze collected patterns: in particular, the center of mass for diffracted intensity was calculated as well as the integrated intensity. The conditions of the experiments were such that a horizontal variation of the center of mass of the spot is proportional to a Bragg angle evolution $\Delta\theta$ (in degrees) that can be translated into a lattice spacing evolution Δa, or strain $\Delta a/a$; a vertical variation of the center of mass of the spot is correlated with a change of the lattice orientation $\Delta\omega$ (in degrees) that can be understood as a tilt (perpendicular to the pillar axis) of the single-crystalline pillar. Note that the other twist component, with an axis aligned with the pillar, could be obtained since rocking curves have been recorded, but is not shown here.

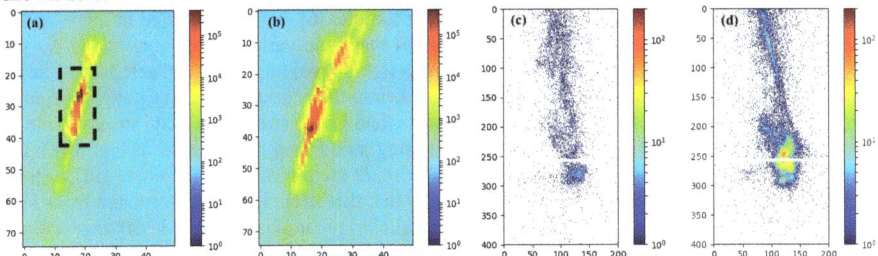

Figure 2: (a) zoom from the full Laue pattern on the 1-1-1 spot collected at the pedestal and (b) at the top of the deformed pillar; the dashed black rectangle is a mask used for analysis. (c) Zoom from the full monochromatic micro-diffraction pattern on 11-1 spot collected at the pedestal and (d) at the top of the pillar. All scales are logarithmic. All axes are in pixels.

Fig. 2 illustrates the shape and intensity of the spots obtained respectively by Laue and monochromatic micro-diffraction at the pedestal (Fig.2 (a) and (c)) and at the top of the pillar (Fig. 2 (b) and (d)), after deformation. It is interesting to observe that in both cases, the shape and

Residual Stresses 2018 – ECRS-10   Materials Research Forum LLC
Materials Research Proceedings **6** (2018) 21-26   doi: http://dx.doi.org/10.21741/9781945291890-4

intensity of the spots are significantly different between the pedestal and the pillar's top, with apparition of peak shift, streaking and splitting that are characteristic of permanent change in lattice strain and orientation, as well as storage of lattice defects: these features are obviously the result of the concentration of plastic deformation at the top of the pillar, as observed in Fig. 1 (c).

Fig. 3 presents the different spatially resolved maps obtained by XRF (Fig. 3 (a)), Laue micro-diffraction (Figure 3 (b)) and monochromatic micro-diffraction (Fig. 3 (c)) on the deformed pillar. The XRF map allows to entirely see the pillar on its pedestal and localize (star symbol) where reference patterns will be taken for the different data analyses. Fig. 3 (b) presents the Laue map of integrated intensity of the 1-1-1 spot (same region as in Fig. 2 (a) and (b)) normalized to the integrated intensity of the same spot at reference position in the pedestal: an unexpected increased normalized integrated intensity is observed in the upper part of deformed pillar, in a region of about 5μm in height. Fig. 3 (c) displays the monochromatic micro-diffraction map of integrated intensity of the 11-1 spot normalized to the integrated intensity of the same spot at reference position in the pedestal. The plotted normalized integrated intensity has been obtained from the summation on the 20 rocking curves, or ϕ-scans, previously described. Again, an increased normalized integrated intensity is observed at top of the pillar; this increase of intensity is more localized than for Laue micro-diffraction, probably because of smaller beam size in monochromatic micro-diffraction. Considering the SEM observations made on the deformed pillar, it seems reasonable to correlate this region of increased diffracted intensity (observed in both maps) to the plastically deformed region at pillar's top.

Fig.4 (a) displays the Laue map of integrated intensity of the 1-1-1 spot, calculated outside the masked region defined in Fig. 2 (a): this region is defined from the pattern at pedestal position, i.e. in a non-deformed region. As a first crude approximation, we consider that the diffracted intensities inside the mask are mainly contributions from the pristine crystal: all intensities outside the mask may then be considered as mainly resulting from plastically deformed regions, inducing shift, streaking and splitting of the 1-1-1 spot. As a result, the obtained map reveals that the most intense region corresponds indeed to the top of the pillar, where plasticity has taken place. Interestingly, this region is more localized than the one revealed previously without any mask (Fig. 3 (b)) and similar to the one revealed by monochromatic micro-diffraction (Fig. 3 (c)). A more in-depth analysis of the Laue spot profile should bring additional information on the origin of observed lattice strain and rotation (incl. lattice defects density) [5]. The monochromatic case, based on a powerful XRD technique [6], has been studied in more details to obtain more quantitative information: Fig. 4 (b) maps the variation in Bragg angle, $\Delta\theta$ (°), deduced from the horizontal shift of center of mass (see above) in patterns collected at a ϕ angle equal to 18.2°. Again here, a reference pattern is taken in the pedestal. The Bragg angle variation is then transformed into lattice spacing variation Δa, and lattice strain $\Delta a/a$, with strain-free InSb lattice being taken as a=6.479 Å [1-2]. The resulting strain map is presented in Fig. 4 (c): the region with largest residual strain is, without surprise, at the top of the pillar where lattice strain of the order of 0.01 is observed. One should recall that this strain has been calculated from 11-1 spot and is thus associated to the lattice planes parallel to pillar axis (see Fig. 1 (d)): this strain is thus smaller than axial strain by a factor of about 1/3 (Poisson ratio) and of opposite sign. A similar approach has been applied to the vertical shift of the center of mass to obtain a map of lattice orientation variation $\Delta\omega$ (°), or lattice tilt, compared to reference in the pedestal. The map shows that the residual tilt is rather small in the pedestal and in all the bottom part of the pillar: it is only the very top that exhibits relatively large residual tilt values (-0.06°) that are footprints of the plastically bent regions (as observed in SEM) but also the tilted regions between cracks at very top of the pillar. The micro-Laue gives access to a quantification of the tilt.

Residual Stresses 2018 – ECRS-10   Materials Research Forum LLC
Materials Research Proceedings **6** (2018) 21-26   doi: http://dx.doi.org/10.21741/9781945291890-4

Overall, these spatially resolved maps are in good agreement with recent results obtained by Laue micro-diffraction combined with compression to investigate the deformation mechanisms in [001]-oriented single crystal MgO cylindrical pillars: a reduction in diffracted intensity was recorded along the vertical direction, moving from the top towards the bottom of the MgO micro-pillars, and associated to gradual reduction of the strain along the pillars height [7]. In our case, both micro-diffraction techniques maps reveal a localized region at the top of the pillar where the integrated diffracted intensity is higher. This can be explained by the fact that in Laue configuration, a broader part of the pink spectrum is selected in the distorted areas. For the monochromatic case, a similar reasoning can be applied considering the summation over the 20 ϕ-scans. Moreover, this region is correlated to the volume where residual lattice strain and tilt were induced by plastic deformation

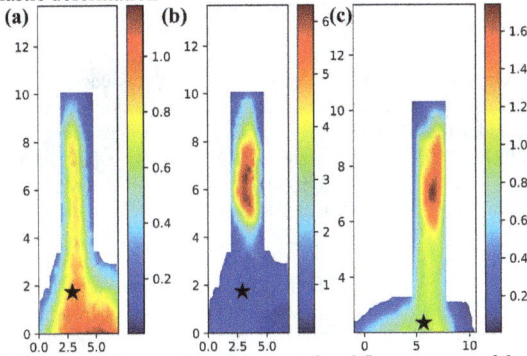

Figure 3: spatially resolved maps from (a) normalized fluorescence of the InSb pillar, (b) normalized integrated intensity of the 1-1-1 spot in Laue micro-diffraction and (c) normalized integrated intensity of the 11-1 spot in monochromatic micro-diffraction. The stars localize the positions for reference patterns. The axes are in μm.

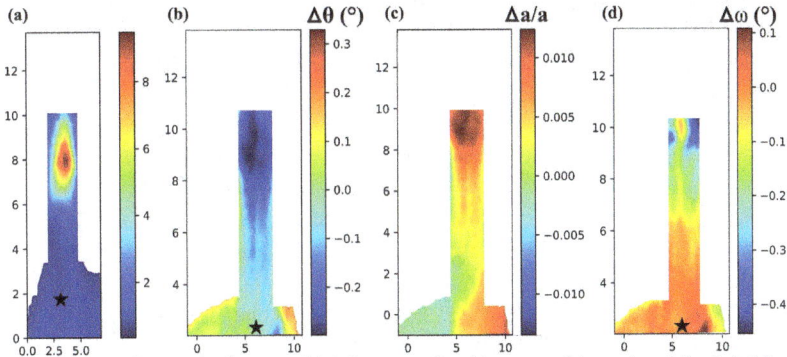

Figure 4: spatially resolved maps of (a) the normalized integrated intensity of the 1-1-1 Laue spot outside a masked region (mask shown on pattern at pedestal in Fig. 2 (a)), (b) the Bragg angle variation Δθ (°) of the 11-1 reflection in monochromatic micro-diffraction and (c) associated lattice strain Δa/a; (d) the variation Δω (°) of lattice tilt in the pillar from 11-1 spot monochromatic micro-diffraction. The stars localize the positions for reference patterns. The axes are in μm.

Materials Research Forum LLC
doi: http://dx.doi.org/10.21741/9781945291890-4

Summary

We reported here the case of one InSb micro-pillar that was deformed via uniaxial cyclic compression up to the plastic regime. 2D spatially resolved maps obtained from Laue micro-diffraction and monochromatic micro-diffraction lead to provide relevant information such as the localization of the plastic activity and increase of strains in a specific region localized at about one third from the top of the pillar and corresponding well to the post-mortem SEM observations. Finally, the two XRD techniques are thus providing correlated but yet complementary information that are useful to better understand the evolution of micro-pillars after mechanical solicitation. This post-mortem benchmarking paves the way for in-situ studies on the same small size objects.

Acknowledgments

The authors gratefully acknowledge Dario Ferreira Sanchez and Carsten Richter respectively from the MicroXAS (SLS, Switzerland) and ID01 (ESRF, France) beamlines for their support during experiments. Guillaume Amiard is also gratefully acknowledged for the fabrication of the pillars. This project has received funding from the European Union's Horizon 2020 research and innovation program under the Marie Skłodowska-Curie grant agreement No 701647, as well as SNSF grant No 200021L_169753, ANR grant No ANR-16-CE93-0006, and by "Investissement d'Avenir" (LABEX INTERACTIFS, ANR-11-LABX-0017-01) and by Nouvelle Aquitaine Region / European Structural and Investment Funds (ERDF No P-2016-BAFE-94/95).

References

[1] L. Thilly, R. Ghisleni, C. Swistak, J. Michler, In situ deformation of micro-objects as a tool to uncover the micro-mechanisms of the brittle-to-ductile transition in semiconductors: the case of indium antimonide, Philos. Mag. 92 (2012), 37-41.
https://doi.org/10.1080/14786435.2012.704422

[2] J.M. Wheeler, L. Thilly, A. Morel, A.A. Taylor, A. Montagne, R. Ghisleni, J. Michler, The Plasticity of Indium Antimonide: Insights from Variable Temperature, Strain Rate Jump Micro-Compression Testing, Acta Materialia, 106 (2016), 283.
https://doi.org/10.1016/j.actamat.2015.12.036

[3] V.L.R. Jacques, D. Carbone, R. Ghisleni, L. Thilly, Counting dislocations in microcrystals by coherent X-Ray diffraction, Phys. Rev. Lett. 111 (2013), 1-5.
https://doi.org/10.1103/PhysRevLett.111.065503

[4] H. Van Swygenhoven, S. Van Petegem, The use of Laue Microdiffraction to study small-scale plasticity, JOM, 62(12), (2010), 36-43. https://doi.org/10.1007/s11837-010-0178-4

[5] X. Chen, C. Dejoie, T. Jiang, C.-S. Ku, N. Tamura, Quantitative microstructural imaging by scanning Laue x-ray micro- and nanodiffraction, MRS Bull., 41(6), (2016), 445-453.
https://doi.org/10.1557/mrs.2016.97

[6] R. Spolenak, W. Ludwig, J. Y. Buffiere, J. Michler, In situ Elastic Strain Measurements-Diffraction and Spectroscopy, MRS Bull., 35 (5), (2010), 368-374.
https://doi.org/10.1557/mrs2010.569

[7] A. Bhowmik, J. Lee, T.B. Britton, W. Liu, T.-S. Jun, G. Sernicola, M. Karimpour, D.S. Balint, F. Giuliani, Deformation behaviour of [001] oriented MgO using combined in-situ nano-indentation and micro-Laue diffraction, Acta Mater. 145 (2018), 516-531.
https://doi.org/10.1016/j.actamat.2017.12.002

Residual Stresses 2018 – ECRS-10　　　　　　　　　Materials Research Forum LLC
Materials Research Proceedings 6 (2018) 27-32　　　doi: http://dx.doi.org/10.21741/9781945291890-5

In-Situ Analysis of Material Modifications During Deep Rolling by Synchrotron X-Ray Diffraction Experiments

Heiner Meyer[1,a]* and Jérémy Epp[1,2,b]

[1]Leibniz Institute for Materials Engineering - IWT, Division Materials Science, Badgasteiner Straße 3, 28359 Bremen, Germany

[2]MAPEX Center for Materials and Processes, University of Bremen, Bibliothekstraße 1, 28359 Bremen, Germany

[a]hmeyer@iwt-bremen.de, [b]epp@iwt-bremen.de

Keywords: Deep Rolling, Synchrotron XRD, In-Situ X-Ray Diffraction, Cold Working, Process Signatures

Abstract. The stress state achieved through elasto-plastic deformation in the deep rolling process has been analyzed with in-situ synchrotron experiment using a roller as tool. For the determination of internal material load and residual stress for specific contact parameters, the material state was analyzed via transmission synchrotron X-ray diffraction for hundreds of measurement positions below the loaded zone. From the individual diffraction rings at each point, the strain and stress fields were reconstructed as 2D mappings of the material modifications. The measured region included material in front of the roller, under the contact point as well as in the already processed near surface region. A 13 mm cylindrical tungsten carbide roller was applied on specimens of AISI 4140 steel in ferritic-pearlitic and quenched and tempered condition to compare the effect of different mechanical properties on the propagation and intensity of the stress and strain fields in the specimen during and after processing. The investigations allowed a comparison of the stress fields for the different material conditions. The results show that the mechanisms of load distribution during the process and the effect of loading stresses on the final material state and on the residual stress distribution are strongly influenced by the initial material state.

Introduction

Residual stresses introduced into the surface through mechanically induced plastic deformation are important aspects of the process effects in deep rolling and roller burnishing. Analysis of surface residual stresses and of depth profiles generated by deep rolling is possible ex-situ after the process through standard methods such as X-Ray Diffraction (XRD), hole drilling techniques and other methods. It is well known that compressive residual stresses may be introduced several hundred micrometers into the material, depending on the process parameters, as shown by Meyer and Kaemmler [1]. These are connected with improved near surface hardness and an increased fatigue resistance for processed parts, as found by Nikitin and Altenberger [2]. The material response to the applied loading stresses and plastic deformation depends on the initial material parameters like yield strength, strain hardening behavior and microstructural stability, but while these parameters may be included in FEM simulations, as done by Sayahi et al. [3], the resulting loading stress and residual stress fields can show deviations from experimentally measured values, or may even be only accessible in-situ during the process. The connection between loading stresses and the resulting residual stresses is however most important to understand the process and its effect on different material states. This approach is used in the determination of process signatures to develop a mechanism-based prediction of the material modifications in metal processing, as proposed by Brinksmeier et al. [4]. The information about loading stresses and the resulting residual stresses from mechanical elasto-plastic deformation can give insight into the material behavior and the modification mechanisms. To actually analyze the stress fields in-situ during processing, diffraction measurements using synchrotron radiation with sufficient penetration depth are an efficient method, as shown by Uhlmann et al. [5] for the analysis of an orthogonal cutting process.

In the present study, in-situ synchrotron XRD experiments were performed during deep rolling of steel grade AISI 4140 (42CrMo4) in two different heat treatment conditions (ferrite-pearlite and quenched + tempered). The aim was to analyze the material behavior under contact during applied pressure and movement to reconstruct 2D maps of stress state and peak width change from the measured diffraction signal. Based on these results, the material reaction during the process was evaluated in terms of cold working and generated residual stresses introduced for the two initial microstructures.

Experimental methods

For the experiments performed on beamline ID11 of the European Synchrotron Radiation Facility (ESRF), a transportable frame was used with a modified deep rolling tool, based on the system of Ecoroll AG, Celle, Germany. The tool head consist of a 13 mm diameter tungsten-carbide cylinder of 15 mm width. Outfitted with a linear drive, specimens of dimension 70 ×20 × 2.8 mm (L × H × W) were processed at a feed rate of v=0.02 mm/s, which can be considered as quasi-static process with low strain rate. Prior to the experiments, the surface was electropolished to a width of 2.8 mm, to reduce influence from heat treatment and preprocessing steps on the measured material state. The cylindrical roller produces a consistent deep rolled state over the whole width of the surface. The force was applied via a hydraulic system at 350 bar which corresponds to a contact force of the tool of 3400 (± 100) N. The setup, as installed in the beamline, is seen in Fig. 1a, while the transmission geometry measurement layout with FReLoN detector position is shown in the sketch in Fig.1b.

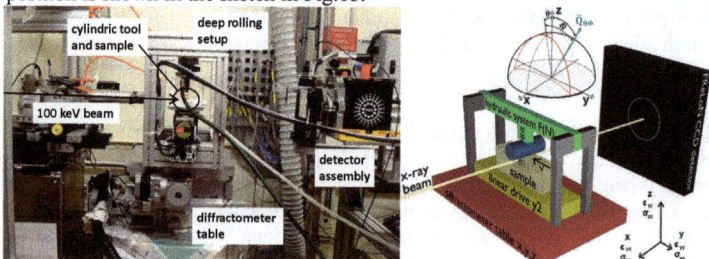

Figure 1. (a) Deep rolling setup installed in beamline ID11 EH3 (b) sketch of system setup and measurement geometry

With the fast-readout camera, full diffraction rings for distributed points inside the measurement window were collected with a 0.05 × 0.05 mm beam size at 100 keV photon energy, with increasing spatial resolution to the contact point, allowing a 2D reconstruction of the material state below the processed surface. The samples are assumed to be isotropic and homogeneous and therefore the samples were moved below the tool during the process, while the scanning was defined in relation to the contact position of the tool. A total of 960 points were measured in a region of 7.8 mm in width and 3.65 mm in depth. The measurement method is comparable to the analysis of static indentation with the same experimental setup in [6].

For the specimens, steel AISI 4140 (42CrMo4) was prepared in two heat treatment states, with a quenched and tempered (QT) state with 506 HV1 hardness, a yield strength of σ_y=1440

Residual Stresses 2018 – ECRS-10   Materials Research Forum LLC
Materials Research Proceedings **6** (2018) 27-32   doi: http://dx.doi.org/10.21741/9781945291890-5

MPa and an ultimate tensile strength of σ_{UTS}=1570 MPa, and a comparatively soft ferritic-pearlitic (FP) state with 188 HV1 with σ_y=420 MPa and σ_{UTS}=730 MPa.

Experimental methods
Using the $\alpha\{211\}$ reflex of the material state, strain value was calculated from the diffraction rings for each of the 960 points in the measurement window, using the fundamental equation of strain as given by Noyan and Cohen [7]. It was determined that only the ε_{yy}, ε_{zz} and ε_{yz} strain components in the respective coordinate system have to be analyzed (Eq.1) in a plane strain approach, since the boundary conditions of the specimen only introduce negligible deviations in this mode, as found by Uhlmann et al [5] for a comparable orthogonal cutting process.

$$\varepsilon_{\theta\delta}(y,z)= \cos^2\theta\sin^2\phi\,\varepsilon_{yy}(y,z)+\cos^2\theta\cos^2\phi\,\varepsilon_{zz}(y,z)+\cos\theta\sin 2\phi\,\varepsilon_{yz}(y,z) \qquad (1)$$

A total number of 72 azimuthal sections were integrated from the diffraction rings with each 5° range, after data treatment with the pyFAI program by Ashiotis et al [8]. These were analyzed for strain and recalculated to stress values on the assumption of plane strain state and using X-Ray elastic constants $\frac{1}{2}\,S_2 = 5.81\ 10^{-5}$ MPa^{-1} for steel specimens [7]. Full-width at half-maximum (FWHM) was determined as the mean value from pseudo-Voigt peak fits.

The error of stress determination in this method ranges from 100 MPa in the ferritic-pearlitic material to around 50 MPa for each stress component in the tempered martensitic structure. The cause for the difference lies in the much larger grains (crystallites) in the ferritic-pearlitic state, which have a detrimental influence on the grain statistic and therefore on the diffraction signal quality.

As a comparable stress value, von Mises equivalent stress σ_{eq} was used to determine the amount and distribution of equivalent loading and residual stresses in the material with respect to its material state, as introduced by Kuznetsov et al. [9] for burnishing processes.

Equivalent stress and FWHM distribution in the processed material
The transmission geometry allows for analysis of the material state in a gauge volume comparable to the width of the specimen and the beam cross section. The equivalent stresses in the ferritic-pearlitic and QT specimens are given in Figure 2. The 2D reconstruction of the equivalent stress fields show that the FP state has a wide distribution of the maximum values from the region directly in contact with the roller to the superposition of loading stress and residual stress in the outgoing material. Calculated equivalent stress values are at a maximum of 800 MPa in depth after plastification of the material, which is a value higher than the determined yield strength of the material measured ex-situ. Since the initial material state would allow further deformation at this value, it can be reasoned that work hardening through increased dislocation density allows for an increase of stress state up to the new increased yield strength of the deformed material, as noted by Altenberger [10]. The increased overall equivalent stress in the material at y=3.3 mm can be attributed to residual stresses in the measured components up to a depth of z=-2 mm, which reach around $\sigma_{eq} = 400$ MPa ($\sigma_{yy} = -70$ MPa). However, a slight superimposition with the loading stress field is still present at this distance.

For the QT specimen, the higher yield strength leads to a concentration of the loading stress to the region directly under the contact point, with a maximum of 1400 MPa in depth, producing a single drop-shaped influence zone of loading stresses in the material. The measured equivalent stress value is not higher than the yield strength in this case, since work-hardening in this material state cannot strongly shift the elastic limit to higher values. This will also be shown in the following section on FWHM behavior. Residual stress values are comparable to the ferritic-pearlitic measurement at around $\sigma_{eq} = 400$ MPa ($\sigma_{yy} = -170$ MPa) in equivalent stress, but at a depth of 0.8 mm more constrained to the surface near region.

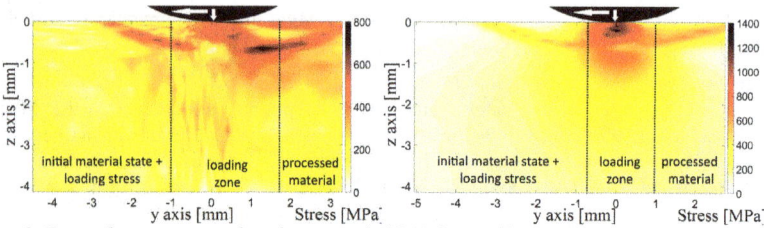

Figure 2. Equivalent stress σ_{eq} distribution at 3400 N deep rolling contact in (a) ferritic-pearlitic and (b) QT material state

Aside from stress evaluation, the full-width at half maximum (FWHM) of the $\alpha\{211\}$ peaks can be used to evaluate the cold-working introduced into the material, giving a qualitative value for the change in crystallite size, microstrains and dislocation density in the specimen introduced by the deep rolling process. To compare the different initial structures, the change of peak width (ΔFWHM) compared to the original state was calculated. Corresponding to the effect measured in the stress fields, the difference between the two material states can be clearly found in the higher affected depth of the plastification in the more ductile ferritic-pearlitic state, where an increase of ΔFWHM can be detected until 2 mm below the surface (Fig. 3a), which corresponds to the residual stress region in Fig. 2a. This shows that these two modifications are directly connected, as already found in ex-situ measurements by Meyer et al. [1]. A maximum increase of FWHM is detected behind the roller, which then decreases again. FWHM increase caused by plastic deformation should remain constant after unloading. Therefore, the reversible increase in peak width at this position is expected to be connected to a superposition of inhomogeneous 1st kind stresses within the gauge volume and possibly inhomogeneous 2nd kind stresses depending on the microstructure, with microstrains (3rd kind stresses) and crystallite size effects. In the case of FP specimen, only small amount of FWHM increase is reversible, while in QT state most local increase is released after unloading (Fig. 3b). Similar behavior depending on the microstructure were observed in in-situ neutron tensile tests by Tomota et al. and attributed to inhomogeneous 2nd kind stresses between different phases [11]. The temporary change in FWHM from the contact is comparable for both material states at a value of around $1.5 \cdot 10^{-2}$ °.

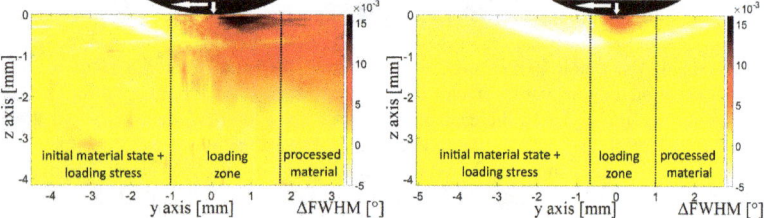

Figure 3. ΔFWHM distribution at 3400 N deep rolling contact in (a) ferritic-pearlitic and (b) QT material state

For the QT state (Fig. 3b), a zone with FWHM decrease compared to the initial state (bright region) starting ahead of the tool can be observed. In the treated region behind the tool, this zone is still present. This behavior is already well-known for initial material state with high hardness and high lattice defect density and is due to dislocation reorganization caused by the introduced mechanical energy, causing dislocation walls, tangles and cells in low energy microstructures, as found by Hoffmann et al [12], which leads to reduced measured FWHM values for deep rolled

Residual Stresses 2018 – ECRS-10   Materials Research Forum LLC
Materials Research Proceedings 6 (2018) 27-32   doi: http://dx.doi.org/10.21741/9781945291890-5

QT specimens, as found from ex-situ experiments of this material state and the study of Perenda et al. [13] on hardened deep rolled parts.

Integrated FWHM and stress progression

Since the effect of deep rolling is restricted to the surface-near region analyzed in this study, the overall effect of stress and FWHM change can be evaluated during the contact sequence if the values are integrated over depth as a function of the y position (Figure 4). For the ferritic-pearlitic material, the effect of loading stress can be first detected around y=-3 mm in front of the roller and is increasing continuously from $\sigma_{eq} = 0.6 \cdot 10^3$ MPa·mm, before contact, derived from the residual stress measurement error, to $\sigma_{eq} = 1.2 \cdot 10^3$ MPa·mm at y = 0 mm directly under the contact point. The corresponding increase in peak width from $\Delta FWHM = 0$ °·mm before contact to a maximum value of $\Delta FWHM = 1.5 \cdot 10^{-3}$ °·mm at y = 0 mm starts 1.2 mm later than the beginning of the loading stress increase but both reach maximum values at the same position. Both components than show more or less stable values in the region of remaining material modifications at y > 2.5 mm, as shown in Fig. 4a. In contrast to this, the QT in Fig. 4b state shows progressive increase in equivalent stresses while the peak width first decreases. This corresponds to the zone of reorganization of dislocation structure due to mechanical energy as described in the previous section. At the position directly below the tool, strong increase of $\Delta FWHM$ take place which can be attributed to superimposition of inhomogeneous 1st and 2nd kind reversible stresses from loading maximum at the same position. After progressive decrease of loading stress influence at y > 1 mm, the integrated FWHM returns back to value close to 0, which confirms the already described phenomena. The equivalent stress maximum below the contact point is followed by a decreasing effect of load stress and the generated residual stress state is visible in the asymmetric decrease of the curve. Since stabilization of the evolution is not reached at y=3 mm, a slight influence of the load stress is still remaining at this distance.

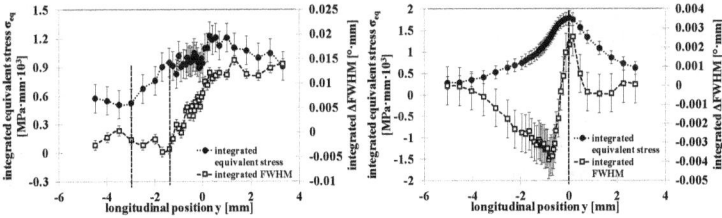

Figure 4. Integrated $\Delta FWHM$ and σ_{eq} over depth in processing direction for (a) ferritic-pearlitic and (b) QT material state

Overall integrated stress state in QT material is higher at around 1.9 MPa·mm compared to 1.2 MPa·mm in the softer FP state, which results from the different contact areas and achievable stress state limited by the local yield strength in each case.

Conclusion

Synchrotron experiments have been performed to analyze the material state of a specimen under deep rolling contact during the process. The local distribution of stresses and peak width could be analyzed during the process, allowing evaluation of loading stresses, generated residual stresses and their superposition in the material. The distribution of stresses is linked to the material properties, where amount and depth of residual stress depends on the yield strength of material. Changes in the width of the analyzed diffraction peaks depend on the initial material state and allowed the evaluation of plastically deformed layer. Since both values show corresponding reactions of the material to the deep rolling contact, an extrapolation of the overall material modifications through the tool contact may be possible depending on material properties and contact parameters. With these results, process signature for deep rolling will be further developed.

Acknowledgement
The funding by the German Research Foundation (Deutsche Forschungsgemeinschaft, DFG) for financial support of subproject C01 within the CRC TRR 136 "Process Signatures" is gratefully acknowledged.

References

[1] Meyer, D., Kämmler, J. (2016). Surface integrity of AISI 4140 after deep rolling with varied external and internal loads. Procedia CIRP, 45, 363-366. https://doi.org/10.1016/j.procir.2016.02.356

[2] Nikitin, I., & Altenberger, I. (2007). Comparison of the fatigue behavior and residual stress stability of laser-shock peened and deep rolled austenitic stainless steel AISI 304 in the temperature range 25–600 C. Materials Science and Engineering: A, 465(1-2), 176-182. https://doi.org/10.1016/j.msea.2007.02.004

[3] Sayahi, M., Sghaier, S., Belhadjsalah, H. (2013). Finite element analysis of ball burnishing process: comparisons between numerical results and experiments. International Journal of Advanced Manufacturing Technology, 67(5-8), 1665-1673. https://doi.org/10.1007/s00170-012-4599-9

[4] Brinksmeier, E., Klocke, F., Lucca, D. A., Sölter, J., Meyer, D. (2014). Process Signatures–a new approach to solve the inverse surface integrity problem in machining processes. Procedia CIRP, 13, 429-434. https://doi.org/10.1016/j.procir.2014.04.073

[5] Uhlmann, E., Henze, S., Gerstenberger, R., Brömmelhoff, K., Reimers, W., Fischer, T., Schell, N. (2013). An extended shear angle model derived from in situ strain measurements during orthogonal cutting. Production engineering, 7(4), 401-408. https://doi.org/10.1007/s11740-013-0471-5

[6] Meyer, H., Epp, J., Zoch, H.W., (2017). In situ X-Ray Diffraction Investigation of Surface Modifications in a Deep Rolling Process under Static Condition. Materials Research Proceedings, 2, 431-436. https://doi.org/10.21741/9781945291173-73

[7] Noyan, I. C., Cohen, J. B. (2013). Residual stress: measurement by diffraction and interpretation. Springer.

[8] Ashiotis, G., Deschildre, A., Nawaz, Z., Wright, J. P., Karkoulis, D., Picca, F. E., Kieffer, J. (2015). The fast azimuthal integration Python library: pyFAI. Journal of applied crystallography, 48(2), 510-519. https://doi.org/10.1107/S1600576715004306

[9] Kuznetsov, V. P., Smolin, I. Y., Dmitriev, A. I., Konovalov, D. A., Makarov, A. V., Kiryakov, A. E., Yurovskikh, A. S. (2013). Finite element simulation of nanostructuring burnishing. Physical Mesomechanics, 16(1), 62-72. https://doi.org/10.1134/S1029959913010074

[10] Altenberger, I. (2005, September). Deep rolling - the past, the present and the future. Proceedings of 9th international conference on shot peening, 6-9.

[11] Tomota, Y., Lukáš, P., Neov, D., Harjo, S., & Abe, Y. R. (2003). In situ neutron diffraction during tensile deformation of a ferrite-cementite steel. Acta materialia, 51(3), 805-817. https://doi.org/10.1016/S1359-6454(02)00472-X

[12] Hoffmann, B., Vöhringer, O., Macherauch, E. (2001). Effect of compressive plastic deformation on mean lattice strains, dislocation densities and flow stresses of martensitically hardened steels. Materials Science and Engineering: A, 319, 299-303. https://doi.org/10.1016/S0921-5093(01)00978-9

[13] Perenda, J., Trajkovski, J., Žerovnik, A., Prebil, I. (2015). Residual stresses after deep rolling of a torsion bar made from high strength steel. Journal of Materials Processing Technology, 218, 89-98. https://doi.org/10.1016/j.jmatprotec.2014.11.042

Residual Stresses 2018 – ECRS-10   Materials Research Forum LLC
Materials Research Proceedings 6 (2018) 33-38   doi: http://dx.doi.org/10.21741/9781945291890-6

Residual Stresses Induced in T Butt Welds from Submerged Arc Welding of High Strength Thick Section Steel Members

G.W. Sloan[1, a, *], S. Pearce[1, b], V.M. Linton[2, c], X. Ficquet[3, d], E. Kingston[3, e], I. Brown[1, f]

[1]School of Mechanical Engineering, University of Adelaide, North Terrace, Adelaide, Australia

[2]Executive Dean Faculty of Engineering and Sciences, University of Wollongong, New South Wales, Australia

[3]Veqter Ltd, Unicorn Business Park, Whitby Road, Bristol, BS4-4EX, UK

[a]glen.sloan@bigpond.com.au, [b]susan.pearce@adelaide.edu.au, [c]valerie.linton@uow.edu.au, [d]xavier.ficquet@veqter.co.uk, [e]ed.kingston@veqter.co.uk, [f]ian.bee13@gmail.com

Keywords: Residual Stress, High Strength Steel, Quenched and Tempered, Neutron Diffraction, Deep Hole Drilling, T Butt Welds

Abstract. The overall residual stress profiles of a weld fabricated structures such as a submarine pressure hull fabricated using high strength welds joining quench and tempered steels is complex as they are the net profiles from the build-up or relaxation of the residual stresses induced into the welds and structural members from each step in the fabrication process. The University of Adelaide with its research partners has attempted to increase the knowledge of residual stresses in this type of structure by undertaking a comprehensive residual stress measurement program using test pieces representing each stage of the fabrication of a submarine pressure hull including full scale test pieces. The primary methods of measuring subsurface residual stress profiles was the neutron diffraction method with the Deep Hole Drilling technique (DHD) to validate key results. The X ray diffraction method and Incremental Centre Hole Drilling methods were used for determining surface residual stresses. This paper presents an overview of the results for residual stress profiles resulting from a full penetration double sided T Butt weld joining a T Frame to a thick flat plate by Submerged Arc Welding. The test piece used a specific pass placement strategy that placed the last pass each side on the base plate. The results presented show that the actual profile is considerably different than those for established fracture mechanics residual stress profiles in that these are over conservative. The discussion will include comparison of results to other measurement results undertaken on similar test pieces but undertaken with an alternate pass placement strategy.

Introduction

Ring stiffened cylindrical structures for the offshore industry and naval platforms are usually fabricated from high strength steels with butt welds and T butt welds. The fabrication method involves manufacturing plate cans from thick plates with the longitudinal direction of initial rolling of the plate coinciding with that for shape rolling into a curve. Depending on the diameter 3 or more plates will be needed to be welded together to form a cylindrical can. The ring frames are also fabricated from high strength steel and are welded into the can via double sided T butt welds undertaken in the 1G down hand orientation. The same also applies for rigid bulkheads in the structures to resist buckling of the structures when subjected to sea pressure. These types of structures operate in environments where Stress Corrosion Cracking (SCC) is a frequent problem for crack initiation in service as is Hydrogen Assisted Cold Cracking (HACC) during manufacture. In the case of naval platforms they are also subjected to a high stress low cycle fatigue regime throughout their service life. Therefore an understanding of residual stress profiles is essential for determining service life and structural integrity safety regimes for the welds and the structural components. While standards such as BS 7910 [1] present residual stress profiles for T butt welds they are limited to the weld toe region only and they assume that the last weld pass is on the base plate with the residual stress level being at tensile yield. No profiles are given for the region under the weld or in the base plates away from the welds and how much the existing profiles in the components are altered by adjacent welding. Hence there is a huge gap in knowledge of residual stress profiles throughout a fabricated ring stiffened cylindrical structure. In 2004 the University of Adelaide and its research partners, ISIS Rutherford Appleton Laboratories, Australian Nuclear Science and Testing Organization, ASC Pty Ltd, and the Commonwealth of Australia, implemented a program to address these gaps [2], [3]. The program was structured to look at the progressive build up or superimposition of residual stress profiles throughout the fabrication process for these types of structures and addressed both current and potential future weld pass placement strategies for the T butt welds. The areas of primary interest were the base plate both beneath the weld and away from the weld region and also the weld. The material used throughout the entire program was BIS 812 EMA quenched and tempered plate with a minimum yield strength of 690 MPa [2] and a Poisson's ratio of 0.3. This yield value was used to normalise all the results. The chemical composition is given in ref [2]. The Young's modulus for the material varies between 208 to 213 GPa. A value of 210 GPa was used for the calculations of strain and stress. The test piece the subject of this paper was that for a weld of a heavy section frame to a flat plate with the welding being undertaken by the Submerged Arc Welding process (SAW).

Experimental Methodology
The flat T Butt test plate was designed to be representative of the welds joining stiffening frames to submarine flat plate rigid bulkheads and also of the pressure hull ring frames to curved shell plate welds but eliminating in this later case the variable of the pre-existing residual stresses from the curved shell plate prior to welding of the frame to shell which is in line with the logic presented in the previous section [2], [3]. These were normally undertaken using the SAW welding process in an automatic welding set up welding side 1 completely, back-gouging the root and then welding side 2 fully [2] to achieve a full penetration T butt weld. All welding was undertaken in the 1 G orientation. The weld cap is a tapered weld, approximately 5^0 to the base plate with the last pass on the base plate. This is a narrower taper than used in BS7910 [1]. The sample was 1 metre long. The brackets indicated in Fig 1 were welded to the base plate and the T Frame before the test weld of the T Frame to the base plate was undertaken to prevent distortion / rotation of the base plate relative to the T Frame and therefore eliminate a variable. All residual stress measurements were undertaken as close as possible to the mid length of the test piece. The welding of the web to base plate was undertaken using a maximum heat input of 1.9 kJ/mm [2], [3] using Mil 120S-1 wire with OP121 TT flux. As SAW is an automatic process the heat input was constant thereby eliminating a variable. Reducing variables is a key approach factor to understand results.

The primary method selected for determining residual stresses in the subsurface regions of the fabricated test piece was the Neutron Diffraction (ND) method [3]. This was chosen as it is totally non-destructive and therefore allowed the subsequent options to use other destructive methods to verify/ validate the results or could be repeated if necessary. Five scan lines were undertaken as detailed in Fig 2 as detailed in [2]. The scan lines concentrated on obtaining strain and stress data across the plate and excluded the zone immediately under the weld due to the large beam path lengths for the longitudinal strain direction, which increase the risk of no effective measurements due to beam scatter, even with the sample tilted to 30^0 which was done.

Residual Stresses 2018 – ECRS-10   Materials Research Forum LLC
Materials Research Proceedings 6 (2018) 33-38   doi: http://dx.doi.org/10.21741/9781945291890-6

The gauge volumes for the neutron diffraction were 4x4x20 mm for the transverse and normal directions and 4x4x4 mm for the longitudinal direction. The test piece was laser scanned and the resultant 3 D model used to generate the sample plan in the computer program SScanSS developed by the Open University [4]. This sample plan was then used to drive the scanning table at Engine X at the Rutherford Appleton Laboratories UK. It has been optimized to reflect available beam time as opposed to full allocation.

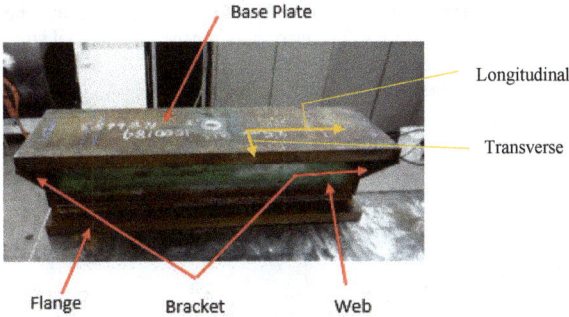

Fig. 1 Showing the design of the Flat T butt test plate and the principal elements and residual stress directions. The longitudinal rolling directions for the plate components is also that for welding.

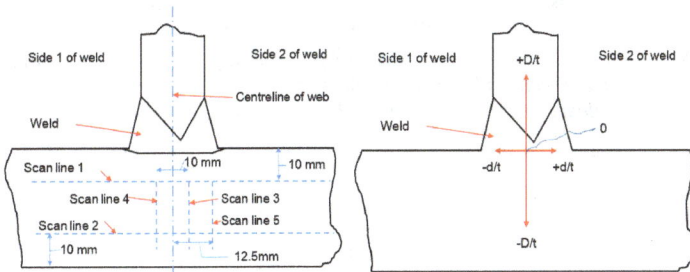

Fig. 2 Showing details of the scanlines for the neutron diffraction testing on the left and the normalized datum's used for result locations on the right. D is the normalized depth through thickness i.e. Depth /base plate thickness t and d is the normalized distance from centerline of web d/ base plate thickness.

In order to validate the neutron diffraction work one scan line was tested by the iDHD/DHD technique. The iDHD/DHD technique is described in [5]. The location chosen was in line with Scan Line 5 of Pearce [2] as this is closest to the weld toe side 2 and therefore could be reviewed against BS 7910 profiles and other publications. As noted in Sloan [3] the gauge volumes for the ND and DHD techniques are slightly different but in this case the DHD measurements were directly centered on Scan line 5 and not offset as occurred in [3] therefore permitting easier comparison of results noting difference in gauge volumes.

Residual Stresses 2018 – ECRS-10   Materials Research Forum LLC
Materials Research Proceedings **6** (2018) 33-38   doi: http://dx.doi.org/10.21741/9781945291890-6

Results

The results presented in this paper will use the datum configuration as presented in Fig 2. extended up into the weld above the top of the base plate and is in line with those used in [3]. The results of ND scans for scan lines 1 and 2 Pearce [2] are shown superimposed in Fig 3. As can be seen in Fig 3 Scan line 1 closest to the web side was undertaken fully across the area of interest of the base plate whereas for Scan line 2 the measurements started at the through thickness scan line 3 and extended out onto side 2 of the weld. The longitudinal residual stresses are still tensile at 10 mm depth but revert to compressive at 25 mm depth. The opposite has been found for the transverse stress profiles. The error bars for the longitudinal stress scan line 1 under the weld reflect the issue of excessive beam path length.

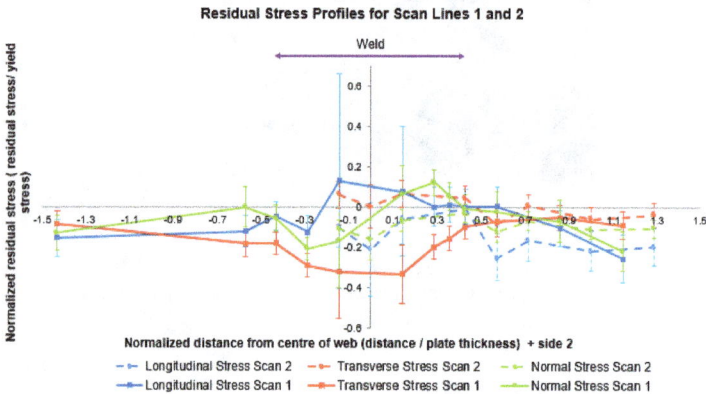

Fig. 3 Shows the plots of residual stress across the base plate for scan lines 1 and 2, see Fig 2, and Fig 1 for the principal directions in a normalized stress and distance d format. The large error bars for longitudinal stress under the weld reflect the excessive beam path lengths.

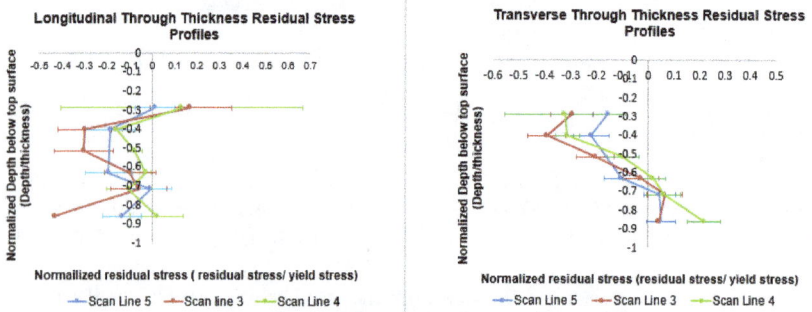

Fig. 4 Details the results for the longitudinal and transverse residual stress profiles for the scan lines identified in Fig 2 in a normalized stress and Depth D format.

The residual stress profiles for the normal direction are not shown in Fig 4 as they cannot be compared with results from the DHD/iDHD measurements undertaken on the test piece with the assistance of Veqter Ltd as it is effectively a 2 dimensional method. They can however be found in Pearce [2]. The results of the DHD/iDHD measurements are shown in Fig 5 below. You will observe that the results for transverse residual stress stop at the intersection with the web

Residual Stresses 2018 – ECRS-10   Materials Research Forum LLC
Materials Research Proceedings 6 (2018) 33-38   doi: http://dx.doi.org/10.21741/9781945291890-6

interface. Above this point the stresses determined by the DHD / iDHD technique become the longitudinal and the normal direction relative to the welding as described in Sloan [3]. The iDHD results closely match the DHD results and so there has not been significant impact of plasticity at or near the weld. The results are totally different from the profiles which are generated from [1], i.e. no yield level stresses at the weld toe.

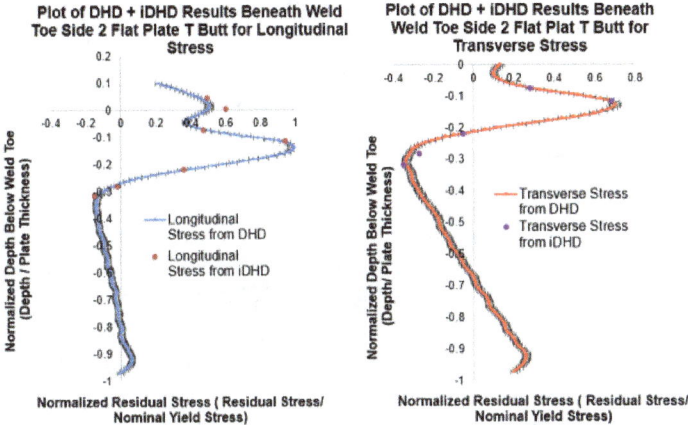

Fig. 5 Detailing the longitudinal and transverse residual stress profiles for the DHD and iDHD measurements of residual stress. The errors by this method are small. The normal stress direction in the base plate is not possible for this method as it is effectively a 2D method. The results are significantly different from the profiles generated from [1].

Discussion

The results presented above clearly highlight that the residual stress profiles detailed in [1] are conservative. The residual stress levels immediately under the weld toe are not tensile yield as per [1] and they change to compressive levels very quickly i.e. between 0.2 and 0.3 D/t which is also different than the predictive profiles of [1]. This was also the case for the results in [3] with a different weld pass placement strategy i.e. last pass on the web member.

In Fig 6 the results for the ND measurements [2] are superimposed on the DHD / iDHD results figure 5. These show very close correlation between results from the two different methods noting the slightly different gauge volumes, however there is that zone beneath the main portion of the T butt weld where there is no ND data which needs to be addressed by other alternative methods such as the Contouring method or Ultrasonic method.

Conclusions

The information presented above shows that the residual stress profiles detailed in [1] are very conservative and may lead to unnecessary weld repairs being implemented which may themselves effect integrity but will certainly effect cost, schedule and life of structure. Extensive further work needs to be undertaken in this field to highlight the extent of conservatism and provide a structured way forward particularly the zone immediately under the T butt weld. Critical methods to explore are those that can be undertaken on full scale structures, either in service or large scale test pieces, which do not involve cutting and therefore relaxing of residual stresses, and therefore the research trend should be towards methods such as ultrasonics provided that they are validated against known results.

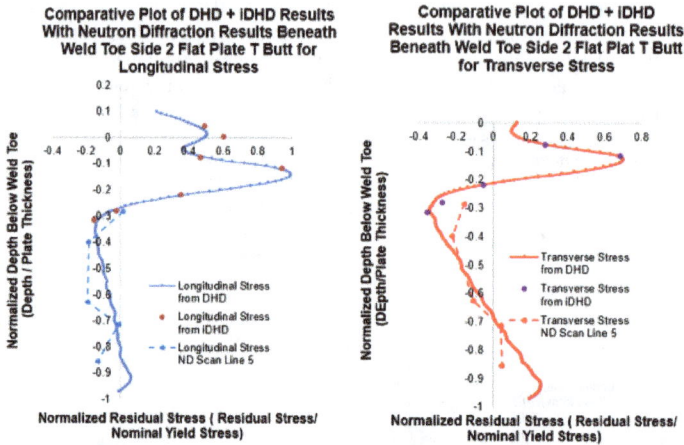

Comparative Plot of DHD + iDHD Results With Neutron Diffraction Results Beneath Weld Toe Side 2 Flat Plate T Butt for Longitudinal Stress

Comparative Plot of DHD + iDHD Results With Neutron Diffraction Results Beneath Weld Toe Side 2 Flat Plat T Butt for Transverse Stress

Fig. 6 Shows an overlay of the ND results as shown in Fig 4. for the longitudinal and transverse directions with those achieved via the DHD/ iDHD method for Scan Line 5.

Acknowledgments

The author wishes to acknowledge the support from the University of Adelaide, ASC Pty Ltd, and the Commonwealth of Australia, and Veqter Ltd. Most importantly the author wishes to acknowledge his parents whose estate funded the DHD work.

References

[1] BS 7910 "Guidance on methods of assessing the acceptability of flaws in metallic structures", 1999 with amendments 2000, 2009

[2] S. Pearce "Breakdown of residual stresses in highly restrained thick section steel welds", Doctoral thesis, University of Adelaide 2009.

[3] G Sloan 'The influence of welding variables on the residual stress profiles around T butt and MMAW butt welds in thick section high strength steel", Master's thesis, University of Adelaide 2015.

[4] J. A. James and L. Edwards "Application of robotic kinematics methods to the simulation and control of neutron beam line positioning systems", Nuclear Instruments and Methods in Physics Research A, 2007. https://doi.org/10.1016/j.nima.2006.11.033

[5] Information on http://www.vqter.co.uk

Residual Stresses 2018 – ECRS-10   Materials Research Forum LLC
Materials Research Proceedings **6** (2018) 39-44   doi: http://dx.doi.org/10.21741/9781945291890-7

In-Situ Synchrotron X-Ray Diffraction Studies in the Chip Formation Zone During Orthogonal Metal Cutting

Jens Gibmeier[1,a] *, Dominik Kiefer[1,b], Rafael Hofsaess[1] and Norbert Schell[2,c]

[1] Karlsruhe Institute of Technology (KIT), Institute for Applied Materials (IAM-WK), Engelbert-Arnold-Str. 4, 76131 Karlsruhe, Germany

[2] Institute of Materials Research, Helmholtz-Zentrum Geesthacht (HZG), Max-Planck-Str. 1, 21502 Geesthacht, Germany

[a]Jens.Gibmeier@kit.edu, [b]Dominik.Kiefer@kit.edu, [c]Norbert.Schell@hzg.de

Keywords: Dry Metal Cutting, Turning, In-Situ X-Ray Diffraction, Synchrotron Radiation

Abstract. In the field of metal cutting a very important parameter is the chip formation zone, where the workpiece, the cutting tool and the chip are in contact. The transient processes inside this zone decisively define the properties of the workpieces near surface zone, e.g. residual stress distribution or the local microstructure. With the development of a special designed turning device we established a methodology to gain insight into the chip formation zone of a workpiece during the turning process with high spatial and temporal resolution. Using the dedicated device we were able to access the strain evolution during a cutting process of aluminum alloy AW-5754 with a measuring rate of 10 Hz inside the chip formation zone.

Introduction

Surface integrity is an important factor for service behaviour of machined workpieces, which is influenced by the chosen cutting parameters and by the tool wear. In this regard the residual stresses, surface quality and the local microstructure are essential parameters for the assessment. For low carbon steels and aluminium alloys built-up edge (BUE) formation is often observed on the rake face of the cutting tool, in particular in the low cutting speed regime and for example in micro machining. Due to the BUE the effect of cutting is displaced from the initial cutting edge towards the cutting edge of the BUE. For the cutting tool the adhering stagnant layers are very important for the wear behaviour. We recently found that the BUE can act as a protective layer and can improve the wear behaviour and thus the tool lifetime. Further, we found that because of detoriation of the surface roughness due to partial breaking of particles off the BUE, surface texturing of the cutting tool can be used to stabilise the BUE [1]. However, the fundamental relationships for BUE formation are not well-understood. In particular these transient processes hamper the prediction of chip forming operations by means of finite element simulations. A valuable step towards an improved process comprehension and for the sustainable improvement of simulation models for proper process prediction can be made if one succeeds in gaining real time insights in chip forming processing using technical relevant machining parameters. A vital tool in this regard is the application of synchrotron X-ray diffraction to monitor the microstructure and strain evolution in the chip formation zone with high temporal resolution. Having knowledge of these highly time-dependent evolutions allow for both, gaining knowledge about the mechanisms during machining in this zone, like local strain formation and texture development, which can further be used to optimize cutting processes, and process simulation through direct validation of simulations with real-time data from local synchrotron X-ray diffraction experiments. Up to know, there have not been a lot of investigations in the field of chip formation mechanism and the complex thermo-mechanical processes during cutting. First results in the field of in-situ cutting were presented by Uhlmann [2] and later by Brömmelhoff [3] analyzing the stress state, the local texture and the microstructure in the shear zone during orthogonal cutting. However, the temporal resolved data were recorded in the very low cutting speed regime, while the chosen process parameters are far from conventional used cutting parameters (e.g. cutting speed etc.)

Here, we report on the establishment of a methodology to get an insight into the chip-tool-workpiece contact zone with high spatial and temporal resolution for cutting conditions near to conventionally used cutting parameters for orthogonal cutting. Angular dispersive high energy synchrotron X-ray diffraction in transmission geometry was performed at beamline P07B@Petra III during turning of aluminium alloy Al-Mg3 (AW-5754) using a specially designed turning rig in the cutting speed regime between 36 - 200 mm/min, which are cutting conditions that are prone to BUE formation.

Experimental
Orthogonal cutting was realized on hollow cylindrical samples made of Al-Mg3 (AW-5754) with a wall thickness of 2 mm, see Fig. 1 for dimension, which were completely turned off using a turning device that was specifically designed for this purpose (see Fig. 2 left). The turning rig was adjusted in the synchrotron radiation beam at P07B@PETRAIII at the German Electron Synchrotron in Hamburg, Germany. The device was specifically built to investigate both, the built-up edges and the chip formation zone. Hence, the tool is mounted at a fixed position and the workpiece realizes the feed [4]. Using this set-up multiple experiments were performed with variations of the cutting speeds v_c. A full list of the turning geometries and parameters is given in Table 1 (experiments A-D). As cutting tool we used uncoated cemented carbide inserts with 6 wt.% Co. The hollow cylindrical sample is mounted on the turning lathe with the 50 mm shaft in a three-jaw chuck.

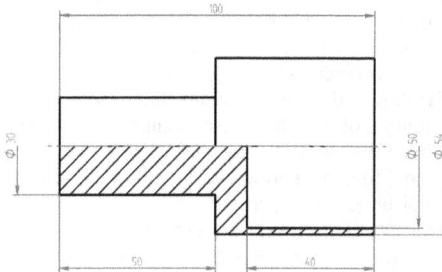

Fig. 1: Technical drawing of the hollow cylindrical sample with 2 mm wall thickness.

Table 1: Turning process angles and further cutting relevant parameters.	
wall thickness t [mm]	2
cutting speed v_c[m/min]	(D)36;(A)150; (B) (C)200
feed rate v_f [mm/min]	25
feed length l_f [mm]	25
clearance angle α [°]	90
wedge angle β [°]	8
rake angle γ [°]	-8
entering angle κ [°]	45
indexable insert type	SNMA120408
clamp mounting type	PSSNR2020K15

The device was adjusted in the primary synchrotron radiation beam in such a way that the chip formation zone in the workpiece is irradiated at an inclination angle of 45° to the feed axis, see also Fig. 2. At P07B@Petra III a fixed photon energy of 87.1 keV ($\lambda = 0.1425$ Å) is provided by a single bounce monochromator (SBM). The beam size was set to 0.2 x 0.2 mm² for this first approach using a cross slit aperture, while the spot center was placed in a distance of 0.2 x 0.2 mm from the cutting edge towards the hollow cylindrical samples (Fig. 2 right). For diffraction pattern detection a 2D flat panel detector of type PerkinElmer XRD 1622 was used. By this means diffraction rings of the aluminum lattice planes {111}, {200}, {220} and {311}

Residual Stresses 2018 – ECRS-10   Materials Research Forum LLC
Materials Research Proceedings 6 (2018) 39-44   doi: http://dx.doi.org/10.21741/9781945291890-7

were recorded with a measuring frequency of 10 Hz and were evaluated using the open-source software *Fit2D* [5].

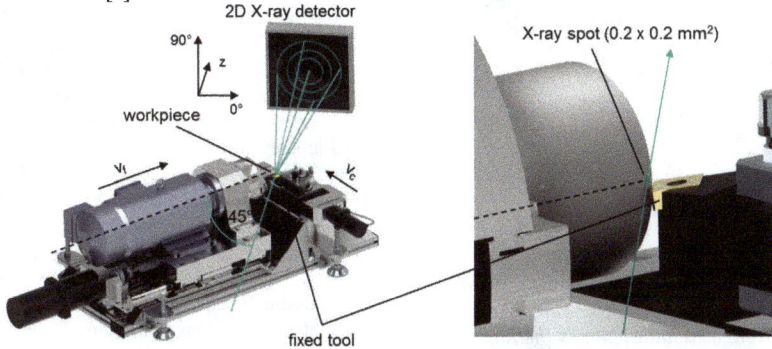

Fig. 2: (a) Experimental setup scheme with turning device and 2D PerkinElmer flat panel detector. (b) Detailed view on the irradiated process zone.

For the strain evolution analysis pie slices with an azimuthal angle range of 10° (± 5°) were defined and integrated for the 0° (normal to cutting plane) and 90° (parallel to cutting plane) direction (see Fig. 2 left and Fig. 3) to increase statistics. Thus, one dimensional intensity vs. 2θ datasets were determined that were further evaluated using a MATLAB script. Peak fitting was done using Pseudo-Voigt fit functions. From the calculated peak positions lattice spacings for the investigated interference lines were calculated on the basis of Bragg's law. The strain evolution for each direction (0° and 90°) was referenced to the first frame monitored; hence, strain changes were determined. The {hkl}-plane specific strain evolutions were averaged according to Daymond [6] using a texture coefficient of 1 for this first approach. The {hkl} specific elastic constants are calculated from single crystal elastic constants [7] using the Kröner model.

Results and Discussion

In Fig. 3 the extracted 1D diffraction profile is given for the 0° direction (experiment C) as an example. In the 2θ range of 2 to 8° the five measured Al diffraction lines are indexed. An analysis of the {222} Al diffraction plane was not performed due to the insufficient counting statistics. The relative intensities of the peaks do not show clear indication of texture or coarse grains, due to the fast rotating workpiece, which is in this investigation of great advantage. Due to the rotating workpiece always closely occupied diffraction rings were recorded. However, this blocks the possibility to monitor local texture evolution. Local texture can only by assessed for the final state, when the workpiece stops rotating.

In Fig. 4 and Fig. 5 the resulting strain evolution for the experiments A, B and C in normal (0°) and parallel (90°) direction to the cutting surface is plotted. In all cases the start of the cutting process is marked by a steep increase in tensile strains in a time period of approx. 5 s. The end of the cutting processes is reached when the feed ends and the strain evolution decreases exponentially. Investigation of the cutting inserts indicated that for all experiments presented here, built-up edges were formed, as intended by the choice of cutting parameters. We distinguished the experiments in ´stable´ (A, partially B) and ´instable´ (C, D) processes. Here, instable machining is characterized by vibrations due to a change in chip formation (i. e. type, conditions) resulting in a split up of the lattice strain courses in 0° direction.

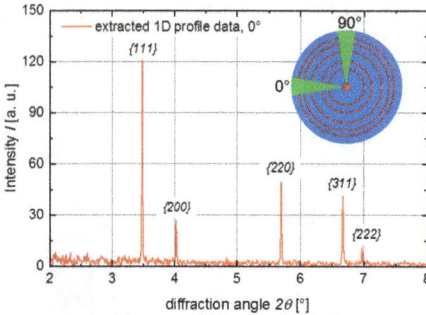

Fig. 3: Extracted 1D diffraction profile from monitored Debye-Scherrer rings of experiment C. Investigated lattice planes {hkl} are indexed.

In this regard, experiment A (Fig. 4 left) shows a stable cutting process, where a plateau is reached at about 10 s with 0.17 % strain for the average value. After approx. 60 s the average lattice strain falls in both directions to about 0.05 % residual strain after stop of processing. Experiment B starts analogously to A as a stable process with a plateau for the average strain of about 0.18 % in both directions. At about t = 55 s a steep rise of the lattice strain evolution in the normal and parallel direction can be observed (Fig. 4, Fig. 5 middle). Additionally a split up of the strain evolution between the evaluated reflections {hkl} in 0° direction can be noticed. This effect can be explained by a change in the cutting behavior. It is assumed that the formation of a built-up edge leads to a change in the chip shape geometry affecting the process stability due to a significant temperature increase leading to maximum average lattice strains of up to 0.45 %. In [4] a similar effect based on a steep temperature rise during the cutting of C45E was observed, which was due to a change in the chip shape geometry changing from spiral towards flow chips during continuous machining. Here, this significant temperature increase in the process zone lowers the warm yield strength of the workpiece resulting in higher plastic deformation, which is in correlation with the higher, residual lattice strain and the split up of lattice strain courses for the different {hkl} lines.

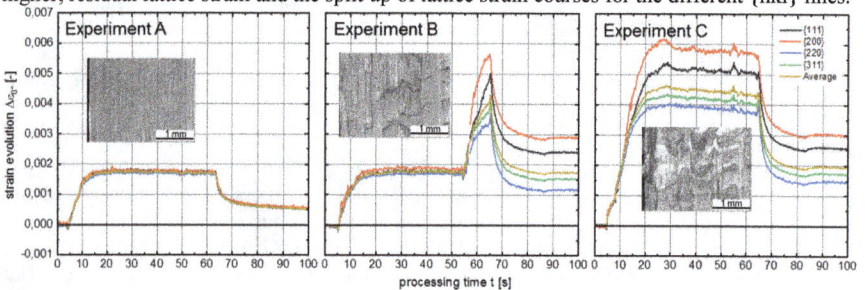

Fig. 4: Strain evolution vs. processing time for the investigated lattice planes and resulting average lattice strain according to [6] in 0° direction for experiments A (left), B (middle) and C (right). Micrographs of the cutting surfaces after the process are shown as thumbnails.

The increase in plastic deformation predominantly occurs in 0° direction, which is promoted by the larger constraint in this direction in the underlying material. The occurring differences in the lattice strain courses for different {hkl} planes are characteristic for fcc metals after plastic deformation and result due to plastic anisotropy through intergranular strains [8]. The split up of the lattice strain courses for different {hkl} lines is a clear indicator for the magnitude of plastic deformation induced during processing and in the present case much more meaningful as the peak widths of the recorded diffraction lines. The data indicate that experiment C (Fig. 4 and 5, right) depict an instable processing right from the start of the cutting.

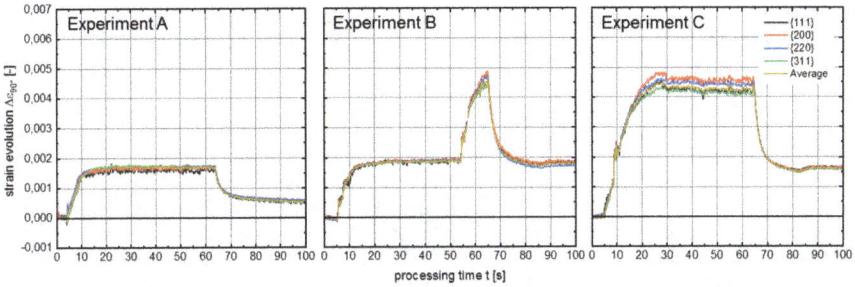

Fig. 5: Strain evolution vs. process time for the investigated lattice planes and resulting average lattice strain according to [6] in 90° direction for experiments A (left), B (middle) and C (right).

The upper average total strain plateau during the cutting is strongly increased to about 0.45 %. The reasons for this are the same as given for explanation of experiment B. It also can be seen, that the split of reflexes is, here again, only pronounced for the 0° direction. This is due to the higher strain restriction of the normal than to the parallel strain component by decreased temperature dissipation in the normal direction. The micrographs given in Fig. 4 show the workpiece surfaces after the process. In the stable case (experiment A) a homogeneous ground surface is depicted, whereas in the instable experiments B and C the cutting surface shows highly deformed areas confirming the higher plastic deformation in these cases.

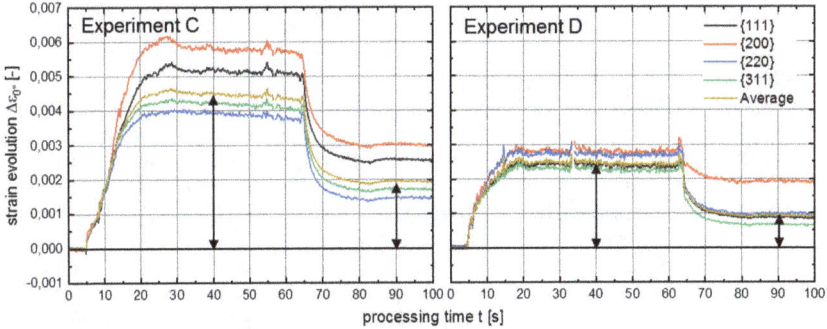

Fig. 6: Lattice strain evolution vs. process time for experiments C (left) and D (right). Arrows marking the averaged lattice strain differences during and after the cutting processes.

In Fig. 6 a direct comparison of experiments C and D is shown for the 0° direction, i.e. the strain component normal to the cut surface. Both processes show a split up of the lattice strain evolution, during and especially after cutting. It can be directly seen that the strain maximum during and after the processing as well as the degree of the split up is more pronounced for experiment C. In both cases the cutting is characterized by an instable processing leading to increased plastic deformation marked by the degree of split up between the lattice planes, especially for the {200} diffraction lines. For experiment C (v_c = 200 m/min) with increased cutting speed the average process induced lattice strain is about 0.2 % higher than for experiment D (v_c = 36 m/min) but the average residual lattice strain for B is about 0.2 % and for C about 0.1 %. The lower lattice strain differences in the final state (compared to the process) as well as the bigger split up in C indicate higher temperature in the process zone during cutting. This can be explained by a larger decrease of the materials yield strength leading to a higher degree of plastic deformation compared to experiment D.

Summary
Real time monitoring of a turning process was successfully carried out using a special developed turning lathe and high energy synchrotron radiation. Thereby the lattice strain evolutions normal and parallel to the cutting surface inside the chip formation zone were investigated for orthogonal cutting using technical relevant cutting parameters of Al-Mg3 (AW-5754) with a measuring frequency of 10 Hz. From the experimental results determined for multiple diffraction lines {hkl} during machining conclusions of the quality of the machining process and the resulting surface can be derived.

The determined lattice strain evolutions clearly indicate changes in the cutting behavior and the chip shape geometry leading to instable cutting processes. The instability induces high temperature increases in the cutting zone, which cause significant plastic deformation especially in the direction normal to the cut surface. As a consequence a split up of the lattice strain courses for different {hkl} planes occurred, which are caused by plastic anisotropy through intergranular strains. The split up effect is a clear indicator for the degree of plastic deformation.

Acknowledgements
The authors gratefully acknowledge DESY for granting beamtime at P07B@PETRAIII and furthermore the technical support provided by DESY and HZG. The authors would also like to thank the WALTER AG for the provision of the cutting inserts.

References
[1] J. Kümmel, D. Braun, J. Gibmeier, J. Schneider, C. Greiner, V. Schulze, A. Wanner, Study on micro texturing of uncoated cemented carbide cutting tools for wear improvement and built-up edge stabilisation, J. Mat. Proc., 215 (2015), 62-70. https://doi.org/10.1016/j.jmatprotec.2014.07.032

[2] E. Uhlmann, R. Gerstenberger, S. Herter, T. Hoghé, W. Reimers, R. V. Martins, A. Schreyer, T. Fischer, In situ strain measurement in the chip formation zone during orthogonal cutting, Production Engineering, 5 1 (2011) 1-8. https://doi.org/10.1007/s11740-010-0266-x

[3] K. Brömmelhoff, S. Henze, R. Gerstenberger, T. Fischer, N. Schell, E. Uhlmann, W. Reimers, Space resolved microstructural characteristics in the chip formation zone of orthogonal cut C45E steel samples characterized by diffraction experiments, J. Mat. Proc., 213 8(2013) 2211-2216. https://doi.org/10.1016/j.jmatprotec.2013.06.016

[4] J. Kümmel, Detaillierte Analyse der Aufbauschneidenbildung bei der Trockenzerspanung von Stahl C45E mi t Berücksichtigung des Werkzeugverschleisses, Dissertation (2015).

[5] A. P. Hammersley, FIT2D: An Introduction and Overview, ESRF Internal Report ESRF97HA02T (1997).

[6] M. Daymond, The determination of a continuum mechanics equivalent elastic strain from the analysis of multiple diffraction peaks, J. Apl. Phys. 96 8 (2004) 4263-4272. https://doi.org/10.1063/1.1794896

[7] A. G. Every, A. K. McCurdy, Low Frequency Properties of Dielectric Crystals-Second and Higher Order Elastic Constants in: Landolt-Börnstein – Group III Condensed Matter, 29a, Springer Verlag, Heidelberg, (1992).

[9] A. J. Allen, M. Burke, W. I. F. David, S. Dawes, M. T. Hutchings, A. D. Krawitz, C. G. Windsor, Effects of Elastic Anisotropy on the Lattice Strains in Polycrystalline Metals and Composites Measured by Neutron Diffraction, ICRS 2 (1989) 78-83. https://doi.org/10.1007/978-94-009-1143-7_10

Residual Stresses 2018 – ECRS-10   Materials Research Forum LLC
Materials Research Proceedings 6 (2018) 45-50   doi: http://dx.doi.org/10.21741/9781945291890-8

Development of Residual Stresses During Laser Cladding

A. Narayanan[a], M. Mostafavi[b], M. Pavier[c] and M. Peel[d*]

[1]University of Bristol, Department of Mechanical Engineering, Queen's Building, University Walk, Bristol, BS8 1TR, United Kingdom

[a]a.narayanan@bristol.ac.uk, [b]m.mostafavi@bristol.ac.uk, [c]martyn.pavier@bristol.ac.uk, [d]matthew.peel@bristol.ac.uk

Keywords: Cladding, Synchrotron X-Ray Diffraction, Stress-Measurement, Rail

Abstract. Laser cladding rail steel with a hard-wearing martensitic stainless-steel coating is a possible technique for improving the track durability of rail networks. However, the cladding process induces significant residual stresses in the clad material, due to the thermal mismatch between the two materials and the shape changes during the martensitic phase transformation. Predictions of the residual stress remain poorly verified as the process is complex and measurements made on final clad parts can be influenced by multiple parameters.

A cladded and heat-treated rail section was subject to sequential laser-pulses representative of the actual cladding process. The thermal cycle of these pulses is much simpler than real clads, easing the task of validating the component parts of simulations. Synchrotron X-ray diffraction was used to determine the phase selective residual stresses around the heated region before and after each pulse. In this manner it was possible to determine the change in stress due to a pulse and the degree of relaxation that is possible due to a neighbouring thermal cycle.

Introduction

A typical piece of rail sees many train wheels passing over it each day. The contact forces involved are high, and involve a mixture of dynamic and static loading, the combination of which can precipitate crack formation and wear towards the surface of the rail. To mitigate against this, it is possible to coat a section of generic rail steel (the substrate material) with a thin layer of a harder, more damage resistant alloy (the clad material) using laser cladding. This involves depositing a powdered form of the clad material onto the surface of the substrate and using a laser to melt it and form a coating. The process involves high, localized thermal gradients occurring over a short time-span and can therefore induce residual stresses in the clad-substrate system. Residual stresses interact with the applied loads to create more complex stress states, the result of which can cause damage at loads that would normally be considered to be within safe limits [1,2] and it is important to be able to understand and predict this interaction.

Martensitic stainless steel is a potential clad but stress development during cooling is complicated by the volumetric and shear strains that occur during the martensitic phase transformation. These can be difficult to account for in simulations [3–5], particularly when the transient temperatures are themselves hard to model. The thermal excursion has been simplified by using a single, tightly-controlled, laser pulse and then measured the residual stresses between pulses. This has been performed to provide a simple analogue of the heat input to the material while minimising the practical difficulties of powder deposition in an X-Ray beamline. If later finite element analysis (FEA) can predict the stress state accurately while incorporating effects of phase transformations, it can be extended for the full laser cladding simulations. Synchrotron X-ray diffraction using 2D imaging provides excellent spatial resolution, far in excess of lab X-ray and neutron diffraction, and simultaneous measurement of multiple phases.

Residual Stresses 2018 – ECRS-10   Materials Research Forum LLC
Materials Research Proceedings **6** (2018) 45-50   doi: http://dx.doi.org/10.21741/9781945291890-8

Materials and Methods

Clad samples were cut from a previously clad rail, 250 mm in length, resulting in 2mm thick slices with the dimensions shown in Figure 1. The clad had been applied by laser cladding in two layers to a depth of 2 mm, before being ground down to provide a smooth surface finish with a clad approximately 1.2 mm thick. The clad consists of a martensitic stainless steel (hereafter referred to as MSS, and whose composition is detailed in Table 1), while the substrate was rail steel grade 260 (also known as UIC 900A) [6]. The use of pre-clad rail ensured the microstructure of the martensite was typical of laser-deposited material. Pre-existing residual stresses were relieved, so far as possible, by tempering the samples at 600°C for two hours.

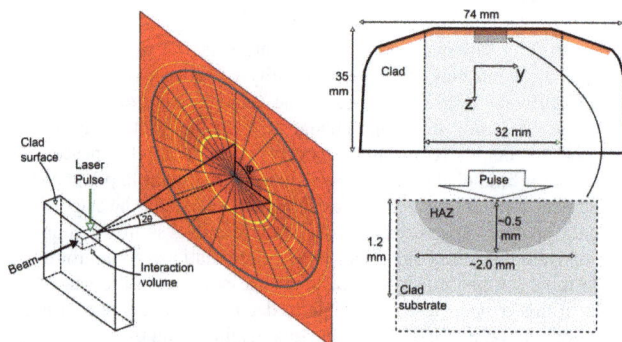

Figure 1. Illustrations of the sample and setup. The orientation of the sample (left) with respect the incoming beam and detector with the diffraction angle 2θ and azimuthal angle φ defined. The sample (right) is cut from a larger rail and was positioned with the laser above it so that the hot zone lies within the clad. The nominal heat affected zone (HAZ) is shown but is only a guide.

Table 1 Composition of MSS clad material including main alloying elements

Element	C	Mn	Si	Cr	Ni	Mo	Fe
Wt %	0.04	0.8	0.6	13.0	4.1	0.5	Balance

The experiment was performed by positioning a 500 W laser above the specimen and focusing it on the clad surface. The laser produced a pulse of set dimensions of 1.4 mm diameter with a set duration of 0.015 s. This would theoretically generate an axisymmetric molten zone on the upper surface, inducing austenitisation to a greater depth followed by martensite formation during the rapid cool. The width of the hot zone was approximately equal to the thickness of the clad slice and about half the depth of the clad (Figure 1). The temperature of the rear surface (facing beam) was measured using a FLIR T650sc thermal imaging camera operating at 30 Hz.

Stresses were measured using synchrotron X-rays on the ID31a beamline at the European Synchrotron Radiation Facility (ESRF). A monochromatic beam of wavelength 0.0173 nm was focussed onto the sample with a spot size of 40 x 15 µm. The interaction volume covered the full thickness of the sample. Due to the wide laser spot size it is assumed the temperature is uniform throughout this volume. The detector was a Pilatus Cadmium-Tellurium (CdTe) 2D detector that allowed for individual photon counting, resulting in patterns with very low noise and high sensitivity to small phase fractions. Tilts were corrected using an image from a Ceria powder sample. Each image was integrated azimuthally into 36 patterns (See Figure 2). It is notable that blank regions between detector sections mean that some reflections are missing or distorted. The

Residual Stresses 2018 – ECRS-10   Materials Research Forum LLC
Materials Research Proceedings **6** (2018) 45-50   doi: http://dx.doi.org/10.21741/9781945291890-8

measurement was from the initial clad and α (i.e. martensite) dominates with only small amounts of retained γ phases. The number of counts away from peaks (i.e. the background) is extremely low and even small peaks are easily resolved and fittable. Peaks were fitted using a pseudovoigt function, which was found to accurately match the peak shape. Fitting has been limited to non-overlapping peaks i.e. ferritic (martensite): (200), (211), (220) and austenite: (220), (222), (311).

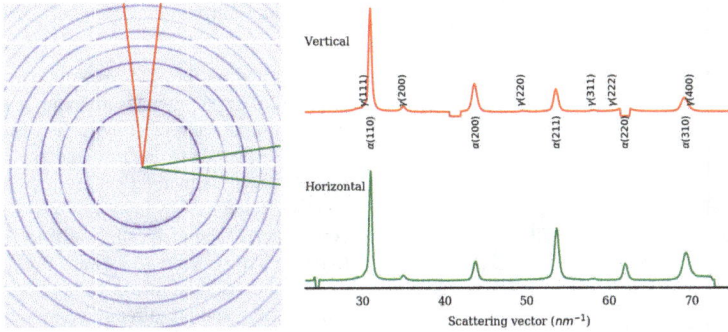

Figure 2. A sample image obtained using the Pilatus detector (left) and the patterns derived from the results (right) with the peaks families labelled. The patterns show the sqrt(I) to emphasise smaller peaks.

The sample had three pulses applied to it sequentially and separated laterally (centre-to-centre, y-direction) by 1 mm. The purpose of this was to mimic the effect of creating a continuous clad. The residual stress field before and after each pulse was measured over a 4 x 1 mm region, entirely within the clad, with a grid spacing of around 100 μm in the z-direction and 140 μm in the y-direction. The low diffraction angle (<10°) means the scattering vectors for all peaks lie close to the plane of the clad slice. The strain from the peak shift Δq for any (hkl) peak is given by

$$\varepsilon_\phi = \frac{\Delta q}{q_0^{hkl}} = p_{11}\sigma_{11} + p_{12}\sigma_{12} + p_{22}\sigma_{22}$$

where q_0^{hkl} is the peak position in the absence of a stress and p_{ij} are the stress factors. These incorporate both the direction (i.e. azimuthal angle, φ) and the (*hkl*) dependent diffraction elastic constants ($S_1, 1/2S_2$). Ignoring the slight out-of-plane component of the scattering vector, assuming alignment of the lab and sample coordinate systems, and setting φ=0 for the vertical detector element gives

$$p_{ij} = \begin{cases} \frac{1}{2}S_2 h_i^2 + S_1 \ if \ i = j \\ 2.\frac{1}{2}S_2 h_i h_j \ if \ i \neq j \end{cases} \quad where \ h = \begin{bmatrix} \cos \varphi \\ \sin \varphi \end{bmatrix}$$

The diffraction elastic constants were obtained using the Isodec software without texture [7]. For each measurement and each phase there are 3 (*hkl*) per spectra and 36 spectra at different φ angles. The stress components can be determined by solving the resulting set of linear equations.

47

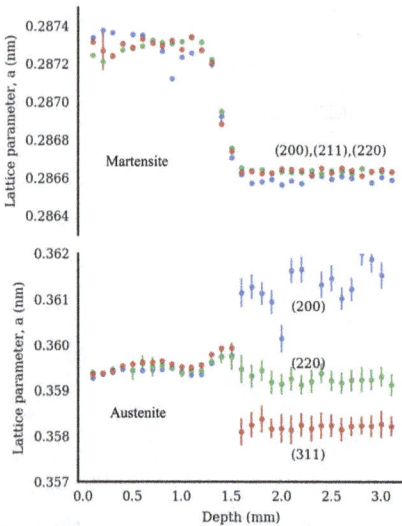

Figure 3 The lattice parameter of the martensite and austenite measured as a function of depth below the clad surface near to the edge of the sample. The clad finishes about 1.2. The low volume fraction of austenite in the substrate (>1.5 mm) makes values beyond this depth unreliable.

Results

Accurate values for q_0^{hkl} were determined as a function of depth by measurements at the edge of the initial sample (Figure 3). The stress state is expected to be low and the location means most remaining stresses should be locally relieved. Since q^{hkl} was broadly constant over the clad (<1.2mm) q_0^{hkl} was set to be the simple average for each (hkl).

Plots of residual stress components in the z-direction and y-direction (normal and transverse in the original rail) are shown in Figure 4 for the martensite phase. The four maps show the initial stress state and after 1, 2 or 3 pulses, respectively. The initial sample has uniform compressive stress in both directions (<-200 MPa). The austenite is uniformly in tension in both directions (<200 MPa) except on the very upper layer where it is unstressed or slightly compressive.

A laser pulse results in a region of high tensile stress broadly corresponding to the material heated above the austenite transformation temperature, and hence reformed martensite upon cooling. The maximum stress in the z-direction occurs around 0.5 mm below the surface – this stress component tends to reduce towards the upper surface as expected from boundary conditions. This stress component (tensile in the martensite, compressive in the austenite) probably doesn't reach zero due to interphase stresses, which balance each other but do not have to be zero at the surface. In the y-direction, the highest tensile stresses are located closer to the upper surface but are of similar magnitude. Compressive stresses are seen at greater depths with the highest magnitude lying just below the tensile region. The residual stress in the austenite (Figure 5) is a mirror image of the martensite with compression in the hottest region and tensile stresses at greater depths. The austenite stress appears more hydrostatic in character with similar magnitudes in both directions. The measured tensile stress at around 0.5 mm depth is perhaps unrealistically high (700-900 MPa) but then this retained austenite will be widely distributed between martensite laths, so high interphase stresses are likely.

Sequential pulses are remarkably consistent. The size and shape of the heat affected zone, and the magnitude of stress components within it, appears independent of the number of pulses and the new zone seems to overwrite the zone next to it. There is a modest tempering effect on the neighbouring pulses, which then appear smaller in extent and magnitude. This is clearer in line plots showing the stress components as a function of depth at the nominal position of each pulse in the final sample (Figure 6). The first and second pulses are almost identical in stress indicating that additional pulses, further away from the first pulse, have little additional tempering effect. The third pulse systematically shows larger magnitude stresses, although not in all locations.

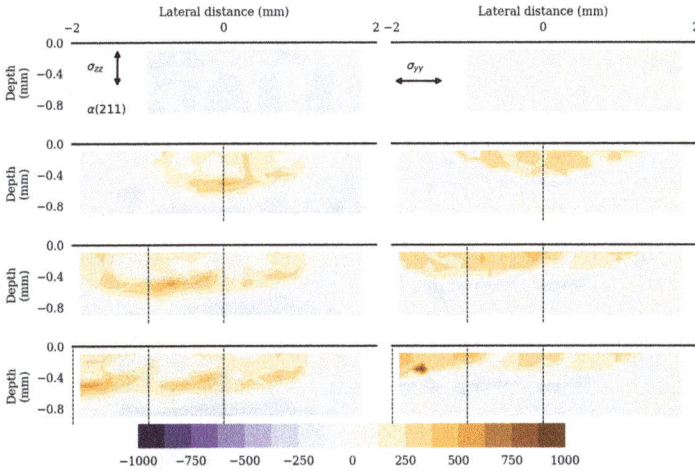

Figure 4. Plots of the z and y components of residual stress (MPa) in the martensite, as measured with the (211) reflection, for sequential laser pulses. The dotted lines show nominal pulse centres.

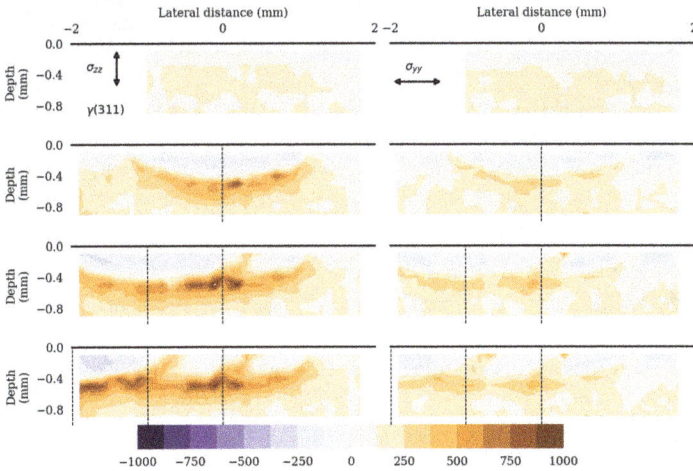

Figure 5. Plots of the z and y components of residual stress (MPa) in the austenite, as measured with the (311) reflection, for sequential laser pulses. The dotted lines show nominal pulse centres.

Conclusions

An experiment was performed to investigate how laser welding changes the residual stress state. A series of three adjacent welds was made to simulate the effect of running a longer track of laser clad. The combination of excellent beam focussing, and a high-speed, low noise detector allows excellent spatial resolution and stress measurement, even in a minority phase.

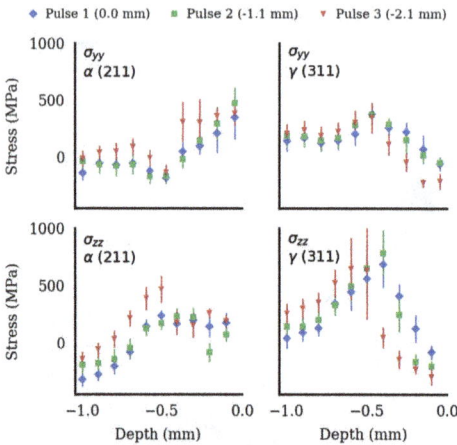

Figure 6. Stress profiles for two orthogonal stress components along the centreline of initial weld (same location of pulse 1); results include the effect of adjacent welds with stress plotted after pulse 1 (blue diamonds), pulse 2 (green squares and pulse 3 (red triangles) and error bars show 95% confidence intervals.

Results have shown that forming a spot weld on the surface of a martensitic clad on UIC 900A substrate produces large tensile stress in the weld zone. This stress is uniaxial towards the surface but becomes biaxial at approximately 0.4 mm depth. The stress components exceed 600 MPa in some regions but the location of maxima and minima is different for each component. The stress state is highly consistent and repeatable in both the ferritic and austenitic phases.

Overlaying multiple pulses starts to approximate a continuous clad more typical of genuine cladded samples. The data suggest that sequential pulses does cause slight tempering of prior adjacent welds. The effect is modest but exceeds the uncertainty in the measurements. Whether this effect is significant enough to need to be included in simulations remains to be seen.

Acknowledgements

This research was funded by EPSRC research grant EP/M023044/1, with equal contributions from EPSRC and the Rail Safety and Standards Board (RSSB). The authors would like to thank the Rail Safety and Standards Board (RSSB) for its support in conducting this research.

References

[1] A. Mirzaee-Sisan, C. E. Truman, D. J. Smith, and M. C. Smith, Eng. Fract. Mech. 74, 2864 (2007). https://doi.org/10.1016/j.engfracmech.2006.12.033

[2] G. A. Webster and A. N. Ezeilo, Int. J. Fatigue 23, Supplement 1, 375 (2001). https://doi.org/10.1016/S0142-1123(01)00133-5

[3] H. Dai, J. A. Francis, H. J. Stone, H. K. D. H. Bhadeshia, and P. J. Withers, Metall. Mater. Trans. A 39, 3070 (2008). https://doi.org/10.1007/s11661-008-9616-0

[4] J. Altenkirch, J. Gibmeier, V. Kostov, A. Kromm, T. Kannengiesser, S. Doyle, and A. Wanner, J. Strain Anal. Eng. Des. 46, 563 (2011). https://doi.org/10.1177/0309324711413190

[5] R. A. Ainsworth, J. K. Sharples, and S. D. Smith, J. Strain Anal. Eng. Des. 35, 307 (2000). https://doi.org/10.1243/0309324001514431

[6] J. W. Ringsberg, A. Skyttebol, and B. L. Josefson, Int. J. Fatigue 27, 702 (2005). https://doi.org/10.1016/j.ijfatigue.2004.10.006

[7] T. Gnäupel-Herold, J. Appl. Crystallogr. 45, 573 (2012). https://doi.org/10.1107/S0021889812014252

Residual Stresses 2018 – ECRS-10   Materials Research Forum LLC
Materials Research Proceedings 6 (2018) 51-55   doi: http://dx.doi.org/10.21741/9781945291890-9

Round Robin Study on Residual Stresses Using X-Ray Diffraction for Shot-Peened Tool Steel Specimens

Jonas Holmberg[1,2,a*], Mikko Palosaari[3,b], Seyed Hosseini[1,c], Henri Larjosuo[3,d] and Pär Andersson[1,e]

[1]Swerea IVF AB, Argongatan 30, Mölndal SE-431 53, Sweden

[2]University West, Dep. Of Engineering Science, Trolhättan SE-461 86, Sweden

[3]Stresstech OY, Tikkutehtaantie 1 FI-40800 Vaajakoski, Finland

[a]jonas.holmberg@swerea.se, [b]mikko.palosaari@stresstech.com, [c]seyed.hosseini@swerea.se, [b]henri.larjosuo@stresstech.com, [e]par.andersson@swerea.se

Keywords: X-Ray Diffraction, Residual Stress, Round Robin, Measurement System Analysis

Abstract. Residual stress measurements using x-ray diffraction is a well established method used within the industrial and academic community to verify the performance of different processes for metallic materials. The measurement gives an absolute value of the stress state which can be used to design and optimize the process route to induce beneficial compressive residual stresses and avoid detrimental tensile stresses. Investigating the uncertainty and accuracy of the measurement system, operator and the material is therefore of high relevance both from an industrial and scientific point of view. Round robin testing is an important way to quantify the uncertainties that could affect the quality of the measured results and hence how a process is optimized and tuned. Such an investigation allows the operator to understand and reduce variations. Current round robin test includes results from five different laboratories using comparable equipments located in Sweden, Finland, Germany and United States. This work focuses on five shot-peened tool steel specimens produced with identical process settings. Additionally, an investigation of the repeatability of the system, influence of the operator, variations within the specimen, and the long time stability of the specimens has been measured.

Introduction

Residual stress measurements using x-ray diffraction is a well established and an essential method to study and verify the results from different processing operations of crystalline materials. This method is today a common practice both within the industrial and academic community. The absolute stress value allows studying how individual process steps affect the material. This opens up the opportunity of process optimization in such way to generate beneficial compressive residual stresses or minimize detrimental tensile stresses. This was for example shown by Matsumoto et al. who studied the influence of cutting parameters for case carburised bearing steels with a hardness of 58-62 HRC [1]. That investigation clearly showed the difference between a ground and a hard turned surface as well as how different cutting tools and parameters affected the stresses. Another example was shown by Kristoffersen et al. [2] who investigated the effect on the residual stresses from the induction hardening process parameters of a AISI 4340 steel. Residual stress measurements are also frequently used to study the impact from shot peening, which is strongly connected to the fatigue performance [3].

Considering that many components today are loaded close to their physical limitations, determination of residual stresses becomes even more important, where the reproducibility and accuracy of the measurement becomes crucial. To capture the variation and scatter in the measurements, round robin activities are an important tool. Such an investigation evaluates the uncertainty and accuracy of the measurement system, operator and the material. A round robin investigation is therefore an important tool to be used both from an industrial and scientific point of view. Previously, one of the authors of this paper conducted a round robin study of carbon steel specimens under tensile stresses. That investigation showed that due to the stress state in the specimens, different source of variations could become more critical [4].

The French Association for residual stress analysis (GFAC) has been working with round robin analyses over a certain period of time, including several laboratories, different equipments and test materials [5]. A part of that investigation was presented for an Inconel 690 material which showed that the laboratories had a repeatability and a reproducibility of about 45 MPa and 85 MPa respectively. Connected to the same round robin study, Francois et al. [6] also evaluated a shot peened steel specimen, which they reported a standard deviation of the raw data from the different laboratories of 36 MPa. However, by analyzing the raw data with the same software and calculation settings the standard deviation was reduced to 18 MPa. Hence, it was clearly demonstrated the importance to correctly evaluate the raw data. On this topic Gibmeier et al. presented the results of a round robin for depth profiles of shot peened 42CrMo4 specimens [7]. That investigation showed even greater variation of the measured results from the different participating laboratories with standard deviation for the surface stress of 80 MPa which was lowered to 40 MPa after uniform evaluation, i.e. using same software.

A part of being an important way to quantify the uncertainties, round robin testing could also be a method to qualify new operators. This in turn could be compared to a measurement system analyses (MSA). However, when performing this there are a many factors that may affect the measurement, of which several are listed in the Measurement Good Practice Guide nr. 52 [8]. Some are related to the geometrical constrains of the specimen, and some are connected to the surface conditions, distribution of the stresses in the material, texture, and presence of coarse grains or high stress gradients. Other important aspects relate to how the measured data is analyzed, such as material properties used when calculating the stresses as well as which Peak-fitting method that is used. Then there are a number of operator related factors that may greatly influence the result such as alignment of specimen relative equipment, selected measurement parameters such as exposure time measurements strategy and equipment alignment/calibration strategy. Evidently there are a great number of factors that needs to be considered when measuring the residual stresses using x-ray diffraction, and interestingly there are limited amount of published research work of round robin activities and how they are conducted. To better assess the quality of the as-obtained results, a deeper understanding of the large scatter in the repeatability and reproducibility of for example the results reported by GFAC is therefore necessary.

The aim of this investigation was to demonstrate the measurements uncertainties both in terms of variations from the measurement system, specimen itself as well as how different laboratories perform. For future repeatability measurements the long time stability of the specimens has also been evaluated with the intention to investigate if there is a decay or degradation with time.

Material

Several test specimens were manufactured from a cold work tool steel which was heat treated to a hardness of 45 HRC. The specimens were shot peened (full coverage) using steel balls with a size specified by SAE S170. All test specimens were shot peened at the same occasion using same parameters. Figure 1 illustrates one of the specimens (1033), and the two main measurement directions.

Residual Stresses 2018 – ECRS-10   Materials Research Forum LLC
Materials Research Proceedings 6 (2018) 51-55   doi: http://dx.doi.org/10.21741/9781945291890-9

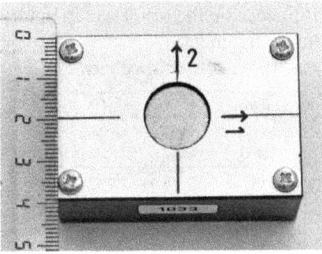

Figure 1.   Overview of specimen 1033 defining measurement direction 1 and 2.

Experimental

This study has been divided into three major subparts; i) measurement system accuracy (MSA), ii) long time repeatability of test specimens and iii) round robin study. Throughout all measurements, a chromium x-ray tube was used and the modified χ method was employed as measurement principle. In this method the specimen is rotated about the χ axis, plane normal to that containing ω and 2 ω, further described in the standard EN-SS 15305:2009 [9]. Further, the (211) diffraction plane was deployed (156.4°). The residual stresses was calculated using Hook's law, assuming elastic strain theory, with a Young's modulus of 211 GPa and 0.3 in Poisson's ratio. For the MSA analyses only one portable X-ray system was used, XStress 3000 G2R from Stresstech, with a 2 mm collimator and 5 ψ-angles in the interval of ±45°. The long time repeatability measurements were performed with eight different Xstress 3000 diffractometers, including G2R, G3 and Robot systems by eight operators using 1-3 mm collimators. For the round robin study the different laboratories used XRD systems such as Xstress 3000 G2R, Xstress 3000 G3 and as well as stationary systems Mahle, Bruker. The round robin measurements were repeated three times (in two direction 1 and 2). It should be noted that aside from performing multiple measurement in the same setup, the samples were even repositioning for each measurement. The specimen was aligned according to the indicators on the specimens, see Figure 1 (arrows on the specimen). The stresses (normal and shear) has been determined by a $\sin^2(\psi)/\sin(2\psi)$ fit of the determined peak positions (or the strain values calculated from them).

Results and discussion

Measurement systems analysis

The intention with this MSA was to study how individual equipment performed in order to compare this in relation to the round robin investigation. In this case one of the participating laboratories performed the measurements with a X3000 G2R system. This investigation was three folded to study variation from a) *the specimen*, b) *the measurement system* and c) *the operator*. In a) measurements were performed at five different locations on the specimen with a displacement by 1.5 mm. Table 1 shows the results, where the standard deviation was 6.5 MPa, or 1.1 %. The deviation, dev, is the results from the peak fitting using cross correlation and constant background. For b) the measurements were repeated without moving the specimen or altering the measurement settings. The variation from the equipment, shown in Table 1, resulted in a standard deviation of 2.6 MPa or ±0.4 %. For c) two operators were included, where measurements were performed by repositioning the specimen between each measurement (removing and repositioning the specimen in the specimen holder). Each operator measured five times, with the target to measure the stresses at the centre of the specimen. The result in Table 2 shows comparable values and a standard deviation of 1.8 MPa and 3.1 MPa for operator 1 and 2.

The MSA results show that the major factors contributing the variation come from the specimen itself and from raw data calculation of the peak fitting according to Table 1-2. In this case, cross correlation with a constant background was employed for peak fitting resulting in a standard deviation of 7.1 MPa. The total measurement uncertainty from this system, calculated as the root mean square of the different factors, a), b), c) and peak fitting, resulted in a value of 10.2 MPa. This result is in line with the results presented by François et al. [6]. That investigation further identified that the operator had a strong influence on the result which in turn also will be influenced by variation in the specimen and specimen alignment.

Table 1. Results from sample and equipment repeatability measurements on specimen 1033.

a) Specimen influence					b) Equipment repeatability				
Position	Direction 1 [MPa]	Dev [MPa]	Direction 2 [MPa]	Dev [MPa]	Measurement no.	Direction 1 [MPa]	Dev [MPa]	Direction 2 [MPa]	Dev [MPa]
1033_1	-611.4	1.9	-614.9	3.1	#1	-585	7	-590	3.8
1033_2	-618.2	2.4	-605.1	2.9	#2	-588.6	7.7	-592.2	3.1
1033_3	-601.4	2.9	-617.5	4.3	#3	-586.1	7	-585	3.4
1033_4	-609.4	2.9	-609	4.5	#4	-583.4	7.1	-589.7	3.2
1033_5	-604.3	4.7	-602.3	3.1	#5	-585	6.5	-589.6	4
Average [MPa]	-608.9	2.96	-609.8	3.6	Average [MPa]	-585.6	7.1	-589.3	3.5
Std. Dev. [MPa]	6.5		6.4		Std. Dev. [MPa]	1.9		2.6	
Std. Dev. [%]	1.1		1.1		Std. Dev. [%]	0.3		0.4	

Table 2. Results from operator influence for five repetitions and alignment on specimen 1033.

c) Operator influence										
Meas. no.	Operator	Direction 1 [MPa]	Dev [MPa]	Direction 2 [MPa]	Dev [MPa]	Meas. no.	Operator	Direction 1 [MPa]	Dev [MPa]	Direction 2 [MPa]	Dev [MPa]
#1	1	-607.1	2.8	-612	2.8	#1	2	-601.4	5.4	-611.5	3.1
#2	1	-605.9	3	-611.4	4.1	#2	2	-602.5	4.2	-609.3	2.7
#3	1	-604.1	4.1	-608.6	3.2	#3	2	-603.6	4.1	-607	3.5
#4	1	-603.8	2.5	-610.5	1.1	#4	2	-603.4	4.4	-606.1	4.1
#5	1	-602.7	3.8	-609.7	3.2	#5	2	-607.2	4.8	-603.5	3.5
Average [MPa]		-604.7	3.2	-610.4	2.9	Average [MPa]		-603.6	4.6	-607.5	3.4
Std. Dev. [MPa]		1.8		1.4		Std. Dev. [MPa]		2.2		3.1	
Std. Dev. [%]		0.3		0.2		Std. Dev. [%]		0.4		0.5	

The purpose of the MSA was to elucidate the influence from other factors when performing the round robin investigation hence the results are therefore considered as generalized for this specimen. However, for a more geometrically complex test specimen, or a specimen where the stress distribution varies, the operator alignment procedure becomes critical which will be reflected in the repeatability. It is further suggested that this type of approach is adopted for new measurement objects in order to retrieve a general value of the expected uncertainty. This result should then be presented along with the measured absolute value, which might not be the case today since these systems typically only give the operator the deviation from the peak fitting.

Long time repeatability of the specimens

Two specimens (1021 and 1022), fabricated in the same batch as the round robin specimens, have additionally been used to analyze the long time repeatability by one of the participant laboratories. Roughly once a week, the specimens were measured with different diffractometers, in total by eight different operators. During 2.5 year period the specimen 1021 was measured 154 times and 106 times for specimen 1022. The results of the long term repeatability are shown in Figure 2 where the measured stress values in direction 1 are plotted as a function of measurement date. The error bars on the graph are the measured 1-sigma deviation of the fitting routine for each measured residual stress value. The average for all of the values were; σ_{1021} = -593 ± 10 MPa, and σ_{1022} = -595 ± 10 MPa, where the uncertainty is one sigma standard deviation.

A linear function was fitted on the results and a slope can be observed, -4.7 MPa /year for specimen 1021 and -1.7 MPa / year for specimen 1022. The reason for this gradual change of specimen 1021 might either be due to wear of the specimen in combination with the rapidly changing residual stress profile of these specimens. It might also be due to drift in the equipment. The maximum compressive stress is located below the surface and the residual stress profiles in this case showed a positioning at about 50 μm below surface. If the surface wears down the measured stresses might be affected. This could be an important factor to consider when designing and producing new reference specimens. It might be a good approach to generate a profile that has similar stress state in the surface and below the surface that take height for any future wear of the specimen. However, further measurements need to be carried out to see if this trend continues. For the other sample the change is less than 1% change per year.

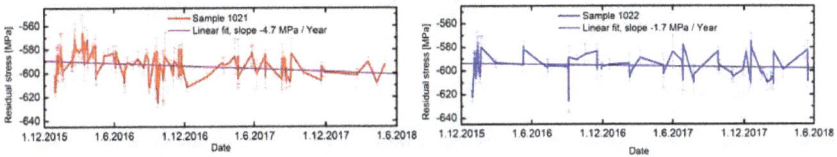

Figure 2. Long term repeatability of specimens 1021 and 1022.

Round robin measurements

The results from the round robin testing have been compared for each specimen and measurement direction in Figure 3. The results are the average values for the three measurements performed for each specimen by laboratory A-E and equipment _1-_4. Respectively, the error bars show the standard deviation for the three measurements. The dotted lines represent the minimum and maximum values including the standard deviation. The summarized result for these specimens lies within the interval (-)554 – (-)611 MPa with a average value of -590.4 MPa and a standard deviation of 10 MPa. Results from the different laboratories show that laboratory D measures the lowest average stress for all specimens while laboratory C measures the highest stresses. However, individual equipments within both laboratory D and E measures both lower and higher results compared to the other laboratories. This means that the individual equipments rather than the laboratories are generating the different results.

The average results for each specimen from the round robin measurements are summarised in Table 3 which show that the values from all laboratories and equipment is in the interval (-)583.9 – (-) 596.8 MPa with a standard deviation of 11.1 MPa as maximum and a 9.1 MPa on an average which corresponds to 1.9 % as maximum and 1.5 % on an average.

Table 3. Result table of the average values from the round robin testing

Specimen	Direction	Mean [MPa]	Std. Dev [MPa]	Std. Dev [%]	Direction	Mean [MPa]	Std. Dev [MPa]	Std. Dev [%]
1028	1	-595.5	7.8	1.4	2	-590.4	7.4	1.2
1029	1	-585.6	9.4	1.5	2	-583.9	10.1	1.7
1030	1	-595.1	7.7	1.4	2	-587.7	11.1	1.9
1031	1	-586.6	7.9	1.3	2	-590.9	11.0	1.8
1032	1	-591.4	9.4	1.5	2	-596.8	9.3	1.5

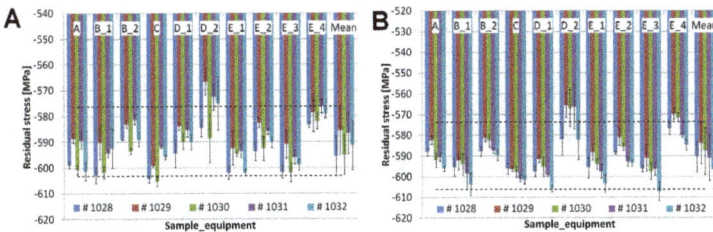

Figure 3. Result of the round robin surface residual stress measurements of the five shot peened specimens in direction 1 (A), and direction 2 (B).

Conclusions

These results clearly show the systems-operator performance, the influence of the test specimen which in turn contributes to the performance of the round robin testing. It could however be difficult to exactly determine how different factors influence on the performance since many of them are linked. The following conclusions were drawn from these investigations:

- The measurement system analysis resulted in a total system uncertainty of 10.2 MPa for this shot peened specimen. The peak fitting of raw diffraction data and variation in surface stress distribution are the most influential on the measurement uncertainty for the system.
- The long term stability of the specimens had a slight decay over the years of -4.7 MPa/year for one specimen and -1.7 MPa / year for the other specimen.
- The round robin investigations showed very good repeatability with an average standard deviation of ±9.1 MPa corresponding to ±1.5 % for these measurement specimens.
- The round robin testing concluded that individual equipment rather than different laboratories influence the variation in the measured result.

References

[1] Y Matsumoto, F Hashimoto, G Lahoti, Surface Integrity Generated by Precision Hard Turning. CIRP Annals - Manufacturing Technology 48 (1999) 59–62. https://doi.org/10.1016/S0007-8506(07)63131-X

[2] H Kristoffersen, P Vomacka, Influence of process parameters for induction hardening on residual stresses. Materials & Design 22 (2001) 637–644. https://doi.org/10.1016/S0261-3069(01)00033-4

[3] S Wang, Y Li, M Yao, R Wang, Compressive residual stress introduced by shot peening. Journal of Materials Processing Technology 73 (1998) 64–73. https://doi.org/10.1016/S0924-0136(97)00213-6

[4] S B Hosseini, B Karlsson, T Vuoristo, K Dalaei, Determination of Stresses and Retained Austenite in Carbon Steels by X-rays - A Round Robin Study. Exp. Mech. 51 (2011) 59–69. https://doi.org/10.1007/s11340-010-9338-2

[5] F Lefebvre, E Wasniewski, M François, External Reference Samples for Residual Stress Analysis by X-Ray Diffraction. Advanced Materials Research 996 (2014) 221–227.

[6] M François, F Convert, S Branchu, French round-robin test of X-ray stress determination on a shot-peened steel. Experimental Mechanics 40 (2000) 361–368. https://doi.org/10.1007/BF02326481

[7] J Gibmeier, J Lu, B Scholtes, Round Robin Test on the determination of residual stress depth distribution by X-ray diffrsction, Mat. Sci. Forum 404-407, 2002, pp. 659-664. https://doi.org/10.4028/www.scientific.net/MSF.404-407.659

[8] M E Fitzpatrick, A Fry, P Holdway, Determination of residual stresses by X-ray diffraction. Measurement Good Practice Guide No 52 (2005).

[9] European Committee for Standardization, Non-destructive Testing – Test Method for Residual Stress analysis by X-ray Diffraction, EN-SS 15305:2009.

Residual Stresses 2018 – ECRS-10
Materials Research Proceedings 6 (2018) 57-62

Materials Research Forum LLC
doi: http://dx.doi.org/10.21741/9781945291890-10

Convergence Behavior in Line Profile Analysis Using Convolutional Multiple Whole-Profile Software

Masayoshi Kumagai[1,a,*], Tomohiro Uchida[1], Kodai Murasawa[2,b],
Masato Takamura[3,c], Yoshimasa Ikeda[3,d], Hiroshi Suzuki[4,e],
Yoshie Otake[3,f], Takayuki Hama[5,g], Shinsuke Suzuki[2,h]

[1] Faculty of Engineering, Tokyo City University, Japan

[2] Faculty of Science and Engineering, Waseda University, Japan

[3] RIKEN Center for Advanced Photonics, RIKEN, Japan

[4] Materials Sciences Research Center, Japan Atomic Energy Agency, Japan

[5] Graduate School of Energy Science, Kyoto University, Japan

[a]mkumagai@tcu.ac.jp, [b]murasawa@akane.waseda.jp, [c]takamura@riken.jp,
[d]yoshimasa.ikeda@riken.jp, [e]suzuki.hiroshi07@jaea.go.jp, [f]yotake@riken.jp,
[g]hama@energy.kyoto-u.ac.jp, [h]suzuki-s@waseda.jp

Keywords: Neutron Diffraction, Line Profile Analysis, Microstructure, CMWP

Abstract. The convergence behavior of the parameters related to microstructural characteristics a–e was studied during optimizations in a common line profile analysis software program based on the convolutional multiple whole profile (CMWP) method. The weighted sums of squared residual (WSSR) was a criterion of the optimization. The parameters b and d, which are related to the dislocation density and to the crystallite size, respectively, strongly affect the line profile shape. Therefore, the distributions of WSSRs on the space parameters b and d were first observed. The variation trajectory of parameters b and d during iterative calculations with several values of parameter e was then observed, along with the variations when all of the parameters were variable. In the case where only three parameters were variable, we found that a smaller initial value of e should be chosen to ensure stability of the calculations. In the case where all parameters were variable, although all of the results converged to similar values, they did not precisely agree. To attain accurate optimum values, a two-step procedure is recommended.

Introduction

The mechanical properties of materials depend on microstructural features such as dislocations. Therefore, the characterization of such features is important in materials and mechanical engineering. Microstructural features can be observed by several methods. One approach is X-ray/neutron diffraction line profile analysis (LPA). In this method, microstructural features are characterized through inverse analysis of diffraction line profiles. LPA is a nondestructive measurement, which enables its use in *in situ* measurements during, for example, tensile tests. Pulsed neutron sources in large proton accelerator facilities are suitable for *in situ* measurements because of the high flux of the neutron beam and the time-of-flight (TOF)-type diffractometer used in these facilities. In addition, whole diffraction profiles can be obtained within a short measurement time.

Several line profile methods have been developed and reviewed in a book [1]. The modified Williamson–Hall/Warren–Averbach method [2] is one of the common methods in the 2000s. However, in this method, each diffraction peak is fitted individually even though whole profiles can be obtained via the measurements with a TOF diffractometer. For the efficient analysis of

Residual Stresses 2018 – ECRS-10
Materials Research Proceedings 6 (2018) 57-62

Materials Research Forum LLC
doi: http://dx.doi.org/10.21741/9781945291890-10

measured profiles, whole-profile analysis methods [3–6] are better than that individual-peak methods, especially for TOF measurements.

The convolutional multiple whole profile (CMWP) fitting method and its software [6] is a common method. The software is convenient approach to whole-profile analysis, and users can obtain parameters related to a sample's microstructure. However, because many parameters are optimized during the CMWP fitting process, users must consider how well to optimize these parameters. Therefore, we studied the convergence behavior in LPA using CMWP software to understand the phenomenon and propose a suitable procedure.

Theorem

In the CMWP method, each measured diffraction pattern is fitted by a theoretical diffraction pattern calculated with five fitting parameters (anisotropy parameter q, variance of the lognormal crystallite size distribution σ_{LN}, effective outer cutoff radius of dislocation R_e^*, dislocation density ρ, and crystallite size L_0) through iterative calculations. These five parameters correspond to the parameters a, b, c, d, and e in the CMWP software as follows:

$$q = a \tag{1}$$

$$\sigma_{LN} = \frac{c}{\sqrt{2}} \tag{2}$$

$$\rho = \frac{2}{\pi \left(b_{Burgers} d\right)^2}$$

$$R_e^* = \frac{\exp(-\frac{1}{4})}{2e}$$

$$L_0 = \frac{2}{3} \exp\left(\frac{5}{4} c^2 + b\right)$$

Experimental

The measured diffraction line profile for 780-MPa-grade steel [7] was used in this study. The profiles covering the scattering vector region from 3.75 to 12.5 nm^{-1}, which includes six peaks (Fig. 1), were used. The diffraction indices of the six peaks of αFe in this region are 110, 200, 211, 220, 310, and 222. The other common analysis conditions are shown in Table 1. The background of the line profile was fixed throughout this study.

Initially, weighted sums of squared residuals (WSSRs) at several values of parameters b and d were obtained. The other parameters—a, c, and e—were fixed as arbitrary values.

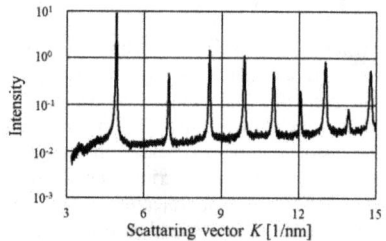

Fig. 1 Measured diffraction profile.

Table 1 Analysis conditions

Latice constant [nm]	0.28665
C_{h00}	0.284
Fit limit	1×10^{-9}
Use weights	No
Fit peak pos	Yes
Instrum. Profiles	LaB$_6$

We next performed optimizations with i) three parameters (b, d, and e) and ii) all parameters varied with four arbitrary initial values; the trajectories of b and d were drawn on the contour maps of WSSRs in b–d space.

Results

Figure 2 shows the WSSRs contour map at each b and d value. The contour colors were interpolations of calculated values shown as plots in the figure. Regions where the WSSR is smaller than in other regions (hereafter referred to as valleys) exist, similar to an inverse

Residual Stresses 2018 – ECRS-10   Materials Research Forum LLC
Materials Research Proceedings **6** (2018) 57-62   doi: http://dx.doi.org/10.21741/9781945291890-10

proportion in d–b space. The parameters b and d are inverse proportional to line profile broadening. Because the parameter b is proportional to crystallite size and profile broadening is inverse proportional to the crystallite size. And the parameter d is inverse proportional to dislocation density and line profile broadening is proportional to the dislocation density. The tendencies of the WSSRs contour map among three graphs are similar; however, the depths and positions of valleys differ among the three plots. The value of WSSR was smallest at $b \approx 4$ and $d \approx 100$ with $e = 0.001$ in the three maps. Thus, the optimum values are reasonably expected to coincide with these values.

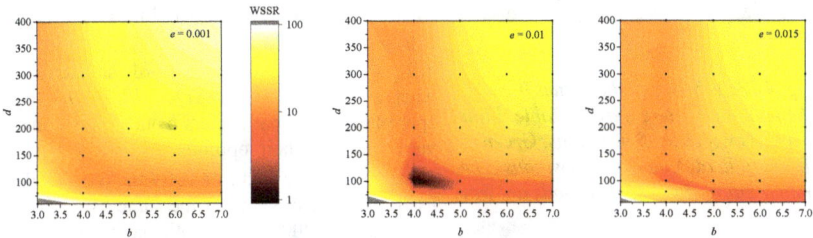

Fig. 2 WSSR contour maps at each value of b and d; parameter e was fixed at a) 0.001, b) 0.01, and c) 0.015. The plots indicate the calculated conditions for b and d. Colors are interpolated from the plots.

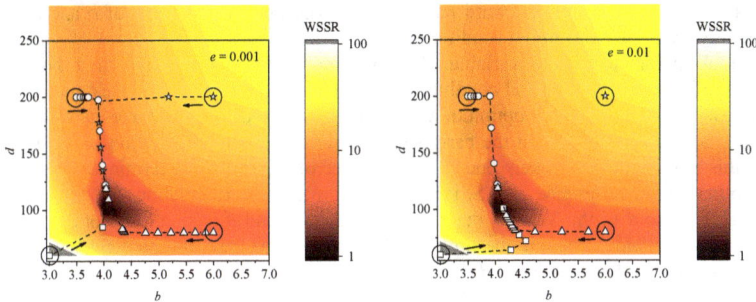

Fig. 3 Trajectories of values b and d as parameters b, d, and e are varied. Initial values of e were 0.0001 and 0.01, b and d were started from cicled points. The contour maps of the backgrounds are from Fig. 2b.

In the case of parameters b, d, and e as variables

Fig. 3 shows the trajectories of the values of b and d in the WSSR contour maps (the contour map in Fig. 2b was used as the background) when b, d, and e were variable and four initial values for b and d were arbitrarily chosen. Plots and lines on the maps show the trajectories of the b and d values during iterative calculations. When parameter e was larger for a calculation with initial values $b = 6$ and $d = 200$, the iterative calculations stopped after 36 times and the values of the parameters did not change. On the contrary, when the initial value of e was smaller ($e = 0.001$), parameters b and d changed with every the initial values. This tendency was also observed for the other values of e. These results suggest that the value of e should be set smaller ($e \le 0.001$) although the value $R_e^* = 389$ nm obtained from eq. (4) is too large in reality.

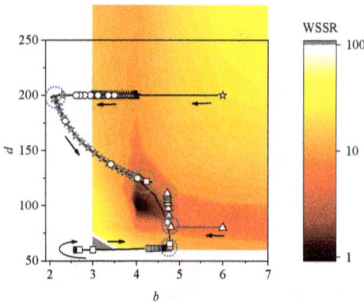

Fig. 4 Trajectories of the values b and d when all parameters were variable. Plots are shown every 5 iterations. Green circles indicate converged positions. The contour maps of the backgrounds are from Fig. 2b.

Murasawa *et al.* indicated that small values of b and d should be chosen [7]. In that study, e was set to 0.1 ($R_e^* = 38.9$ nm) because the value is reasonable as effective radius for a dislocation. Therefore, small values of the parameters b and d were appropriate in that study.

In the case of all parameters as variables

Fig. 4 shows the trajectories of the b and d values in optimizations with four initial values for b and d when all parameters were variable. When the initial values of b and d were large (marked as ☆ in Fig. 4), parameter b varied toward smaller values, whereas d did not vary. When b reached approximately 2, parameters b and d simultaneously varied toward their optimum values. When initial values of $b = 3.5$ and $d = 200$ (marked as ○) were used, the path was almost the same as the case $b = 3.5$ and $d = 200$. When both parameters b and d were small (marked as □), parameter b initially varied toward smaller values and then increased, whereas parameter d varied little. Parameter d then varied toward larger values as parameter b varied. When $b = 6$ and $d = 60$ (marked as △), parameter b varied toward smaller values and stopped at $4.5 \leq b \leq 5$, which corresponds to a position on the trace where b and d were small. Then, only parameter d varied toward larger values. The optimized values of b and d differed substantially from the other initial conditions.

When all parameters were variable, the converged values reached from a given initial condition differed when the initial condition was changed. However, in all cases, the parameters almost reached their assumed optimum values. In a case where only three parameters were variable, calculations stopped in the early steps of iteration without variation of the parameter values; this termination was likely due to a low degree of freedom. The termination of iterations is controlled by the difference in WSSRs before and after each calculation (ΔWSSRs). Even if the slope of contour of the WSSRs due to the values of b and d is small, the other parameters must change to avoid exceeding the limit of ΔWSSRs; in this case, the parameter e is the only variable parameter other than b and d.

Discussion

The converged values are summarized in Table 2. All of the WSSRs were similar to 0.111 and sufficiently small. In the case of initial conditions $b = 6$ and $d = 80$ (△), the parameters converged to different values than those reached under other initial conditions, as shown Fig. 4. Not only parameters b and d but also the other parameters, a, c, and e, were different.

Figure 5 shows typical microstructural characteristics calculated from the CMWP parameters using Eqs. (1)–(4). Although the differences are not fatal, the values of all of the parameters differed substantially, especially in the case of the aforementioned initial conditions. However, avoiding such initial conditions is better from the viewpoint of obtaining precise optimum values. When three parameters were variable, the parameters converged and agreed, independent of the initial values of b and d except in the case of a large initial value e. Therefore, before carrying out optimizations with all parameters variable, a pre-calculation should be performed with small values of e and arbitrary values of b and d. Then, a full calculation should be performed with all parameters variable using the three converged values obtained via the pre-calculation. In actuality, a full calculation was conducted with all parameters variable using the averaged converged values obtained from the calculation in which only three parameters were variable (Table 2). We thus obtained the following optimum values: $a = 1.57$, $b = 4.10$, $c = 0.688$, $d = 124$, and $e = 5.43 \times 10^{-3}$; the WSSR was 0.111. These values agree with the other results in Table 2 with the exception of the unsuitable initial condition. In the procedure, roughly optimum values are obtained by pre-calculation and then precise solutions are obtained by full calculation.

Table 2 Optimized results obtained with several initial values of b and d in the analyses with all parameters or only three parameters variable.

Initial condition			Converged value					
b	d	Symbol	a	b	c	d	$e \times 10^{-3}$	WSSR
6	200	☆	1.57	4.09	0.693	123.9	5.40	0.111
3.5	200	○	1.57	4.10	0.692	123.9	5.40	0.111
3	60	□	1.58	4.23	0.611	121.9	5.93	0.111
6	80	△	1.70	4.71	0.046	112.6	8.11	0.112
Average of results only b, d, e varied			1.96 (fixed)	4.05	0.7 (fixed)	118.8	4.49	0.114

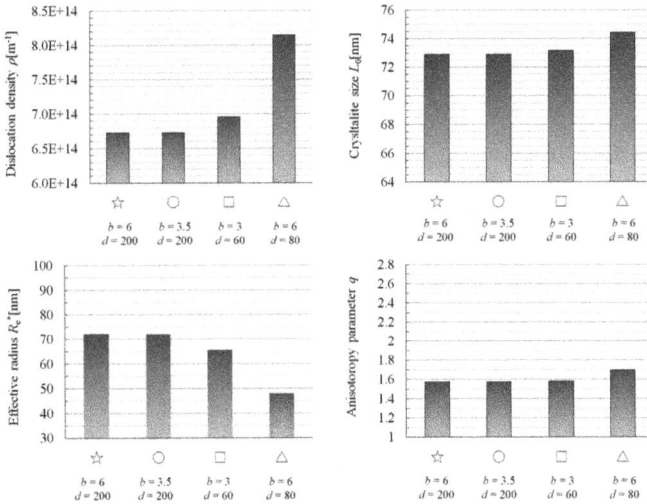

Fig. 5 Values of microstructural characteristics calculated from obtained CMWP parameters a–e in Table 2.

Fig. 6 shows the number of iterations in each calculation. The total iteration numbers of the pre- and full-calculations are approximately half the number of iterations of the calculations with all parameters simultaneously variable. An exception occurred when the initial values were $b = 6$ and $d = 80$, which were unsuitable initial values; in this case, both iteration numbers were the

Residual Stresses 2018 – ECRS-10
Materials Research Proceedings 6 (2018) 57-62

Materials Research Forum LLC
doi: http://dx.doi.org/10.21741/9781945291890-10

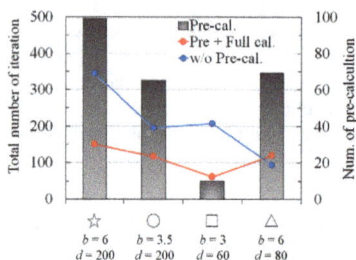

Fig. 6 Comparison of the number of iterations.

same. The two-step procedure is therefore effective in terms of both reliability of the converged values and time efficiency.

Summary

The convergence behavior in LPA using CMWP software was studied. Five parameters related to microstructural characteristics, $a–e$, were optimized using CMWP software. We observed the distributions of WSSRs on the space parameters b and d and obtained WSSR contour maps in $b–d$ space. We then showed the variation trajectory of parameters b and d under several initial sets of conditions where only the three parameters b, d, and e were variable and where all parameters were variable. In the case where only three parameters were variable, the selection of a smaller value of e tended to improve the stability of the calculations. In the case where all parameters were variable, although all of the results converged to similar values, they did not precisely agree. To obtain accurate optimum values, a two-step procedure is recommended. The procedure can lead to optimum solutions, with the additional advantage of time efficiency.

The neutron diffraction experiment was performed as a proposed program (2015A0294) at MLF, J-PARC. We thank Dr. Harjo, Dr. Kawasaki, Ms. Saori Iwata, and Mr. Hayato Komine for their experimental assistance.

References

[1] E. J. Mittemeijer and U. Welzel, in Mod. Diffr. Methods, Wiley-VCH (2013), pp. 87–126. https://doi.org/10.1002/9783527649884.ch4

[2] T. Ungár and A. Borbély, Appl. Phys. Lett. 69, 3173 (1996).

[3] P. Scardi, M. Leoni, and Y. H. Dong, Eur. Phys. J. B 18, 23 (2000). https://doi.org/10.1007/s100510070073

[4] G. Ribárik, T. Ungár, and J. Gubicza, J. Appl. Crystallogr. 34, 669 (2001). https://doi.org/10.1107/S0021889801011451

[5] P. Scardi and M. Leoni, Acta Crystallogr. Sect. A Found. Crystallogr. 58, 190 (2002). https://doi.org/10.1107/S0108767301021298

[6] G. Ribárik, Modeling of Diffraction Patterns Based on Microstructural Properties, Eötvös Loránd University, 2008.

[7] K. Murasawa, M. Takamura, M. Kumagai, Y. Ikeda, H. Suzuki, Y. Otake, T. Hama, and S. Suzuki, Mater. Trans. 59, 1135 (2018). https://doi.org/10.2320/matertrans.M2017380

Residual Stresses 2018 – ECRS-10
Materials Research Proceedings 6 (2018) 63-68

Materials Research Forum LLC
doi: http://dx.doi.org/10.21741/9781945291890-11

High-Resolution Neutron Diffraction Setting for Studies of Macro- and Microstrains in Polycrystalline Materials

Pavol Mikula[1,a,*], Massimo Rogante[2,b], Jan Šaroun[1,c] and Miroslav Vrána[1,d]

[1]Nuclear Physics Institute ASCR, v.v.i., Husinec-Řež 130, 25068 Řež, Czech Republic

[2]Rogante Engineering Office, I-62012 Civitanova Marche, Italy

[a]mikula@ujf.cas.cz, [b]main@roganteengineering.it, [c]saroun@ ujf.cas.cz, [d]vrana@ ujf.cas.cz

Keywords: Neutron Diffraction, Bent Perfect Crystals, Bragg Diffraction Optics, Polycrystals, Strains

Abstract. On the basis of our previous experience [1-4], a unique three axis high-resolution experimental setting for nondestructive strain measurements which is based on neutron Bragg diffraction optics with cylindrically bent perfect crystals is presented. The use of focusing in real and namely, in momentum space, from *FWHM* of diffraction lines the three axis setting provides the $\Delta d/d$ resolution (d-lattice spacing) of about 4×10^{-3} for bulk samples. It permits studies not only macrostrain components resulting from angular shifts of diffraction peaks but also estimations of microstrains in a plastically deformation region by means of profile-broadening analysis. The feasibility of the experimental setting is demonstrated on low carbon steel shear deformed steel wires.

Introduction

Non-destructive X-ray and neutron diffraction techniques for studies of internal strain fields in polycrystalline materials have been successfully used for many years [5-17]. At present, the investigations of residual strains/stresses are usually carried out at the dedicated double axis diffractometers (strain scanners) with a bent perfect crystal (BPC) focusing monochromator situated on the first axis, a sample situated on the second axis and with a position sensitive detector (PSD). With respect to the experimental conditions, the BPC crystal is optimally bent which results in a highly collimated beam (often called quasiparallel beam) reflected by the polycrystalline sample [12,15-17]. However, the $\Delta d/d$ resolution of these dedicated scanners derived from the *FWHM* of the diffraction lines is sufficiently high for small sample gauge volumes but rarely better than 8×10^{-3} for bulk samples. Important thing is that the dedicated scanners operate with open beams without the necessity of Soller collimators. A further way, how to increase the resolution which would permit to investigate an influence of microstrains on the diffraction profile, namely, in the case of plastically deformed samples, is the use of a third axis set-up when employing a third BPC crystal on the third axis as an analyzer. The first attempts of the use of the three axis set-up were described two decades ago [1,2]. The drawback of such a set-up in comparison with the conventional scanners consists in using the step-by-step analysis (by rocking the BPC analyzer). Therefore, the effective measurements could be carried out within a reasonable measurement time at high-flux neutron sources. From the point of view of luminosity, some improvements can be done by an employment of the BPC monochromator at a rather small Bragg angle, while the resolution can be optimized by a suitable choice of the BPC analyzer and its thickness, as it was used in the present case (see Fig. 1).

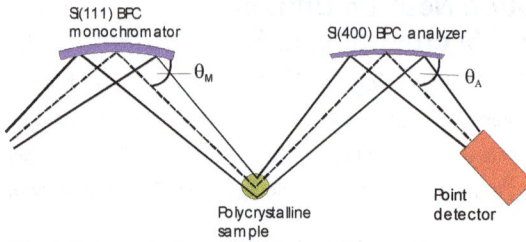

Fig. 1. Schematic diagram of the experimental performance with the sample in the vertical position.

Experimental details

The set-up shown in Fig. 1 was experimentally realized on the three axis neutron optic diffractometer installed at the medium power research reactor LVR-15 situated in Řež. Si(111) and Si(400) single crystals had the dimensions of 200x40x4 mm^3 and 20x40x1.3 mm^3 (length x width x thickness), respectively. The monochromator Si(111) had a fixed curvature with a radius R_M of about 12 m. The curvature of the analyzer Si(400) was changeable in the range from R_M=36 m to R_M=3.6 m and finally set for the optimum radius of curvature of R_A= 9 m, where a best resolution was found. For a practical demonstration of the feasibility of using the three axis set-up for diffraction line analysis we used α-Fe(110) nondeformed as well as deformed wires with accumulated shear deformation, as a result of rolling with the shear of the metal ingot and conventional wire drawing. Due to the deformation the diameters of the samples are not the same, however, but they are in the vicinity of 5 mm. The samples were already studied in detail elsewhere [18] and in this case several of them were used for the feasibility studies of the set-up. The description of the nondeformed as well as with accumulated shear deformation samples – (low carbon steel - Grade 08G2S GOST 1050) is shown in Table1. Table 2 describes the chemical composition of the steel.

Table 1. Description of deformation of the low-alloyed steel samples. Percentage in the fourth column shows the reduction degree in drawing deformation.

Sample number	ϕ [mm]	Shear def. [%]	Drawing def. [%]
1	5.10	0	0
2	4.28	8	23.2
3	5.35	0	0
4	4.28	16.6	23.2
5	5.57	0	0
6	4.28	23	23.2

Residual Stresses 2018 – ECRS-10   Materials Research Forum LLC
Materials Research Proceedings 6 (2018) 63-68   doi: http://dx.doi.org/10.21741/9781945291890-11

Table 2. Chemical composition of low-alloyed structural steel grade 08G2S GOST 1050 element

Element	C	Mn	Si	S	P	Cr	Ni	Cu	N2
wt %	0,08	1,87	0,82	0,020	0,022	0,02	0,02	0,02	0,007

Experimental results of the diffraction profiles - Radial components

The steel samples were situated on the second axis of the diffractometer in vertical position. The width of the incident beam was 8 mm and therefore, the whole volume of the sample was irradiated (within the whole diameter). The height of the incident beam was 20 mm. From the introduced Fig. 2, we can detect the following features: The peak intensities and *FWHM*s related to the deformed samples differ very little. This is brought about by the fact that the diameter of the deformed samples was equal and that the shear deformation has a negligible effect. Very close values of *FWHM* point out on the fact that the lattice deformation in the radial direction was basically brought about by drawing deformation. The integrated intensity under the peak profiles related to the nondeformed samples N.1, N.3 and N.5 primarily corresponds to the irradiated volume of the sample, which is naturally maximum for the sample N. 5.

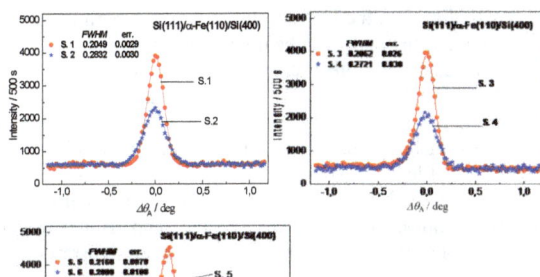

Experimental results of the diffraction profiles - Axial components

In the next step the steel samples were situated on the second axis of the diffractometer in the horizontal position. The obtained results are shown in Fig. 3. In comparison with the previous case, for the samples in the horizontal position their irradiated volume is much smaller and

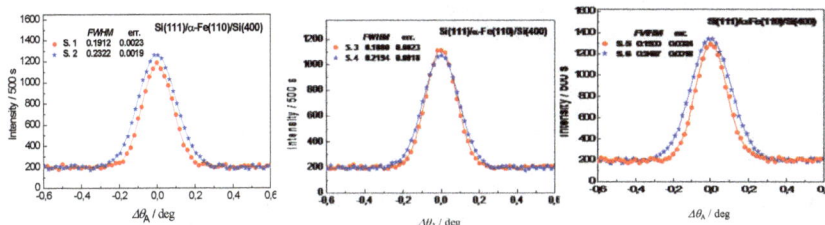

Fig. 3. Diffraction profiles related to the samples N.1-N.6 situated in the horizontal position for measurement of the axial component as analyzed by the bent perfect Si(400) analyzer.

correspondingly neutron signal was smaller. In this case, it was found that the resolution was dependent on the width of the incident beam impinging the sample and therefore, we used the slit width of 5 mm. It can be seen from Fig. 3 that the resolution represented by *FWHM* was practically the same for all nondeformed samples N.1, N.3 and N.5. Small differences in *FWHM* can be seen for plastically deformed samples. It points out the fact that the lattice deformation was basically brought about by drawing deformation and influencing the radial strain component and much less the axial strain component.

Summary

The presented neutron diffraction results obtained on the samples of low-alloyed quality structural steel (Grade 08G2S GOST 1050) document the feasibility of the unconventional three axis set-up for studies of some properties of polycrystalline samples in the plastic deformation region. It can be seen from the Tab. 3 that though the *FWHM*-effects resulting from the applied

Table 3. Summary of the FWHMs as calculated in ($\Delta d/d$)- scale

Sample number	N.1	N.2	N.3	N.4	N.5	N.6
$FWHM(\Delta d/d)$ Radial [10^{-3}]	4.09	5.66	4.12	5.44	4.32	5.61
$FWHM(\Delta d/d)$ Axial [10^{-3}]	3.82	4.64	3.76	4.38	3.80	4.97

deformation on the samples are very small, thanks to the high resolution of the experimental set-up, they are clearly measurable. In particular, such results can be used as an additional support to complement the information achieved by using the other characterization methodologies. It should be pointed out that contrary to the conventional double axis strain scanner, the three axis setting provides a high resolution for bulk samples e.g. of the diameter of 5-10 mm (for the radial component) while for conventional scanner such diameter itself introduces to the resolution an uncertainty in *FWHM* of the diffraction profile at least of 1×10^{-2} rad. Thanks to the high resolution property of the three axis set-up, it can be, of course, used also for measurements of residual elastic deformation macrostresses, however, less efficiently in comparison with the conventional two axis neutron diffraction measurement [18] which has not so high resolution requirements.

Acknowledgement

Measurements were carried out at the CANAM infrastructure of the NPI CAS, v.v.i. in Řež supported through MŠMT project No. LM2015056. The presented results were also supported in the frame of LM2015074 infrastructural MŠMT project "Experimental nuclear reactors LVR-15 and LR-0" as well as by the Czech Science Foundation GACR through the project No. 16-08803J. The authors would like to thank Ms. Michalcová for the help in measurements and evaluations of the data.

References

[1] J. Kulda, P. Mikula, P. Lukáš and M. Kocsis, Utilisation of bent Si crystals for elastic strain measurements, Physica B 180&181 (1992) 1041-1043.

[2] M. Vrána, P. Lukáš, P. Mikula, and J. Kulda, Bragg diffraction optics in high resolution strain measurements, Nucl. Instrum. Methods in Phys. Research A, 338 (1994) 125-131. https://doi.org/10.1016/0168-9002(94)90172-4

[3] B.S. Seong, V. Em, P. Mikula, J. Šaroun, M.H. Kang, Unconventional Performance of a Highly Luminous Strain/Stress Scanner for High Resolution Studies, Materials Science Forum, 681 (2011) 426-430. https://doi.org/10.4028/www.scientific.net/MSF.681.426

[4] P. Mikula, M. Vrána, J. Šaroun, B.S. Seong, W. Woo, Double Bent Crystal Monochromator for High Resolution Neutron Powder Diffraction, Powder Diffraction, 28, Issue S2 (2013) S351-S359. https://doi.org/10.1017/S0885715613000912

[5] I.C. Noyan and J.B. Cohen, Residual stress: measurement by diffraction and interpretation, 1st edn; 1987, New York, Springer-Verlag. https://doi.org/10.1007/978-1-4613-9570-6

[6] M.T. Hutchings, and A.D. Krawitz, (Eds.), 1992, "Measurement of residual and applied stress using neutron diffraction", NATO ASI Series, Applied Sciences, Vol. 26, Kluwer Academic Publisher, 1992.

[7] M. R. Daymond, M. A. M. Bourke, R. B. von Dreele, B. Clausen, and T. Lorentzen, Use of Rietveld refinement for elastic macrostrain determination and for evaluation of plastic strain history from diffraction spectra, J. Appl. Phys. 82 (1997) 1554–1562. https://doi.org/10.1063/1.365956

[8] G.A. Webster (ed.), Polycrystalline materials—determination of residual stress by neutron diffraction, ISO/TTA3, ISO/Technology, Trends Assessment, Geneva, Switzerland, 2001.

[9] V. Stelmukh, L. Edwards, J.R. Santisteban, S. Ganguly, and M.E. Fitzpatrick, Weld stress mapping using neutron and synchrotron x-ray diffraction, Materials Science Forum, 404-407 (2002) 599-604. https://doi.org/10.4028/www.scientific.net/MSF.404-407.599

[10] M.T. Hutchings, P.J. Withers, T.M. Holden and T. Lorentzen, Introduction to the characterization of residual stress by neutron diffraction, 1st ed.; 2005, London, Taylor and Francis.

[11] P. Mikula, M. Vrána and P. Lukáš, Power of Bragg Diffraction Optics for High Resolution Neutron Diffractometers for Strain/Stress Scanning, In IAEA-TECDOC-1457 "Measurement of residual stress in materials using neutrons", IAEA Vienna, 2005, pp. 19-27.

[12] T. Pirling, G. Bruno, and P.J. Withers, SALSA, Advances in Residual Stress Measu-rement at ILL, Materials Science Forum, 524–525 (2006) 217–22. https://doi.org/10.4028/www.scientific.net/MSF.524-525.217

[13] T.M. Holden, H. Suzuki, D.G. Carr, M.I. Ripley and B. Clausen, Stress measurements in welds: Problem areas, Mater. Sci. Eng. A, 437 (2006) 33–37. https://doi.org/10.1016/j.msea.2006.04.055

[14] P. J. Withers, Comptes Rendus Physique, Mapping residual and internal stress in materials by neutron diffraction, 8 (2007) 806–820.

[15] P. Mikula, M. Vrána, Ľ. Mráz and L. Karlsson, High-Resolution Neutron Diffraction Employing Bragg Diffraction Optics - A Tool for Advanced Nondestructive Testing of Materials, In Proc. of ASME Conference on Engineering Systems Design and Analysis, 7 to 9 July 2008, Haifa. ISBN 0-7918-3827-7. https://doi.org/10.1115/ESDA2008-59174

[16] R.C. Wimpory, P. Mikula, J. Šaroun, T. Poeste, Junghong Li, M. Hoffman nand R. Schneider, Efficiency Boost of the Materials Science Diffractometer E3 at BENSC: One Order of Magnitude Due to a Double Focusing Monochromator, Neutron News, 19 (2008) 16-19. https://doi.org/10.1080/10448630701831995

Residual Stresses 2018 – ECRS-10   Materials Research Forum LLC
Materials Research Proceedings **6** (2018) 63-68   doi: http://dx.doi.org/10.21741/9781945291890-11

[17] W. Woo, Z. Feng, X. Wang and S.A. David, Neutron diffraction measurements of residual stresses in friction stir welding: A review, Science and Technology of Welding and Joining, 16 (2011) N.1 23-32.

[18] M. Rogante, P. Mikula, P. Strunz, A. Zavdoveev, Residual stress determination by neutron diffraction in low-carbon steel wires with accumulated shear deformation, Proc. 7th Int. Conf. "Mechanical Technologies and Structural Materials" MTMS2017, Split, Croatia, 21-22 Sept. 2017, S. Jozić, B. Lela, Eds., Croatian Society for Mechanical Technologies, Split, Croatia (2017), pp. 111-115.

Residual Stresses 2018 – ECRS-10
Materials Research Proceedings 6 (2018) 69-74

Materials Research Forum LLC
doi: http://dx.doi.org/10.21741/9781945291890-12

Stress Measurements of Coarse Grain Materials using Double Exposure Method with Hard Synchrotron X-Rays

Kenji Suzuki[1,a,*], Takahisa Shobu[2,b] and Ayumi Shiro[3,c]

[1]Faculty of Education, Niigata University, Ikarashi-2-no-cho, Nishi-ku, Niigata, 950-2181, Japan

[2]Japan Atomic Energy Agency, 1-1-1, Koto, Sayo-cho, Hyogo, 679-5148, Japan

[3]National Institutes for Quantum and Radiological Science and Technology, 1-1-1, Koto, Sayo-cho, Hyogo, 679-5148, Japan

[a]suzuki@ed.niigata-u.ac.jp, [b]shobu@sp8sun.spring8.or.jp, [c]shiro.ayumi@qst.go.jp

Keywords: X-Ray Stress Measurement, Coarse Grain, Double Exposure Method, Hard Synchrotron X-Ray

Abstract. The double exposure method (DEM) is proposed herein for X-ray stress measurement of coarse grain materials. A diffraction angle was obtained from an incident beam and a spotty diffracted beam. Each X-ray beam was measured by an area detector on a linear motion stage on a 2θ-arm. To verify the effectiveness of the DEM, the residual stress of a plastically bent specimen and the residual stress distribution of an indented specimen was measured. The results obtained with the DEM were similar to the results of simulations using the finite element method confirming that the DEM is useful for X-ray stress measurements of coarse grain materials.

Introduction

There are three X-ray stress measurement methods that use area detectors: the 2D method [1], the cosα method [2], and the direct-method [3]. However, these methods cannot be applied to the stress measurement of coarse grain materials. Therefore, a diffraction spot trace method (DSTM) has been developed and employed for strain scanning coarse grain materials [4]. However, as a rotating-slit system is indispensable, the DSTM involves a complicated experimental system. The details of the rotating-slit system were described in a previous study [5].

Herein, we propose a double exposure method (DEM), which does not require a slits system and is therefore suitable for the stress measurement of coarse grain materials.

Double Exposure Method

The difficulties in the stress measurement of coarse grain materials are as follows:

1. The diffraction pattern from a coarse grain is spotty.
2. The diffraction angle cannot be determined from the diffraction centre because the diffraction centre depends on each grain.
3. The measured diffraction angle of the coarse grain is affected by the divergence of the incident beam.

These problems are solved by using an area detector, a DEM, and synchrotron radiation. Usually, a diffraction angle is measured by a diffractometer or by using a tangent, but these methods cannot be applied to coarse grains. In the DEM proposed herein, the diffraction angle is determined from two lines consisting of the incident X-ray beam and the diffracted beam.

Figure 1 shows the configuration and coordinate system of the DEM. As shown in Fig. 1 (a), the specimen is set on a sample stage, and the area detector (PILATUS) is mounted on a linear motion stage on a 2θ-arm. The diffraction spots from transmitted X-rays are measured by an area

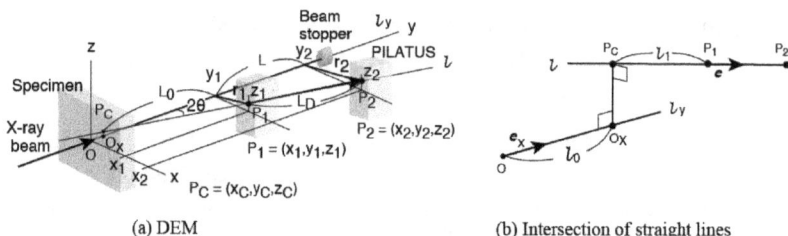

(a) DEM   (b) Intersection of straight lines

Fig. 1 Configuration and geometry of double exposure method (DEM).

detector. The straight line of the incident beam, l_y, is determined by the area detector and is defined as the y-axis.

When a diffraction spot is measured with the area detector, the straight line of the diffraction spot is given by the two points $P_1(x_1,y_1,z_1)$ and $P_2(x_2,y_2,z_2)$. The radii r_1 and r_2 are given by

$$r_1 = \sqrt{x_1^2 + y_1^2}\,, \quad r_2 = \sqrt{x_2^2 + y_2^2}\,. \tag{1}$$

The diffraction angle 2θ is obtained from the following relation.

$$2\theta = \arctan\left(\frac{r_2 - r_1}{y_2 - y_1}\right). \tag{2}$$

The straight line l is the diffracted beam and passes through P_1 and P_2. The unit vectors of the straight lines l_y and l in Fig. 1 (b) are given by

$$\boldsymbol{e_X} = (0,1,0)\,, \quad \boldsymbol{e} = \frac{(x_2 - x_1, y_2 - y_1, z_2 - z_1)}{\|L_D\|}\,. \tag{3}$$

As shown in Fig. 1 (b), the line segment between the points P_C and O_X is perpendicular to both straight lines; therefore, the inner products become zero.

$$\boldsymbol{e} \cdot (\boldsymbol{OP_C} - \boldsymbol{OO_X}) = \boldsymbol{e_X} \cdot (\boldsymbol{OP_C} - \boldsymbol{OO_X}) = 0\,. \tag{4}$$

The position vectors are given by

$$\boldsymbol{OP_C} = \boldsymbol{OP_1} - l_1\boldsymbol{e}\,, \quad \boldsymbol{OO_X} = l_0\boldsymbol{e_X}\,. \tag{5}$$

By substituting Eq. 5 into Eq. 4, the lengths l_0 and l_1 are obtained as follows:

$$l_0 = y_1 - \frac{\cot^2 2\theta}{L}[(x_2 - x_1)x_1 + (z_2 - z_1)z_1] \tag{6}$$

$$l_1 = \frac{\cos 2\theta}{L \sin^2 2\theta}[(x_2 - x_1)x_1 + (z_2 - z_1)z_1] \tag{7}$$

Finally, the position of the diffracted grain, $P_C(x_C,y_C,z_C)$, can be obtained from Eq. 5 and 7.

$$x_C = x_1 - \frac{\cot^2 2\theta}{L^2}[(x_2 - x_1)x_1 + (z_2 - z_1)z_1](x_2 - x_1) \tag{8}$$

$$y_C = y_1 - \frac{\cot^2 2\theta}{L}[(x_2 - x_1)x_1 + (z_2 - z_1)z_1] \tag{9}$$

$$z_C = z_1 - \frac{\cot^2 2\theta}{L^2}[(x_2 - x_1)x_1 + (z_2 - z_1)z_1](z_2 - z_1) \tag{10}$$

Synchrotron X-Ray Measurements and Test Specimens
The stress measurements using the DEM were carried out at SPring-8 on the beamline BL22XU. A linear motion stage with a stroke length of 800 mm was set on the 2θ-arm, and a PILATUS-300K area detector was mounted on the linear motion stage. The 2θ-arm was moved to $0°$, and the beam centre was measured by the detector at two positions $(0,y_1,0)$ and $(0,y_2,0)$. The 2θ-arm was moved to $26°$, and the diffraction spots were measured at P_1 and P_2. In this experiment, L_D was 791.889 mm, L_0 was 707.989 mm, and L was 711.746 mm.

The energy of the X-rays was 30.034 keV, and the dimensions of the incident beam was 0.2×0.2 mm^2. The effective area of the detector was 83.8×106.5 mm^2, the total number of pixels was 487×619 pixel, and the spatial resolution was 0.172×0.172 mm^2.

The test specimen was made of the aluminum alloy A5052 and was annealed at 623K for 1 h. The mean grain size of the specimen was 27 μm. The Al 331 diffraction was employed, and the stress-free diffraction angle was $2\theta_0 = 25.627°$ ($d_0 = 93.006$ pm). The stiffness of the single crystal c_{ij} was obtained from another study [6]. The diffraction elastic constants were calculated with the Kröner model [7,8] and are $E_X = 71.39$ GPa and $v_X = 0.347$ for the Al 331 diffraction.

Residual Stress of the Plastically Bent Specimen
Diffraction images were measured with the PILATUS at P_1 and P_2, and examples are shown in Fig. 2 (a). The rectangular box in the P_1 image indicates the detected area at P_2. The diffraction spots from the 331 diffraction were used for the stress measurement. Each diffraction image was converted into a binary image, from which the outline of the diffraction spot was extracted.

As shown in Fig. 2 (b), the brightness centre of gravity was calculated, and the coordinates $P_1(x_1,z_1)$ and $P_2(x_2,z_2)$ of the diffraction spots were determined by the matching program. Finally, the diffraction angle 2θ and the diffraction position $P_C(x_C,y_C,z_C)$ were obtained. As the process was very complicated, we developed an expert system that performed the analyses and matching mentioned above.

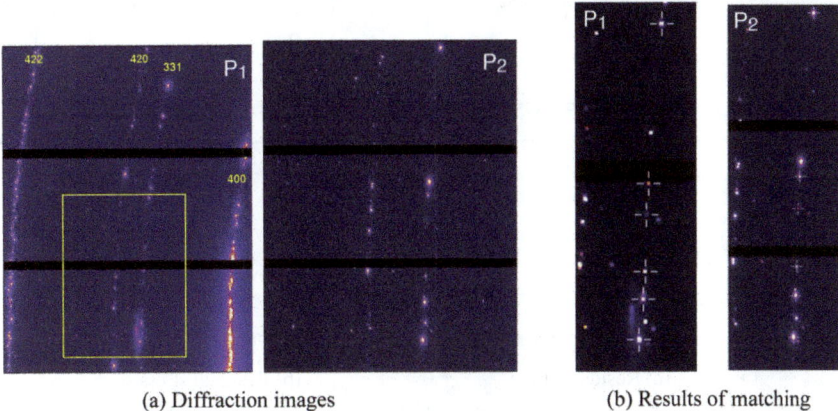

(a) Diffraction images (b) Results of matching

Fig. 2 (a) Diffraction images measured by DEM and (b) matching diffraction spots at P_1 and P_2.

Residual Stresses 2018 – ECRS-10   Materials Research Forum LLC
Materials Research Proceedings **6** (2018) 69-74   doi: http://dx.doi.org/10.21741/9781945291890-12

The plate specimen was 4.7-mm wide and 3-mm thick and was plastically bent. The diffraction angle and position of the diffracted grain were measured using the DEM. Figure 3 (a) shows the measured diffraction angles. The direction along the width of the specimen is for $\chi = 0$ and that of the longitudinal bar axis is for $\chi = 90$. The error bars in the figure indicate the standard deviation. The diffraction angle along the longitudinal bar axis ($\chi=90$) changes due to plastic bending. The plots in Fig. 3 (b) show the positions of the diffracted grains. The rectangle with the black dashed line in Fig. 3 (b) indicates the cross section of the specimen. The vertical red dashed lines correspond to the incident X-ray beam. The position of each coarse grain can be obtained by the DEM as shown in Fig. 3 (b). In the plastic zone, a diffraction ring appears, and the spotty pattern is less pronounced. Therefore, the number of points in the plastic zone decreases compared with the elastic zone.

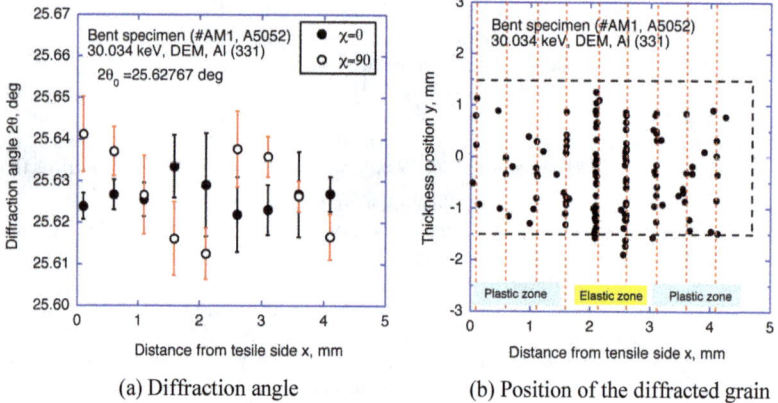

(a) Diffraction angle   (b) Position of the diffracted grain

Fig. 3 Results of the plastically bent specimen.

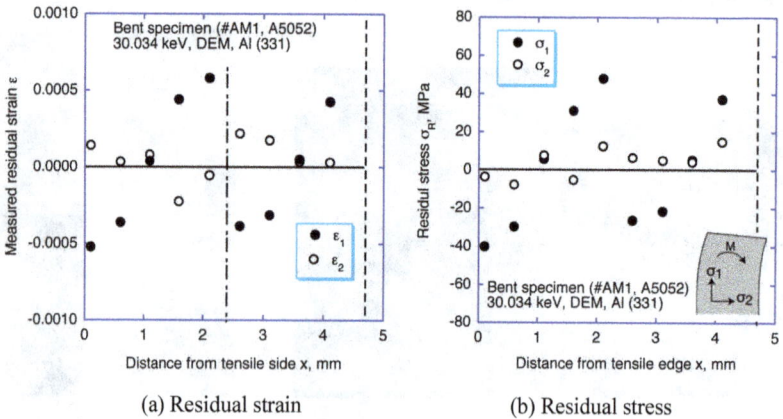

(a) Residual strain   (b) Residual stress

Fig. 4 Residual strain and stress of the plastically bent specimen measured by DEM.

The distributions of the residual strain and stress are shown in Fig. 4 (a) and (b). The strain ε_1 and the stress σ_1 are along the longitudinal axis, while the strain ε_2 and the stress σ_2 are along the

Residual Stresses 2018 – ECRS-10   Materials Research Forum LLC
Materials Research Proceedings 6 (2018) 69-74   doi: http://dx.doi.org/10.21741/9781945291890-12

width of the bar. The stresses were determined from the strains for the bent specimen under the biaxial stress assumption. The compressive residual strain and stress are introduced in the tensile side of bending, and these results show the typical distribution of the plastically bent plate.

Residual Stress Distributions of the Indented Specimen

To assess the effectiveness of the DEM, we measured the residual stress distribution of the indented specimen. A rectangular bar was used to strike the side edge of the plate specimen twice, as shown in Fig. 5 (a). The 1st indentation was tilted to the left, and the 2nd indentation was driven vertically. Figure 5 (b) shows the indented specimen, and the distribution of the residual stress around the asymmetric indentation is interesting.

Figure 6 (a) shows the stress maps of the indented specimen measured by the DEM. The measured area was 8-mm long and 4.7-mm wide. The diffraction angle with respect to the horizontal and vertical directions was measured every 0.2 mm. The incident X-ray beam penetrated through a thickness of the specimen, and the diffraction angles with $\chi = 0°$ and $90°$ were measured.

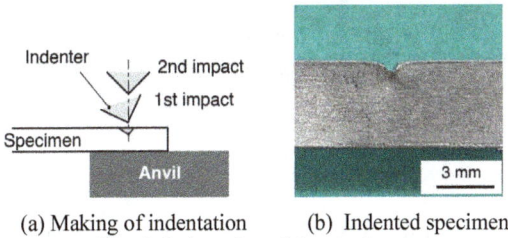

(a) Making of indentation   (b) Indented specimen
Fig. 5 Indentation of the specimen.

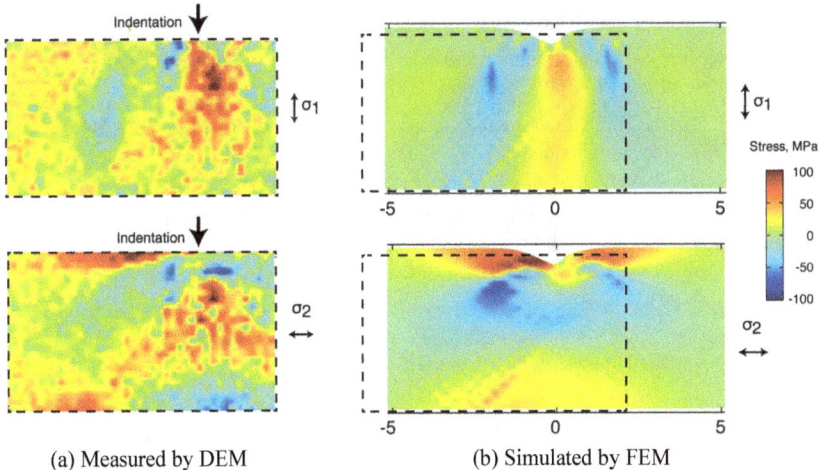

(a) Measured by DEM   (b) Simulated by FEM

Fig. 6 Distribution of the residual stress in the indented specimen.

A plane stress regime was assumed, and the stress and strain were obtained from the measured diffraction angles. For the residual stress σ_1, a large tension is introduced under the indentation, and a compressive region exists around the indentation, as shown in Fig. 6 (a). For the residual stress σ_2, a large tensile region was generated in the left side of the indentation, and compression is introduced under the indentation. A tensile region was introduced near the edge of the anvil.

A simulation of a model with a similar indentation was performed to verify the accuracy of the obtained residual stress distribution. The simulation was performed with the finite element program "Impact" [9], and the results are shown in Fig. 6 (b). The broken line in the figure indicates the area measured by the DEM. When Fig. 6 (a) is compared with Fig. 6 (b), the measured residual stress distribution is similar to the simulated one, indicating that the DEM is useful to measure the stress of coarse grain materials.

Summary

A DEM using hard synchrotron X-rays was proposed. The residual stresses of a plastically bent specimen and the residual stress distribution within an indented specimen were measured, and the accuracy of the method was verified by comparison with simulated data. The results show that the DEM is useful for the stress measurement of coarse grain materials.

Acknowledgements

The authors would like to acknowledge the financial support from the Japan Society for the Promotion of Science, JSPS KAKENHI for Scientific Research (C) 17K06046. The synchrotron radiation experiment on the BL22XU beamline at SPring-8 was performed with the approval of the JAEA (No. 2017A-E10), and the experiment at BL22XU was supported by the "Nanotechnology Platform" of the Japanese Ministry of Education, Culture, Sports, Science and Technology.

References

[1] B.B. He, K.L. Smith, A new method for residual stress measurement using an area detector, in: T. Ericsson, M. Odén, A. Andersson (Eds), Proc. of the 5th international conference on residual stresses (ICRS-5), Linköping, Sweden, 1997, pp. 634-639.

[2] S. Taira, K. Tanaka, Local residual stress near fatigue crack tip, Transactions of Iron and Steel Institute of Japan, 19 (1979) pp. 411-418.

[3] K. Suzuki, Proposal for a direct-method for stress measurement using an X-ray area detector, NDT and E International, 92 (2017) pp. 104-110. https://doi.org/10.1016/j.ndteint.2017.07.012

[4] K. Suzuki, T. Shobu, A. Shiro and S. Zhang, Internal stress measurement of weld part using diffraction spot trace method, Materials Science Forum, 777 (2014) pp. 155-160. https://doi.org/10.4028/www.scientific.net/MSF.777.155

[5] K. Suzuki, T. Shobu, A. Shiro and H. Toyokawa, Evaluation of internal stresses using rotating-slit and 2D detector, Materials Science Forum, 772 (2014) pp. 15-19. https://doi.org/10.4028/www.scientific.net/MSF.772.15

[6] G.N. Kamm and G.A. Alers, Low-temperature elastic moduli of aluminum, Journal of Applied Physics, 35 (1964) pp. 327-330. https://doi.org/10.1063/1.1713309

[7] E. Kröner, Berechnung der elastischen Konstanten des Vierkristalls aus den Konstanten des Einkristalls, Zeitschrift Physik, 151 (1958) pp. 504-518. https://doi.org/10.1007/BF01337948

[8] Information on http://www.rigaku.co.jp/app/kroner/kroner_c.html

[9] Information on https://sourceforge.net/projects/impact/

Residual Stresses 2018 – ECRS-10   Materials Research Forum LLC
Materials Research Proceedings 6 (2018) 75-80   doi: http://dx.doi.org/10.21741/9781945291890-13

Study of Error Analysis and Sources of Uncertainty in the Measurement of Residual Stresses by the X-Ray Diffraction

E.T. Carvalho Filho[1,a,*], J.T.N. Medeiros[1,b], L.G. Martinez[2,c] and V.C. Pinto [1,d]

[1]Federal University of Rio Grande do Norte, Brazil

[2]Nuclear and Energy Research Institute, Brazil

[3]Federal Institute of Rio Grande do Norte, Brazil

[a]eugenioteixeira_@hotmail.com, [b]jtelesforo@yahoo.com, [c]lgallego@ipen.br,
[d]vinicius_c_pinto@hotmail.com

Keywords: Residual Stress, X Ray Diffraction, Repeatability, Reproducibility

Abstract. The aim of this work is to analyze the sources of errors inherent to the residual stress measurement process by X-ray diffraction technique making an interlaboratory comparison to verify the reproducibility of the measurements. For this work were machined specimens with grinding finish, with polishing finish and to be a reference standard an iron powder was used To verify the deviations caused by the equipment, those specimens were positioned and with the same analysis condition, seven measurements were carried. To verify sample positioning errors, seven measurements were performed by positioning the sample at each measurement. To check geometry errors, measurements were repeated for the geometry Bragg Brentano and Parallel Beams. In order to verify the reproducibility of the method, the measurements were performed in two different laboratories and equipments. The results were statistically worked out and the quantification the type A errors that suggests that is a significant difference between the methods and orientation of grooves directions.

Introduction

According with ASTM E2480/2017, generally, *"a single test specimen may be measured more than once and the results combined to produce a test result if the protocol or test method so specifies"* and *"the same combination of operator and apparatus is used to obtain every test result on every specimen"*. In this study four test specimens were measured to investigate the uncertainty sources of XRD residual stress measurements and different laboratories, operators and apparatus – this is the main motivation for this work.

Nowadays, reliability is essential to quantify the residual stresses according to loading rates and mechanical behavior of structural materials. The X-ray diffraction technique is one of the most sensitive techniques for small variations of the crystalline lattice, since the X-ray beam interacts with the interplanar distance and one of the most used non destructive techniques to evaluate stress level in polycrystalline materials [1,2]. In each obtained scan, the chosen diffraction peak position has to be accurately determined, i.e. the 2θ value has to be measured from a very broad and sometimes irregularly shaped peak [3]. The obtained 2θ values are then plotted in a graphic versus d-spacing where the linear regression is performed in order to obtain the slope and intercept values to be used with material elastic constants for the stress calculation [3,4]. The accuracy of the XRD-$\sin^2\psi$ residual stress measurement depends on the minimization of various measurement errors derive from several sources. Materials to be measured, machine limitation, methods of analyses, manpower and software measurement conditions are the main big uncertain sources [3,4,5,6].

Residual Stresses 2018 – ECRS-10   Materials Research Forum LLC
Materials Research Proceedings **6** (2018) 75-80   doi: http://dx.doi.org/10.21741/9781945291890-13

Experimental

Ishikawa diagram is an excellent tool for mapping the possible sources of uncertainties [7]. An Ishikawa diagram was developed for residual stress measurements by XRD (Fig. 1).

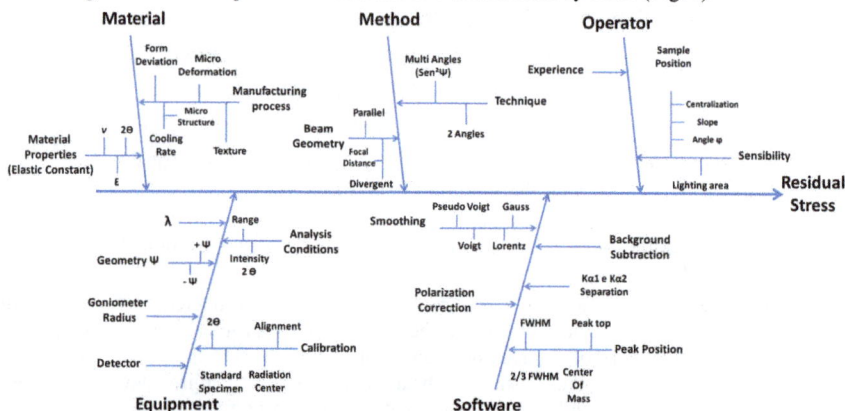

Figure 1 - *Ishikawa diagram (Cause and Effect) to XRD Residual Stress Measurements.*

Considering the sources of uncertainties grouped in the Ishikawa diagram the present work studies the influence of the material manufacturing process, the incident beam geometry, the positioning of the sample and with the collaboration of the laboratory located at the IPEN/USP it was possible to study the reproducibility of the measurements.

For this paper, two specimens were machined from a previously characterized drawn steel bar (AISI 1020). One of the specimens was polished with alumina suspension (0.3 μm). The other specimen was grinded in 4 steps with 0.03 mm of grinding depth each, grinding wheel speed of 3500 rpm and transverse motion of 0.10 mm/stroke.

As SAE HS784 and ASTM E915-10 standards recommend a ferritic steel powder with a grain size> 45 μm and heat treated in a vacuum for relieving stress was used as reference.

The XRD residual stress measurements were made at UFRN in a Shimadzu model 7000 and repeated at IPEN in a Rigaku Ultima IV model, both with the target of CrKα (2.289Å). The XRD settings are shown in Table. 1.

Tabel 1 – *XRD settings of the diffraction measurements for residual stress analysis.*

	UFRN	IPEN
Method	*Iso-inclination* (ψ-constant)	*Iso-inclination*(ψ-constant)
Goniometer radius	200 cm	200 cm
2Θ	156° (211)	156° (211)
Range	150° to 162°	152° to 161°
Step	0.1°	0.1°
Counting time	2 sec.	2 sec.
Speed	3°/min.	3°/min.
Current	30 mA	40 mA
Voltage	30 kV	40 kV
Ψ Tilts	(0; 12.92; 18.43; 22.79; 26.56; 30; 33.21; 36.27; 39.23; 42.13; 45.)	(-60; -48.59; -37.76; -25.65; 0.)

For the analyzes performed in both laboratories a device that restricts the degrees of freedom of the sample positioning in the equipment was built, leading the specimens to be

Residual Stresses 2018 – ECRS-10   Materials Research Forum LLC
Materials Research Proceedings **6** (2018) 75-80   doi: http://dx.doi.org/10.21741/9781945291890-13

always positioned in the same place. To evaluate the influence of the positioning of the samples (analyzing the – method – error), 7 measurements were performed by repositioning the sample at each measurement and to evaluate the repeatability of the equipment, 7 uninterrupted measurements were performed without interfering with the positioning of the samples (analyzing the – equipment – error). In order to evaluate the influence of the machining direction, this same procedure was repeated in the grinded specimen for two conditions: aligning the X-ray beam in the longitudinal direction of the grooves ($\varphi = 0$) and in the transversal direction of the grooves ($\varphi = 90$) of the specimen machined. For the polished sample and the iron powder a random position was initially established and then they were measured following the same procedures as cited above. The standards and technical books advise the use of parallel beam geometry for the calculation of residual stresses to minimize the effects of shape deviation, since the divergent beam geometry (Bragg Brentano) is sensitive to sample positioning and surface irregularities. For an evaluation of the beam geometry the analyzes were performed in both geometries: Pseudo-Parallel Beam and Divergent Beam.

The measurement uncertainty generally includes many variables which can be estimated based on the statistical distribution of the measurement series and they can be characterized by experimental standard deviations. The procedures used to treat the data obtained by the equipments were similar, but the software used by Shimadzu equipment was limited, it did not own some functions present in the software used by Rigaku equipment. The function adjustment for the smoothing and the intensity of the background radiation was determined by the measurement of 5 points on the left and right sides of each diffraction peak, adjusted by the absorption effect. Thereafter, the background radiation is subtracted, the diffraction line separations in $K\alpha1$ and $K\alpha2$ components were performed and thus the position of the 2Θ peak is obtained. After plotting the graph ($\varepsilon_{\varphi\psi}$ versus $sin^2\psi$), a linear function is fitted in which the slope of this line multiplied by the elastic constant of the material gives the value of the residual stress.

Results and discussion

Initially, the average values and experimental standard deviations were calculated for the measurements performed at UFRN (Fig. 2) and IPEN (Fig. 3)

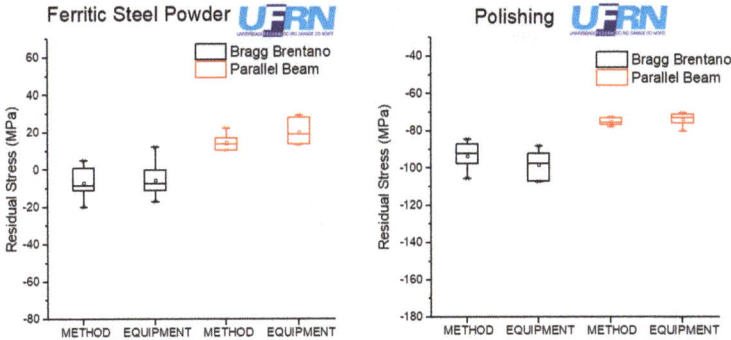

Residual Stresses 2018 – ECRS-10　　　　　　　　　　　　　　　Materials Research Forum LLC
Materials Research Proceedings 6 (2018) 75-80　　　　doi: http://dx.doi.org/10.21741/9781945291890-13

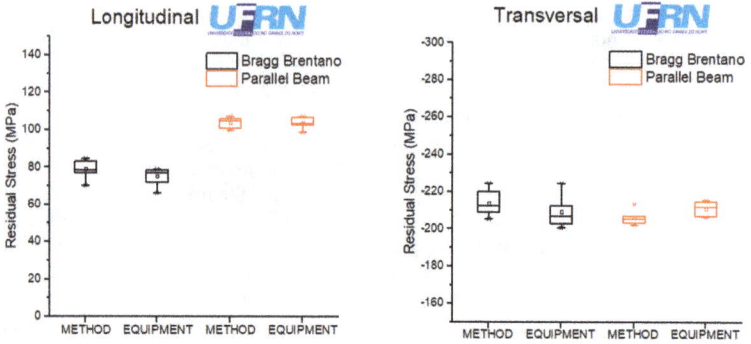

Figure 2 *– Residual stress measurements obtained at UFRN.*

This box chart graphic show four quadrants that represents 25% of the measurements and it's possible to use to study midhinge, range, mid-range, e trimean. The values obtained in the equipment from UFRN showed repeatability, demonstrated by the low values of experimental standard deviations. The values obtained with the iron powder were very close to zero (as expected according to standards and technical books) for both geometries. Although the values are within the acceptable range according to the standards (they can fluctuate between ± 20 MPa), in the geometry of divergent focusing (Bragg Brentano) it was possible to get closer to the true value. It was observed that, in most cases, the experimental standard deviation of the equipment showed to be slightly superior to the deviation of the measurement method, evidencing that the device built to perform the analyzes and the method adopted to carry out the measurements was not interfering in the quality of the presented results.

The measurements performed in the longitudinal direction to the grooves ($\varphi = 0$) presented to be tensile (positive values) while the measurements performed in the transversal direction to the grooves ($\varphi = 90$) presented to be compressive (negative values), indicating a heterogeneous state of stress, which is expected for grinding [8].

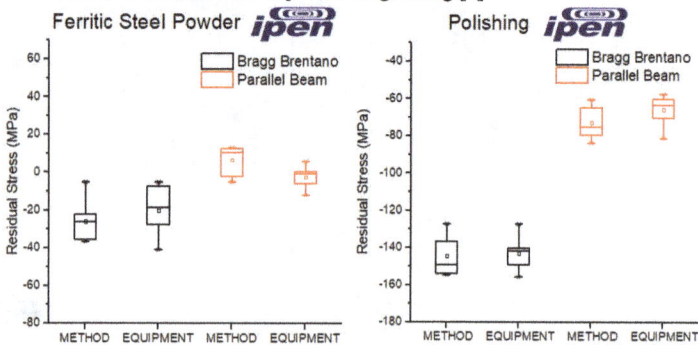

Residual Stresses 2018 – ECRS-10   Materials Research Forum LLC
Materials Research Proceedings 6 (2018) 75-80   doi: http://dx.doi.org/10.21741/9781945291890-13

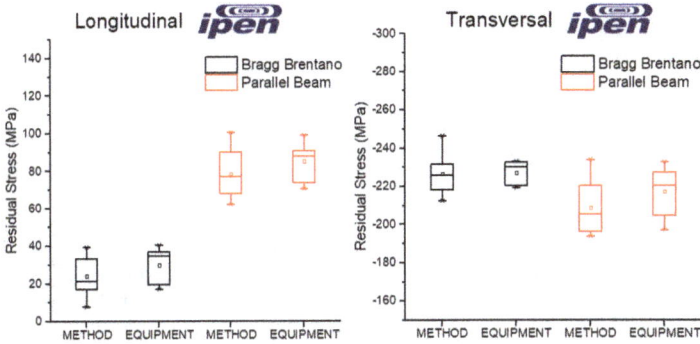

Figure 3 – *Residual stress measurements obtained at IPEN.*

Despite the low values of experimental standard deviations, the values obtained in the IPEN equipment were slightly higher than the values obtained in the UFRN and also demonstrated repeatability. The values of the iron powder were very close to zero for the geometries of pseudo-parallel beams, however for the geometry of divergent beams, the results were slightly above the acceptable by the standards. The aspect of the residual stresses measured in the IPEN equipment also showed to be consonant with those measured in the UFRN, but they diverged somewhat in module. The experimental deviations, analyzing the uncertainty of the method and the uncertainty of the equipment, proved to be very close, corroborating what was obtained with the UFRN equipment.

Given the measured data, using standard experimental deviations, type A (MPa) uncertainties were calculated from repeated observations, in this case, the uncertainty value (u) can be estimated as the standard deviation (s) divided by the number of the performed observations (Table 2).

Table 2 – Type A (MPa) uncertainty calculated from UFRN and IPEN measurements.

		Bragg Brentano		Parallel Beam	
		Method	Equipment	Method	Equipment
Ferritic Steel Powder	UFRN	3.07	3.53	3.80	1.52
	IPEN	4.01	4.60	2.80	2.11
Polishing	UFRN	2.65	2.73	0.74	1.26
	IPEN	3.95	3.35	3.26	8.01
Grinded Specimen	UFRN	1.78	1.74	0.99	1.09
(Longitudinal)	IPEN	4.07	3.56	5.06	3.75
Grinded Specimen	UFRN	2.60	3.06	1.38	1.37
(Transversal)	IPEN	4.23	2.28	5.44	4.73

The results showed that the equipment of the UFRN was more repetitive since the values of the uncertainty of type A suggested smaller values than for the measurements realized in the IPEN. Furthermore, most of the values for the geometry of divergent beams results were higher for both laboratories, indicating a greater susceptibility to the error, since there is a relation between the focal distance of the X-ray source, the specimen positioning and the detector. Therefore, any variation in this distance causes shifting of the peak, leading to a fluctuation in the measurement.

Conclusions

The results from this study suggest that the technique of X-ray diffraction using the geometry of pseudo-parallel beams is accurate and reproducible. The distinct responses between the geometry of pseudo-parallel beams and divergent beams proved to be different, but books and technical standards such as SAE HS784-03 consider that a difference of ± 20 MPa within the uncertainty limit of the technique. The direction of the machining scratches plays an important role in the variation of residual stresses. The X-ray diffraction is considered as a reference technique for validation of other methods of residual stress analysis. However, the results of this study revealed that it is necessary to minimize the sources of uncertainties during the operation of X-ray diffraction measurement, since this technique demonstrated considerable sensitivity to external influences, as demonstrated in the Ishikawa diagram. Hence, for a more detailed study on this quantification of this error, it is necessary to observe these variables inherent to the technique, constructing a worksheet of the measurement uncertainty estimation.

References

[1] P.J. Withers, H.K.D.H. Bhadeshia, Mater. Sci. Technol. 17 (2001) 355. https://doi.org/10.1179/026708301101509980

[2] B.D. Cullity, S.R. Stock, Elements of X-ray Diffraction, 3rd edition, Prentice Hall,

Upper Saddle River, NJ, 2001, p. 435.

[3] M.E. Fitzpatrick, A.T. Fry, P. Holdway, F.A. Kandil, J. Shackleton, L. Suominen, Determination of Residual Stresses by X-ray Diffraction—Issue 2, DTI, 2005, Measurement Good Practice Guide No. 52.

[4] Q. Luo, A. H. Jones, High-precision determination of residual stress of polycrystalline coatings using optimised XRD-$Sin^2\psi$ technique, Surface and Coatings Technology, 205(5)(2010)1403–1408. https://doi.org/10.1016/j.surfcoat.2010.07.108

[5] A T Fry, F A Kandil, A Study of Parameters Affecting the Quality of Residual Stress Measurements Using XRD, Materials Science Forum Vols. 404-407 (2002) pp. 579-584. https://doi.org/10.4028/www.scientific.net/MSF.404-407.579

[6] Czan, A., Zauskova, L., Sajgalik, M., & Drbul, M. (2016). Triaxial Measurement Method for Analysis of Residual Stress after High Feed Milling by X-Ray Diffraction, *Technological Engineering*, *13*(2), 31-33. https://doi.org/10.2478/teen-2016-0019

[7] L. Luca et al., "Study on Identification and Classification of Causes which Generate Welds Defects", Applied Mechanics and Materials, Vol. 657, pp. 256-260, 2014. https://doi.org/10.4028/www.scientific.net/AMM.657.256

[8] PINTO, Vinícius Carvalho. Influence of cutting parameters on the surface integrity of SAE 1045 steel after turning. 2017. 115p. Dissertation (Master in Mechanical Engineering) - Technology Center, Federal University of Rio Grande do Norte, Natal, 2017.

Mechanical Relaxation Methods

Residual Stresses 2018 – ECRS-10　　　　　　　　　　Materials Research Forum LLC
Materials Research Proceedings 6 (2018) 83-88　　　doi: http://dx.doi.org/10.21741/9781945291890-14

A Procedure for Plasticity Error Correction for Determination of Residual Stresses by the ESPI-HD Method

Leonid Lobanov[1,a], Frederico A.P. Fernandes[2,b] and Viktor Savitsky[1,c,*]

[1]E.O.Paton Electric Welding Institute, Kyiv, Ukraine

[2]Federal University of ABC, São Bernardo do Campo, 09606-045, Brazil

[a]holo@paton.kiev.ua, [b]codoico@gmail.com, [c]viktor.savitsky@gmail.com

Keywords: Residual Stresses, Speckle-Interferometry, Hole Drilling, ESPI-HD Method, Plasticity Effect

Abstract. The standard strain gage hole drilling method for residual stress determination assumes a linear elastic behavior of the material. However, if the stress level is high, plastic deformation may occur near the drilled hole, resulting in significant errors in the calculation of residual stress. The application of optical whole-field interferometry methods like speckle-interferometry (ESPI) for determining residual stresses allows to gather more information about the stress state of an object compared to the data from resistance strain gauges. Such advantage is achieved by calculating a significantly larger amount of displacement values induced by the hole-drilling. The ESPI-HD method additionally enables automated data analysis and selection of predefined areas with displacements in the vicinity of the drilled hole to determine the plasticity effect without the need of the material's yield stress. In the present work, an approach is described to determine the plasticity effect without using the yield stress. The approach is based on the fact that plastic deformation in the vicinity of the hole is not uniform. Analysis of stress variation at different sectors near the hole allows retrieving the real stress state. The developed method can significantly reduce errors in the determination of high residual stresses. The influence of materials properties, such as Young and plastic tangent moduli, Poisson ratio and yield stress, on the results obtained by the proposed procedure for plasticity error correction is investigated.

Introduction

Reliability and service life of materials are noticeably dependent on the residual stresses, which a high level can result in crack initiation, decrease of corrosion resistance, etc. Therefore, the development of reliable and informative methods of investigation of the stress state in structures is important. One of the most widely used methods for determination of residual stresses is the hole-drilling method. The procedure involves drilling a small hole at the material surface and the strain is measured using a special strain gauge rosette attached concentrically around the drilled hole. According to the standard procedures [1], calculations are made under the assumption of linear elastic behavior of the material. It is widely known that a drilled hole might act as a stress concentrator that can further induce significant plastic deformation, particularly if the nearby residual stresses are high. It was previously shown that if the stress state does not exceed about 60% of the material's yield stress (σ_Y) the errors induced by yielding around the hole are less than 3%, but the plasticity error increases with increasing residual stresses [2-4]. This is associated with the fact that elastic deformations which for stresses should be calculated, are added to unknown plastic deformation. Various approaches have been proposed to calculate the plasticity effect on the results of residual stresses, mainly by using the material's yield strength [3]. However, determining the exact values of the yield stress require additional experimental efforts and may be a new source of error. The application of optical whole-field interferometry methods for determination of residual stresses allows gathering more information about the residual stress state

Residual Stresses 2018 – ECRS-10   Materials Research Forum LLC
Materials Research Proceedings 6 (2018) 83-88   doi: http://dx.doi.org/10.21741/9781945291890-14

of an object when compared with the data from the resistance strain gauges [5, 6]. This is achieved by calculating a significantly larger amount of displacement (or strain) values induced by the hole-drilling. Electronic speckle-interferometry hole-drilling (ESPI-HD) method allows an automation of the data analysis and selection of predefined areas with displacements in the vicinity of the drilled hole in order to determine the residual stresses [6]. This additionally enables determination of the plasticity effect without the need of the material's yield stress [4].

Determination of residual stresses by the ESPI-HD method

Figure 1 schematically shows a typical ESPI setup [6] for surface displacement measurement. The light from a coherent laser source (1) is split into two parts using a beam splitter (2). Optical configuration for measuring in-plane displacements is based on the two-beam illumination arrangement. The observation direction is normal to the specimen surface. The two parts of the laser light interfere on the object's surface to form a speckle pattern, which can be imaged by a digital camera (7) and further calculations assume that the ESPI device measures only an in-plane component of the surface displacements.

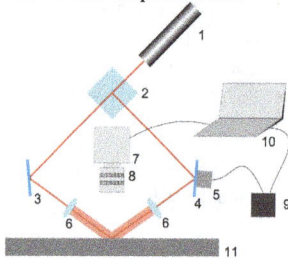

Fig. 1. A schematic presentation of the ESPI setup sensitive to in-plane displacement: 1 – laser, 2 – beam splitter, 3 – mirror, 4, 5 – piezoelectric-driven mirror, 6 – beam expanders, 7, 8 – digital camera with lenses, 9 – piezo controller, 10 – computer, 11 – object.

Fig. 2. The ESPI device for determination of residual stresses installed on a welded specimen

The displacements, caused by the relaxation of residual stresses in the bulk material, due to the drilling of blind holes are then measured using the ESPI method. Measurements are made in the following sequence [6]: the ESPI device is placed at the surface of the object and four-phase shifted speckle patterns are recorded, characterizing the initial state before drilling a hole. A phase map of the initial state is calculated by the four-step phase shifting algorithm. After stress relaxation caused by drilling a blind hole with a diameter of 0.5-2.0 mm ($2r_0$), another set of speckle images, corresponding to the deformed state, is recorded and the resulting phase map is calculated. Computer processing of fringe patterns allows the evaluation of the displacements, u_x, in the laser illuminated area, as well as the calculation of the residual stresses. The proposed ESPI-HD method uses displacements, $u_x(r, \theta)$, at a constant radius $r = 2.5r_0$ around a hole to calculate stresses because of: 1) the mean radius of the strain gauge rosette usually equals to $2.5r_0$; 2) automatic processing of speckle images is affected by the noise due to surface damage from chipping after drilling, as well as electronic noise from the digital camera, speckle-noise, laser radiation noise, etc. According to previous studies [4-6] the dependency of displacements, $u_x(r, \theta)$, due to blind hole drilling, on the residual stresses can be written in the following form [5, 6]:

$$u_x(r,\theta) = F(r,\theta)\sigma_{xx} + G(r,\theta)\sigma_{yy} + H(r,\theta)\tau_{xy} \qquad (1)$$

Residual Stresses 2018 – ECRS-10   Materials Research Forum LLC
Materials Research Proceedings 6 (2018) 83-88   doi: http://dx.doi.org/10.21741/9781945291890-14

where the basic functions $F(r,\theta)$, $G(r,\theta)$ and $H(r,\theta)$ depend on the mechanical properties of the material, hole dimensions, radial position around the blind hole and can be determined from the displacement data of finite-element modeling of the blind hole drilling process [5, 6].

To find the three unknowns σ_{xx}, σ_{yy} and τ_{xy} from a system of linear equations (1) it is sufficient to measure displacements u_x at three points $u_{x,n}(2.5r_0, \theta_n), n = 1 \ldots 3$, which are located on a circle with a center at the drilling point [6]. The system of equations to be solved can adopt $n > 3$ values θ_n to enable a least squares solution.

Influence of the plasticity effect on residual stresses calculation
Mathematical simulation with a finite element method was conducted in order to estimate the influence of plastic deformation on the accuracy of residual stress determination by the ESPI-HD method for different materials properties, such as Young (E) and plastic tangent (Et) moduli, Poisson ratio (ν) and yield stress (σ_Y). FEM simulations were carried out by using the Code-Aster software. Due to the symmetrical geometry of the problem, only one quarter-plate finite element model consisting of 50976 HEXA8 elements was used. The following conditions were used in the simulation: hole diameter and depth of 1 mm, object thickness was 5 mm and object width was 35 mm. Stress-strain curves are presented in Fig. 3.

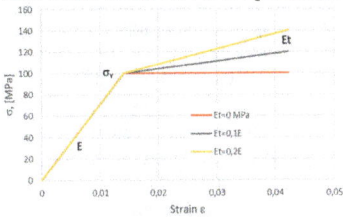

Fig. 3. Stress-strain curves used for FE simulation.

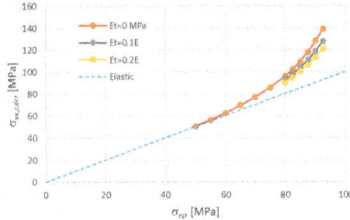

Fig. 4. Influence of plastic tangent module on residual stresses calculation. The dotted line represents an elastic material behavior.

As a result from simulation, displacement values of material surface were then calculated, depending on the stress level in the object. Afterwards, displacements, $u_x(2.5r_0, \theta)$, were selected at different points at the surface located at a distance of $2.5r_0$ from the hole center. Accordingly, the stress values σ_{xx}, σ_{yy} and τ_{xy} were found using Eqs. 1. For simplicity in the simulations only the $\sigma_{xx} = \sigma_{rs}$ stress component was investigated while σ_{yy} and τ_{xy} were assumed to be zero. The displacements $u_{x,calc}(2.5r_0, \theta)$ at the distance of $2.5r_0$ for angles of θ in the range $[0;2\pi]$ were calculated for a given σ_{rs}/σ_Y ratio. Then, the system of equations (Eqs. 1) was composed and the values of $\sigma_{xx,calc}$ were calculated by the least-squares method. Then the residual stress determination error $\eta = (\sigma_{calc} - \sigma_{rs})/\sigma_{rs}$, depending on the ratio (σ_{rs}/σ_Y), was calculated. It is shown that for the ratio $\sigma_{rs}/\sigma_Y > 0.6$, the difference between σ_{rs} and $\sigma_{xx,calc}$ increases (see Fig.4).

Results of FE simulation for different sets of material properties (E, Et, σ_Y, ν) are presented in Table 1. It is possible to state that σ_Y has the largest influence on the plasticity effect, while the Poisson ratio, ν, does not affect the results significantly. Therefore an exact value of the yield stress is very important to achieve an accurate stress correction using the following steps of the procedure, proposed by Beghini et al, 2011 [3]: "1) evaluate the residual stress, by assuming the material as linear elastic ("as-elastic" residual stress); 2) calculate a plasticity effect intensity parameter; 3) re-evaluate the residual stress components from the as-elastic initially found and the plasticity effect intensity". Step 3 uses a relationship derived from FE simulation (Fig 4).

Table 1. Influence of material properties on plasticity effect

N	input data				as-elastic calculation		plast.eff. correction $(\sigma_Y = 100\ MPa)$		plast.eff. correction $(\sigma_Y = 110\ MPa)$	
	σ_{rs}, [MPa]	σ_Y, [MPa]	v	Et	$\sigma_{xx,calc}$, [MPa]	η, [%]	$\sigma_{xx,calc}$, [MPa]	η, [%]	$\sigma_{xx,calc}$, [MPa]	η, [%]
1	85	100	0,33	0	109,45	28,8	85,0	0,0	89,70	5,5
2	85	110	0,33	0	98,91	16,4	81,2	-4,4	85,00	0,0
3	90	100	0,33	0	127,93	42,1	89,9	-0,1	95,43	6,0
4	90	110	0,33	0	110,22	22,5	85,3	-5,2	90,00	0,0
5	85	100	0,27	0	110,38	29,9	85,3	0,4	90,06	6,0
6	90	100	0,27	0	129,63	44,0	90,3	0,4	95,80	6,4
7	85	100	0,30	0	110,17	29,6	85,3	0,3	89,98	5,9
8	90	100	0,30	0	129,34	43,7	90,3	0,3	95,74	6,4
9	85	100	0,33	0,1E	103,87	22,2				
10	85	100	0,33	0,2E	119,12	17,7				
11	90	100	0,33	0,1E	100,05	32,4				
12	90	100	0,33	0,2E	113,22	25,8				

Determination of the plasticity effect by the ESPI-HD method

There is no difference on the results for different sets of angles θ_n, if the assumptions involved in basic theory, described in the ASTM E837 standard [1], are valid. But if the stress/strain relationship for the test material is nonlinear due to yielding or other causes, the calculated stresses may vary. One of the possibilities of the ESPI-HD method is the ability to compute stresses using the displacement data u_x at points of an arc BC (Fig. 5) that could be chosen around the drilled hole (Fig. 5). Variation of length (defined in radians by the angle ψ) and position (defined by the angle α) of the circular arc BC allows to develop a new technique for residual stress calculation, which takes into account the plastic deformation. The arc is defined by the following procedure. A point A with polar coordinates $(2.5r_0, \alpha)$ is marked on the circle. Regarding this point, the arc BC that includes all available N points with angular coordinates θ_n (where $\alpha - \psi \leq \theta_n \leq \alpha + \psi, n = 1, ..., N$) is chosen. The displacements $u_x(\theta_n)$ on the arc BC is then used to calculate the stress $\sigma_{xx,calc}(\alpha, \psi)$ from the system of Eqs 1.

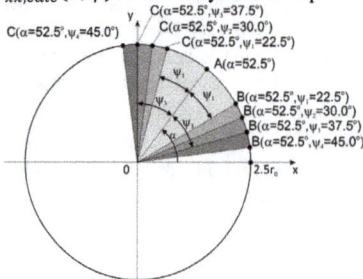

Fig. 5. Location of the arc BC relative to the drilled hole

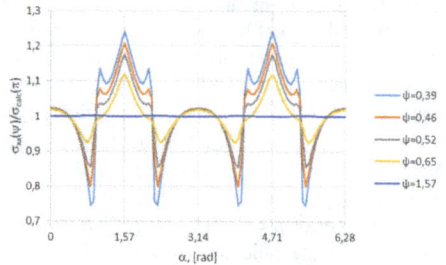

Fig. 6. Dependencies of the ratio, $\sigma_{xx,calc}(\alpha, \psi)/ \sigma_{xx,avr}$ on α calculated by FEM for different values ψ providing for $\sigma_{rs} = 0.8\sigma_Y$

The dependencies of the $\sigma_{xx,calc}(\alpha, \psi)$, calculated using displacement data measured on the arc BC for an angle α, are presented on Fig. 6, for $\sigma_{rs} = 0.8\sigma_Y$, $E = 70\ GPa$, $Et = 0\ GPa$, and $v = 0.33$. It is shown that in the presence of plastic deformation decreasing the length of the arc BC (lower values of ψ) leads to an increase of the relative standard deviation (RSD_ψ), where RSD

Residual Stresses 2018 – ECRS-10   Materials Research Forum LLC
Materials Research Proceedings 6 (2018) 83-88   doi: http://dx.doi.org/10.21741/9781945291890-14

is defined as the ratio of the standard deviation of $\sigma_{xx,calc}(\alpha, \psi)$ to the averaged value $\sigma_{xx,avr} = \sigma_{xx,calc}(\alpha, \pi)$ calculated using all points along the circle. The curves on Fig. 6 show a significant increase in the difference between the extreme and averaged value of the stress $\sigma_{xx,avr}$ when the angle ψ is decreased. This is a consequence of localized areas with plastic deformation. Thus RSD_ψ is one of the parameters that may characterize the plasticity effect. Dependencies of the RSD_ψ on $1/\psi$ for different σ_{rs}/σ_Y ratios are shown in Fig. 7a. It should be noted that RSD_ψ tends to zero in the absence of plastic deformation. A monotonic increase of RSD_ψ with increasing σ_{rs}/σ_Y is observed. Thus, based on the analysis of the $RSD_\psi(1/\psi)$ dependency, a proof about the absence or presence of plastic deformation after drilling a hole in the object is achieved. Fig. 7b shows that the ratio σ_{rs}/σ_Y is uniquely determined by the slope k_ψ of the graph $RSD_\psi(1/\psi)$.

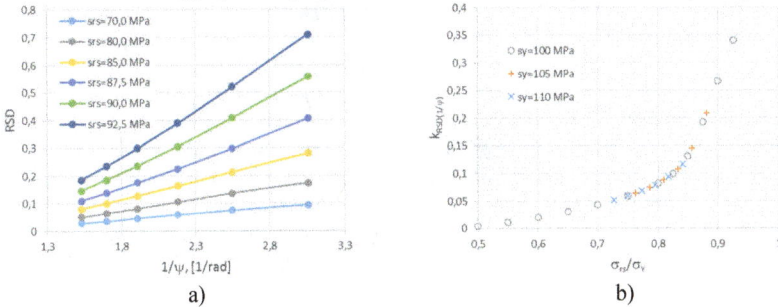

Fig. 7. Dependencies of a) RSD_ψ on $1/\psi$ for different σ_{rs}/σ_Y ratios, and b) k_ψ on the ratio σ_{rs}/σ_Y obtained by FE simulation ($E = 70$ GPa, $Et = 0$ GPa, $v = 0.33$).

The presently developed ESPI-HD method for determination of residual stresses taking into account the plasticity effects includes the following stages:

1. Find the functions $F(2.5r_0, \theta)$, $G(2.5r_0, \theta)$ and $H(2.5r_0, \theta)$ for real experimental conditions (diameter and depth of the hole, thickness of the object, mechanical properties of the material) using linear-elastic FE simulation or from a database.

2. Calculate averaged values of stresses $\sigma_{xx,avr}$ and $\sigma_{yy,avr}$ from Eqs. 1 using all displacement data along a circle with a radius of $2.5r_0$ with the center in the drilling point. If $\sigma_{xx,avr} \gg \sigma_{yy,avr}$, do the next steps.

3. For a set of ψ values in the range $[\pi/6; \pi/2]$ calculate stresses $\sigma_{xx,calc}(\alpha, \psi)$, $\alpha = [0; 2\pi]$.

4. Calculate RSD_ψ for each value of ψ. Build the curve $RSD_\psi(1/\psi)$ and determine the slope $k_{\psi,exp}$ of the curve.

5. Compute the dependency of the slope k_ψ on σ_{rs}/σ_Y from elasto-plastic FE simulation.

6. Determine the plasticity effect by substituting the experimentally obtained $k_{\psi,exp}$ value, calculated in Stage 4, into the dependency of k_ψ on σ_{rs}/σ_Y received obtained using FEM on Stage 5.

7. Calculate the residual stress applying the procedure described by Beghini et al, 2011 [3] and adapted to the displacement-stress form, but using experimentally determined plasticity effect.

The proposed procedure assumes that high residual stress exists only in one direction, which coincides with the laser illumination direction of the ESPI device. Other stress components are considered to be negligible. Such conditions are usual for residual stresses resulting from welding. As an example, residual stresses have been measured by the developed ESPI-HD method considering the plasticity effect in welded joints made of an aluminum alloy AlMg6 (Fig. 2).

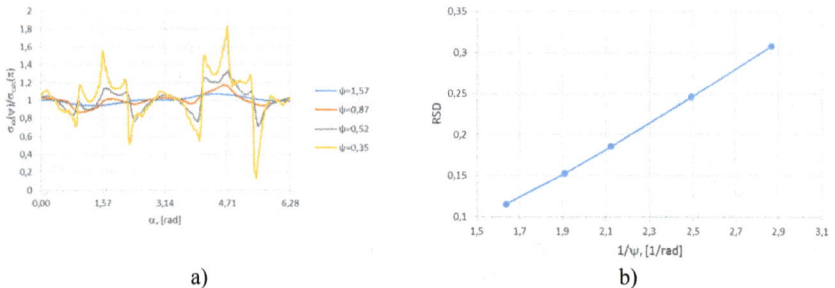

a)   b)

Fig.8. Determination of residual stresses by the ESPI-HD method taking into account the plasticity effect: a) dependence of $\sigma_{xx,calc}(\alpha,\psi)/\sigma_{xx,avr}$ on α for different values of ψ; b) dependence of RSD_ψ on $1/\psi$. Material is AlMg6.

Approach, described in [6], determined that $\sigma_{xx,avr} = 246$MPa. Values of $\sigma_{xx,calc}(\alpha,\psi)$ and RSD_ψ were calculated according to stages 2 to 4 and as a result the $RSD_\psi(1/\psi)$ curve has been found (Fig. 8a and 8b). These graphs are similar to those obtained by mathematical modelling (see Fig. 6 and Fig. 7a) and they clearly show the presence of significant plastic deformation around the hole. FEM simulation has shown that the slope k_ψ of the curve in Fig.8b fits values of the ratio $\sigma_{xx,avr}/\sigma_{rs} = 1.28$. Thus, in the drilling point the residual stresses are $\sigma_{xx} = 192\ MPa$.

Summary
The proposed approach has been approved for measuring residual stresses taking into account the plasticity effect in a welded specimen with high uniaxial stress state. Determination of residual stresses in the welded plate shows that the theoretically predicted influence of high residual stresses on displacement variations is similar to those retrieved experimentally.

Application of the present approach for a biaxial stress state and incremental hole drilling should be further considered as well as experimental validations should be conducted.

References

[1] ASTM E837-13a, Standard test method for determining residual stresses by the hole-drilling strain-gage method, in Annual Book of ASTM Standards, Am. Soc. Test. Mat, 2013.

[2] Lin Y.C., Chou C.P., Error induced by local yielding around hole in hole drilling method for measuring residual stress of materials, J. Mater. Sci. Tech., Vol.11, 1995, pp. 600–604. https://doi.org/10.1179/mst.1995.11.6.600

[3] Beghini M., Santus C., Valentini E., Benincasa A., Experimental Verification of the Hole Drilling Plasticity Effect Correction, Materials Science Forum, Vol. 681 (2011), pp. 151-158. https://doi.org/10.4028/www.scientific.net/MSF.681.151

[4] Lobanov L., Savitsky V., Residual Stresses Determination with Plasticity Effects by Electron Speckle-Interferometry Hole-Drilling Method, Mat. Res. Proc., Vol. 2, 2017, pp. 389-394. https://doi.org/10.21741/9781945291173-66

[5] Baldi, A., A new analytical approach for hole drilling residual stress analysis by full field method, J. Eng. Mater. Technol., Vol. 127, No. 2, 2005, pp. 165–169. https://doi.org/10.1115/1.1839211

[6] Lobanov L., Pivtorak V., Savitsky V. and Tkachuk G., Technology and Equipment for Determination of Residual Stresses in Welded Structures Based on the Application of Electron Speckle-Interferometry, Materials Science Forum, Vol. 768, 2014, pp. 166-173.

Residual Stresses 2018 – ECRS-10
Materials Research Proceedings 6 (2018) 89-94

Materials Research Forum LLC
doi: http://dx.doi.org/10.21741/9781945291890-15

Hole Drilling Residual Stress Evaluations in Cast Iron

Mattias Lundberg[1,a,*], Lennart Elmquist[1,b]

[1]Swerea SWECAST, Tullportsgatan 3, SE-550 02, Jönköping, Sweden

[a]mattias.lundberg@swerea.se, [b]lennart.elmquist@swerea.se

Keywords: Incremental Hole Drilling, Residual Stresses, Grey Cast Iron, Ductile Iron

Abstract. Incremental hole drilling for residual stress measurement are widely used in industry today and is considered to be a cheap and fairly reliable method for stress measurements. Even though the method assumes isotropic material, it has been expanded to orthotropic materials such as composite laminates. With heterogeneous material like grey cast iron, the reliability and accuracy of the method still often fails to provide residual stress data valid for analysis. Cast iron microstructural aspects that complicate the analysis are the graphite and its morphology, variations in matrix structures and casting defects. These features can extend over different length scales and give cast iron highly localised mechanical properties. Global engineering parameters, such as Young's modulus and Poisson's ratio, are used together with the locally measured strains to calculate the residual stresses. Utilizing global material parameters while measuring locally can provide false stress results. Grey cast iron exhibits a non-linear elastic behaviour and the Young's modulus can change significantly and can therefore result in very different calculated residual stresses. Experiments were conducted on cast stress lattices utilizing incremental hole drilling to measure strains. To calculate the residual stresses, global material parameters and standard evaluation procedures in accordance to ASTM E837 were used. Results show that the method is questionable for grey cast iron but can be used for ductile iron. Lack of material properties knowledged are suggested to be the main obstacle for residual stress evaluations on grey cast iron as the accuracy of the method decreases as hole depth approaches 1 mm.

Introduction

Cast irons are widely used in for example heavy duty diesel engines. The material allows for complex components to be cast to near-net-shapes without extensive post machining. A complex shape tends to result in different cooling rates over the cast component. These differences often results in pores at thermal centres and residual stresses from the thermal gradients during cooling. Different thicknesses of the component will result in thermal gradients that lead to residual stresses. Quantification of casting induced residual stresses is therefore important. Tensile residual stresses are detrimental and promote fatigue crack initiation and propagation whereas compressive residual stresses are beneficial for the fatigue strength. Critical load bearing positions on the component could rupture prematurely if the stresses are detrimental. It is therefore important for the industry to have a fast and reliable residual stress measurement method at hand.

Hole drilling and X-ray diffraction for residual stress measurements are common and widely used on a regular basis in industry today. These two techniques are mainly being used on steels, aluminium, titanium and nickel-base materials but also on some cast iron [1,11]. Incremental hole drilling (IHD) is a fast method for stress measurements and allows for collection of data from the first 1 - 2 millimetres below the surface. It is often described as semi-destructive testing and sometimes as non-destructive testing since the hole dimensions are small compared to the component. With IHD it is possible to measure in-plane non-uniform stresses on a macroscopic level. The method is based on strain relaxation as the hole is drilled. Changes in strain are measured by strain gauges attached to the surface. IHD works by drilling the hole in several small steps (increments) and between each step the relaxed strains are recorded. Strains are converted into stresses via one of the available mathematical methods such as Integral method, Kockelmann and Power series. Whilst main part of the published articles concerns the residual stresses from shot peening, welding or machining of the above mentioned material groups, very little has been done on cast iron [7]. There exists a lack of knowledge on how good IHD is for accurate stress measurements on cast iron. Therefore, this study has focused on IHD stress measurement on cast stress lattices in two different classes of cast iron: grey cast iron (GJL) and ductile iron (GJS). With the simple geometry of the cast stress lattice, the residual stresses can be simulated and thought be possible to verified by IHD.

Material and experimental procedure

Stress lattice were cast in grey cast iron class GJL-300 and ductile iron class GJS-500. Thickness were 25 mm, width of side bar were 12 mm, width of middle bar were 68 mm, distance between the bars were 35 mm and the overall length were 295 mm, see Fig 1a). The free graphite in GJL is in the shape of flakes, see Fig. 1b), and GJS as nodules. Young's modulus for GJL and GJS were set to 116 GPa and 185 GPa respectively, whereas the Poisson's ratio were set to 0.3 for both materials. For easier application of strain gauges the surface were machined on a small section across the width of the lattice, as visible in Fig. 1a). Depth of the machining was 600 - 700 μm.

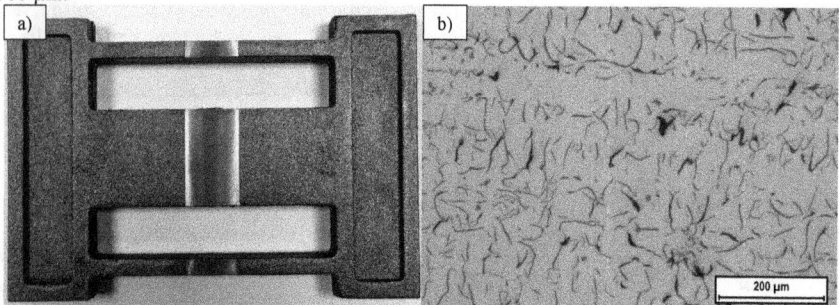

Figure 1: Stress lattice with the machined surface (a) and microstructural image of GJL from centre of the side bar showing type A and type E [12] graphite (b).

Hole drilling equipment used was Restan MTS-3000 from SINT Technology, having a high speed drilling of 350,000 - 400,000 rpm and 1.8 mm diameter inverted cone drill. It is a fully automatic process to ensure good repeatability and is compliant to ASTM E837-13. The method used is based on incremental hole drilling, meaning that the hole is being drilled step by step and between each step the relaxed strains are measured. Increments in steps of 180 μm were used, based on data given by Stefanescu *et al* [13]. With IHD it is possible to achieve good accuracy of the measured stresses [14,15] and therefore thought to be good for investigating cast iron. One hole drilled on GJS was conducted with 90 μm step size. Hole depth were 1.26 mm resulting in z/D = 0.494. Measurements were conducted on one stress lattice of either material. One hole were drilled on each side bar on both lattices and two holes in the middle bar on GJL and one hole on the middle bar on GJS.

Rosette strain gauges from HBM with three strain gauges oriented at 45° to each other having a gauge length of 1.5 mm and a rosette diameter of 5.1 mm were used.

Results

Measured stress distribution in GJL in the thicker middle bar and the thinner side bars overlap each other as illustrated in Fig. 2. Minimum stresses were found closest to the machined surface; -225 MPa for the middle bar and -285 MPa on the side bar. Maximum stresses were found to be 100 MPa for the middle bar and 50 MPa for the side bar located at the bottom position of the drilled hole. Around 700 µm depth all curves kinked and then continued rising.

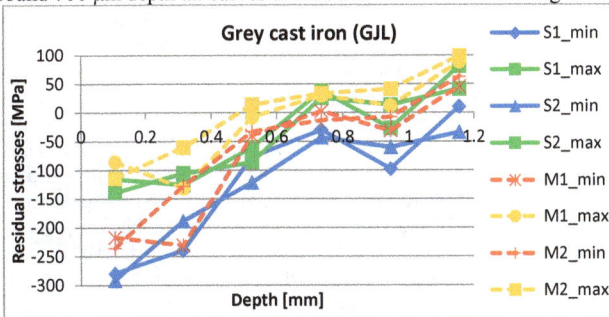

Figure 2: Residual stress distribution in as-cast stress lattice of grey cast iron. S1 and S2 are notations for drilled hole number 1 and 2 in the side bar and M1 and M2 are notations for hole 1 and 2 in the middle bar.

For spheroidal graphite iron the minimum stresses in the side bars were always lower than the middle bar as shown in Fig. 3. For the side bar, there is a trend of increased compressive stresses with depth from -100 MPa at the surface to -250 MPa at 1 mm depth. Maximum stresses are more or less always on the compression side with a few data points between 25 - 45 MPa at 700 µm to 1 mm depth.

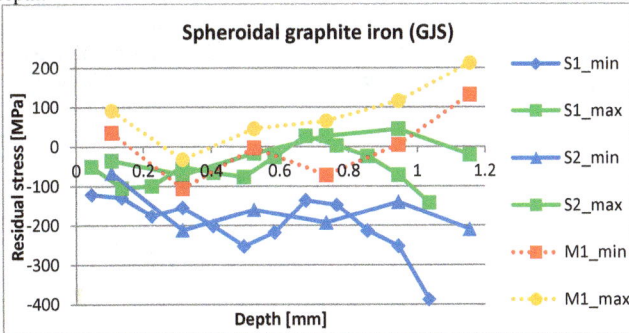

Figure 3: Residual stress distribution for spheroidal graphite iron. S1 and S2 refers to hole number 1 and 2 in the side bar and M1 refers to hole number 1 in the middle bar.

In the middle bar, the minimum stresses distribution shows a concave upward curve with tensile stresses at the surface followed by a decrease in stresses to -100 MPa before changing direction and reaching tensile stresses of 130 MPa. Same features can be said about the maximum stresses, with 100 MPa at the surface followed by its lowest value of -35 MPa at 300 µm and reaching it highest value above 200 MPa at full depth.

Residual Stresses 2018 – ECRS-10   Materials Research Forum LLC
Materials Research Proceedings 6 (2018) 89-94   doi: http://dx.doi.org/10.21741/9781945291890-15

Discussion

When the stress lattices are cast, the side bars will solidify and cool down before the middle bar which has more material. When the middle bar solidify and cool down it want to contract but is hindered by the already solidified side bars. There will be compressive residual stresses in the side bars and tensile residual stress in the middle bar. Direction of the stresses will be along the bar axis. From simulation on GJL [16], the residual stress in the side bars should be around -190 MPa and in the middle bar around 95 MPa. For GJL this is not seen with IHD, no significant differences in residual stresses can be observed. Possible reasons for this could be machining induced plasticity, casting skin or variation in microstructure. One measurement was conducted directly on the as-cast surface on the middle bar after gentle polishing, to smoothen out the roughest patches. This measurement showed the same result as previously measurements and therefore the observed compressive stresses are not just from machining. GJL exhibit a non-linear elastic behaviour and is brittle. Graphite's loadbearing capacity in tension is low and basal planes start to slip below 30 MPa tensile load [17]. The graphite tip acts as stress raises making them perfect points for crack initiation. In compression the graphite supports the matrix and is loadbearing, resulting in a strongly anisotropic stress-strain curve between tension and compression [18]. Stress evaluations of IHD data on GJL material are complex due to its anisotropic load responds. Global parameters derived from tensile testing might not be enough for proper stress calculation and evaluation. Local variations in microstructure are a part of cast iron nature, providing some of its unique properties. Variations in graphite morphology are often seen because of different cooling rates and chemical variations. Microstructure investigation of the side bar revealed type D and type E graphite with some rosettes (type B) and no type A graphite as seen in Fig. 4. Microstructural study conducted prior to this investigation provided somewhat false conceptions regarding the actual graphite morphology in the investigated area. To the authors' best knowledge, no investigations have been conducted about local responses to these kinds of graphite morphologies. GJL matrix consists of different amounts of ferrite/pearlite, which also could lead to misinterpretations of the stresses if local variations are present [19,20]. If the matrix just underneath the attached strain gauges is pearlitic and almost fully ferritic after a few hundredth microns, then it is easy to suspect that material behaviour differs. However, the matrix of the investigated GJL was fully pearlitic and thus not being a concern of irregular loading responds between grains. It is a complex microstructure with different behaviour when investigated over different length scales. Therefore, several holes should be drilled and the results compared, as well as microstructural investigations. Better knowledge from thorough investigations are needed to understand the complex GJL material propertied before adequate stress evaluations of GJL using IHD are really feasible.

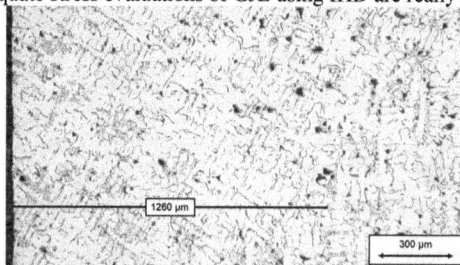

Figure 4: Typical microstructure in the side bar of GJL, left side is the machined surface.

Ductile iron load respond differs from grey cast iron. GJS and construction steel have a close resemblance in stress-strain curves with a clear linear part followed by plastic deformation and

Residual Stresses 2018 – ECRS-10   Materials Research Forum LLC
Materials Research Proceedings 6 (2018) 89-94   doi: http://dx.doi.org/10.21741/9781945291890-15

finally fracture. IHD was developed as a method to measure residual stresses in steels. And since GJS is somewhat similar to steel in its load respond, the IHD should result in plausible and realistic residual stress values for our GJS stress lattice and also seen in our investigation.

General recommendation for IHD measurement state that the measured stresses should be less than 80% of materials yield strength. This recommendation is cleared for GJS but not for GJL. Trying to meet this recommendation for GJL is difficult since the materials non-linear elastic behaviour turns the concept of yield strength to a fundamental issue to deal with for the scientific community. IHD stress calculations assume homogeneous material and material properties, which is not the nature of cast iron and especially GJL and this material seems to be too far away in its properties allowing for good stress measurements utilizing IHD. ASTM E837-13 also specifies that the maximum eccentricity error is 0.02 mm which was found to be difficult to fulfil for GJL, but not for GJS.

Measuring depth in [7] reveals the same tendency as seen in this work that stress evaluations beyond 700 µm depth isn´t really feasible for cast iron, not even GJS. The method as such is physically limited by Saint-Venant principle, meaning that the surface strain response to stresses quickly becomes insensitive and result in larger uncertainties with depth. For cast iron the total measuring depth suggested by the authors is $z/D = 0.3$. Residual stress measurements utilizing IHD on GJL should be done with extreme care and thorough analysis of the results are required for best possible outcome.

Other stress measuring techniques should preferably be conducted parallel to verify the IHD such as electronic speckle pattern interferometry or X-ray diffraction since these techniques captures the local mechanical responds better[21,22].

Conclusions

Feasibility of incremental hole drilling as a residual stress measurement technique on cast iron is still not clearly answered. Stress lattices, cast in grey cast iron and ductile graphite iron, were investigated and the following conclusions could be made:

- Stress measurements on GJS using IHD can be done and returns plausible residual stress values.
- IHD method seems to deteriorate beyond $z/D = 0.3$ depth.
- IHD for stress measurement on GJL gives questionable results.
- More investigation on material behaviour at different length scales are needed.

Acknowledgement

The authors acknowledge all the project members of the project *Optimization of weight and volume intelligent cast components (OLGA)* for material and discussion. The project is founded by the Swedish Energy Agency.

References

[1]   D. Dye, S.M. Roberts, P.J. Withers, R.C. Reed, The determination of the residual strains and stresses in a tungsten inert gas welded sheet of IN718 superalloy using neutron diffraction, J. Strain Anal. Eng. Des. 35 (2000) pp 247–259.
       https://doi.org/10.1243/0309324001514396

[2]   V. Fontanari, F. Frendo, T. Bortolamedi, P. Scardi, Comparison of the hole-drilling and X-ray diffraction methods for measuring the residual stresses in shot-peened aluminium alloys, J. Strain Anal. Eng. Des. 40 (2005) pp 199–209.
       https://doi.org/10.1243/030932405X7791

[3]   D. Kirk, D.G. Birch, Residual stresses induced by peening austenitic ductile cast iron, in: 7th Int. Conf. Shot Peen., 1999: pp. 23–32.

[4]   C.R. Knowles, T.H. Becker, R.B. Tait, Residual stress measurements and structural

integrity implications for selective laser melted ti-6al-4v, Sajie. 23 (2012) pp 119–129. https://doi.org/10.7166/23-3-515

[5]   W. Bouzid Saï, N. Ben Salah, J.L. Lebrun, Influence of machining by finishing milling on surface characteristics, Int. J. Mach. Tools Manuf. 41 (2001) pp 443–450. https://doi.org/10.1016/S0890-6955(00)00069-9

[6]   K. Sasaki, M. Kishida, T. Itoh, The accuracy of residual stress measurement by the hole-drilling method, Exp. Mech. 37 (1997) pp 250–257. https://doi.org/10.1007/BF02317415

[7]   E. Kingston, Residual stress measurements within the nodular cast iron PWR insert of a radioactive waste canister, 2013.

[8]   J.E. Wyatt, J.T. Berry, A.R. Williams, Residual stresses in aluminum castings, J. Mater. Process. Technol. 191 (2007) pp 170–173. https://doi.org/10.1016/j.jmatprotec.2007.03.018

[9]   M. Lundberg, R.L. Peng, M. Ahmad, T. Vuoristo, D. Bäckström, S. Johansson, Residual stresses in shot peened grey and compact iron, HTM - J. Heat Treat. Mater. 69 (2014) pp 38–45. https://doi.org/10.3139/105.110207

[10]   J. Saarimäki, M. Lundberg, J.J. Moverare, 3D residual stresses in selective laser melted Hastelloy X, Mater. Res. Proc. 2 (2016) pp 73–78.

[11]   M. Lundberg, J. Saarimäki, J.J. Moverare, R.L. Peng, Effective X-ray elastic constant of cast iron, J. Mater. Sci. 53 (2017) pp 2766–2773. https://doi.org/10.1007/s10853-017-1657-6

[12]   SS-EN ISO 945-1:2008, 2011.

[13]   D. Stefanescu, C.E. Truman, D.J. Smith, P.S. Whitehead, Improvements in residual stress measurement by the incremental centre hole drilling technique, Exp. Mech. 46 (2006) pp 417–427. https://doi.org/10.1007/s11340-006-7686-8

[14]   G.S. Schajer, Measurement of non-uniform residual stresses using the hole-drilling method. Part II—Practical application of the integral method, J. Eng. Mater. Technol. 110 (1988) pp 344–349. https://doi.org/10.1115/1.3226060

[15]   G.S. Schajer, Measurement of non-uniform residual stresses using the hole-drilling method. Part I—Stress calculation procedures, J. Eng. Mater. Technol. 110 (1988) pp 338–343. https://doi.org/10.1115/1.3226059

[16]   P. Schmidt, L.P. Ru, V. Davydov, M. Lundberg, M. Ahmad, T. Vuoristo, D. Bäckström, S. Johansson, Analysis of residual stress in stress harps of grey iron by experiment and simulation, Adv. Mater. Res. 996 (2014) pp 586–591. https://doi.org/10.4028/www.scientific.net/AMR.996.586

[17]   M.R. Ayatollahi, A.R. Torabi, Tensile fracture in notched polycrystalline graphite specimens, Carbon N. Y. 48 (2010) pp 2255–2265. https://doi.org/10.1016/j.carbon.2010.02.041

[18]   D. Kirk, Ductility and strength properties of shot peened surfaces, 2006.

[19]   M. Lundberg, M. Calmunger, R.L. Peng, In-situ SEM / EBSD study of deformation and fracture behaviour of flake cast iron, in: 13th Int. Conf. Fract., 2013: pp. S12-038.

[20]   T. Sjögren, Influences of the Graphite Phase on Elastic and Plastic Deformation Behaviour of Cast Irons, Institutionen för ekonomisk och industriell utveckling, 2007.

[21]   C. Barile, C. Casavola, G. Pappalettera, C. Pappalettere, Remarks on residual stress measurement by hole-drilling and electronic speckle pattern interferometry., ScientificWorldJournal. (2014) id:487149.

[22]   M. Lundberg, J. Saarimäki, R.L. Peng, J.J. Moverare, Residual stresses in uniaxial cyclic loaded pearlitic lamellar graphite iron, Mater. Res. Proc. 2 (2017) pp 67–72. https://doi.org/10.21741/9781945291173-12

Residual Stresses 2018 – ECRS-10
Materials Research Proceedings 6 (2018) 95-100

Materials Research Forum LLC
doi: http://dx.doi.org/10.21741/9781945291890-16

Residual Stresses in Railway Axles

ŘEHA BOHUSLAV[1,a,*], VÁCLAVÍK JAROSLAV[2,b] NÁVRAT TOMÁŠ[3,c] and DAVID HALABUK[3,d]

[1] GHH-BONATRANS GROUP a.s., Revoluční 1234, 735 94 Bohumín, Czech Republic

[2] Výzkumný a zkušební ústav Plzeň s.r.o., Tylova 1581/46, 301 00 Plzeň, Czech Republic

[3] Faculty of Mechanical Engineering, Brno University of Technology, Technická 2896/2, 616 69 Brno, Czech Republic

[a]bohuslav.reha@ghh-bonatrans.com, [b]vaclavik@vzuplzen.cz, [c]navrat@fme.vutbr.cz, [d]david.halabuk@vutbr.cz

Keywords: Residual Stresses, Railway Axle, Hole-Drilling Method, X-Ray Diffraction

Abstract. The companies producing railway axles are obliged to demonstrate a level of residual stresses in their products based on the specification of standard EN 13261 in which the applicable methods are suggested and the allowable values of residual stresses are specified. An objective of this contribution is to arouse a discussion whether the requirements of the standard can be met by means of the suggested measuring methods - hole-drilling (according to ASTM E837-13 standard) or X-ray diffraction. It especially applies to the ability to determine the residual stress distribution in the depth of 2 mm below the surface. The mentioned issue is demonstrated by means of both measuring methods under various measurement conditions.

Background

The railway axles are exposed to the high operating stress. It especially includes the high-cycle fatigue where the high tension stresses reduce the material fatigue limit. At the same time, unevenly distributed residual stresses may cause the deformation of an axle and increase the stress amplitude originating during the axle rotation. These both values also refer to the quality of heat treatment and the manufacture of the axle. The demonstration of the low level of the residual stress is one part of the existing standards dealing with checking the quality of railway axles EN 13261 [1] and EN 13260 [2].

Both standards require that surface tension residual stresses measured in the depth of 0.1 mm below the axle surface are lower than 100 MPa. The standards also specify three sections where the measurements shall be carried out (Fig. 1). In each section there is requirement for two measurements in the angular separation by 120°. In addition, the standard EN 13261 requires that the mutual deviation of the residual stresses measured at 6 points separated by the angle of 60° in the central section of the axle (a difference of the maximal and minimum values) must be within the interval up to 40 MPa (Fig. 1, section A).

All the tests shall be carried out by means of a strain gauge or an X-Ray method. At the present time a new version of the EN 13261 standard is being prepared where in addition to the conventional strain gauge techniques such as the layer removal method and the sectioning method, the hole-drilling method according to ASTM E837-13a [3] standard is recommended.

Measurement Methods

Hole-Drilling Method. A hole is drilled into the centre of the strain gauge rosette in the steps by means of an end mill or a high speed turbine at the revolutions ranging from 20,000 to 400,000 rpm and the residual stress is calculated from the released strains using calibration constants.

Residual Stresses 2018 – ECRS-10
Materials Research Proceedings **6** (2018) 95-100

Materials Research Forum LLC
doi: http://dx.doi.org/10.21741/9781945291890-16

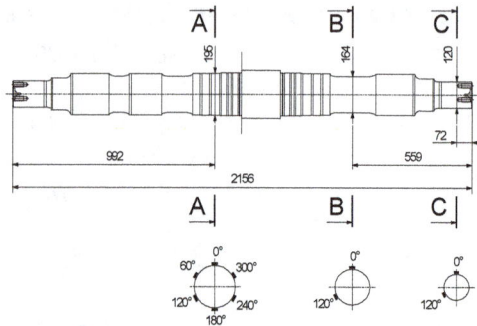

Fig. 1 The points being measured on a railway axle according to the requirement of [1]. The level of the surface residual stresses are demonstrated at three sections (A, B, C). In the centre (A), the evenness of the residual stress is tested at another 6 points by 60° along the perimeter.

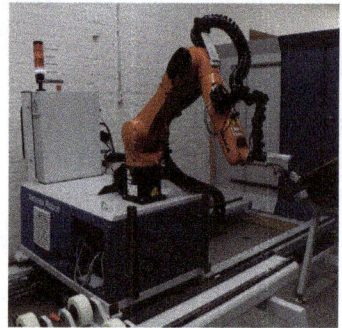

(a) (b)

Fig. 2 The equipment used for measuring residual stresses by means of the hole-drilling method RESTAN (a) and for measuring by means of the X-Ray method (b)

The ASTM E837 standard specifies these constants for the Vishay rosettes of the sizes D 1/32, 1/16 and 1/8 inches. Using the largest rosette it is possible to determine the residual stresses to the depths up to 2 mm. The minimum depth is within the range from 0.025 to 0.1 mm according to a size of the rosette. Thus it would look like the minimum and maximal depths met the requirements of the standard for the tests of the railway axles. However, applying this standard is questionable in several aspects. No other shape of the strain gauge rosette than the one specified in the standard can be used as no calibration constants are specified for it. The hole must be drilled coaxially with the rosette centre with the accuracy of ±0.004D (i.e. 0.016 mm for the largest rosette) which sometimes is difficult to keep. The standard does not state any method for correcting the misalignment. The radial clearance angles on the end face of the cutting tool should not exceed the value of 1°, however, the standard does not give any possibility how to correct the fact that the value would be greater. In the standard the calibration constants are set for a theoretical diameter of the hole being drilled. For the actual diameter of the drilled hole it is recommended to correct them using the quadratic interpolation.

Uncertainty of the Hole-Drilling Method. According to [3] the uncertainty of determination of the uniform stresses by means of the power-series method is ±10%. It is not determined for the

96

Residual Stresses 2018 – ECRS-10   Materials Research Forum LLC
Materials Research Proceedings 6 (2018) 95-100   doi: http://dx.doi.org/10.21741/9781945291890-16

integral method. In [4] an extensive analysis of a measuring error is carried out based on the uncertainties of the most of the input parameters. The main source of the uncertainties specified herein is the eccentricity of the hole being drilled (5 %), stress induced by drilling (5.5 %), a diameter of the hole being drilled and the material constants. When using the integral method, the error is the maximal on the surface due to the influence of the high uncertainty of the input parameters (misalignment, presence of large strain gradients, smaller gauge outputs in the first drilling steps, identification of the zero reference surface) and in the last step of the hole drilling when the uncertainty of the input parameters is increased due to the method lowest sensitivity.

In [5] the effect the hole-bottom fillet radius for the case of using the power-series method is investigated however, the calculated measuring error for the first drilled depth is valid also in case of the evaluation by the integral method. A deviation of the released relative deformation for the radius of 5% of the hole diameter is estimated to be 10%.

The diameter of a railway axle is approx. 200 mm. The derived error when we use the coefficients for flat surface is shown at Fig. 3a for the 1/8 and 1/16 inch rosettes. The released strain and the calculated residual stress are always lower on the curved surface at the surface layer. We derived also the error we would make if we use the correction of the standardized sensitivity coefficients by the quadratic approximation recommended by the standard [3] for the actual diameter of the drilled hole. The resultant error of the released relative deformation is shown at Fig. 3b where the comparison is carried out for the standardized hole diameter of $D_0 = 4$ mm. The approximation of coefficients is carried out for the other diameters. The error increases with the depth being drilled and with a higher deviation from the standardized diameter. The maximal error is in the maximal depth being drilled.

It is obvious from the considerations performed that the uncertainty of determination of the residual stresses on the axle surface and in the depth of 2 mm below the surface is very high and very hard to identify. We estimate that it can reach even 30% and more of the measured value depending on the stress state of the axle. For roll burnished or surface hardened axles the measuring error is increased due to the tri-axial stress state by additional at least 5% especially in bigger depth being drilled [6].

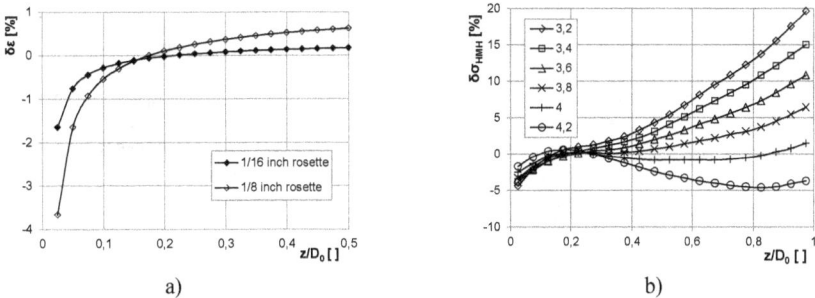

a)   b)

Fig. 3 The dependence of the released relative deformation on the standardized depth below the surface. Due to the curvature of the surface the maximal error for the axle diameter D = 200 mm is reached on the surface (a). Due to the recalculation of the calibration coefficients to the actual diameter of the hole being drilled for the deviation from the hole diameter 4 mm (b) the error is the maximal in the maximal depth being drilled.

Results

An influence of the drilling steps. An example of the distribution of the measured residual stress on the surface-hardened axle is shown at Fig. 4. Three sets of measurements were carried out with a 1/8 inch rosette (HBM RY61-3.2/120S, an electric motor of 20,000 rpm) with three different steps of drilling $\Delta z = 0.025$ mm (96 steps), 0.05 mm (48 steps) and 0.1 mm (24 steps) always to the final depth of 2.4 mm (An evaluation is carried out always for 20 depths).

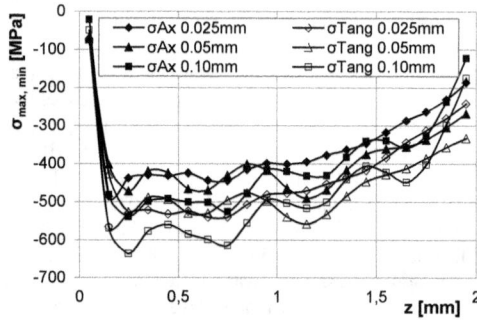

Fig. 4 An influence of the size of drilling steps $\Delta z = 0.025$, 0.05 and 0.1 mm on the distribution of the evaluated residual stresses – by making the step finer the curve becomes smoother and correspond more to the reality.

It is obvious that the resultant residual stress depth profile is not so warped for a finer step of the material drilling which rather corresponds to the reality. For the roughest step specified by the standard [3] the profile is the most warped especially at the place of the maximal depth being drilled. This is an undesirable effect, especially in case we evaluate a deviation of the residual stress in the depth of 2 mm at 6 points in one section of the axle. The hole drilling was carried out to the depth of 2.4 mm, not to the depth of 2 mm as required by the standard [3]. If the drilling is carried out only to the depth of 2 mm, then the dispersion of the residual stress value is very high probably for the reason that when the curve is smoothed, there is a lack of the values of bigger depths. Therefore we recommend carrying out the measurements to the depth bigger by approx. 4 steps than for which the evaluation is carried out.

Stress variation around the axle. It was examined what was the mutual deviation of the evaluated residual stresses in the centre of the axle in the depth of 2 mm. The measurement was carried out with a 1/8 inch rosette again at six points separated by an angle of 60° and the result is shown at Fig. 5a where a double standard deviation of the evaluated stresses is given. The measurement was carried out for 4 axles made of the material A1N ($R_e = 520$ MPa), A4T ($R_e = 580$ MPa), for the surface-hardened axle and for the roll-burnished axle, the material OS (according to GOST 4728). The angular deviation of the residual stress for the axles made from the materials A1N and A4T is low and the requirement of the standard [1] can be reached. The results for the hardened and roll-burnished axle significantly exceed the permissible deviation of 40 MPa. This fact may be caused not only by a big measuring error which is increased with the measured value of the residual stress but also by the fact that the level of the residual stresses in both axles approximates to the yield point of the material. The stresses drawn at Fig. 5a are not corrected to the elastic-plastic stress state as there is no reliable method for integral method.

A reliable method for the correction of the stresses approximating the yield point is derived for the uniform stress distribution through the depth [4]. This correction is presented at Fig. 5b for one measured point for the roll-burnished axle A1N. The profiles of non-corrected and corrected stresses measured both by a small 1/16 inch rosette (serving for measuring the residual stress on the axle surface), and by a big 1/8 rosette for determining the values in the depth of 2 mm are plotted here. The depth of the roll-burnished layer is approx. 5 mm and a drop of the

Residual Stresses 2018 – ECRS-10   Materials Research Forum LLC
Materials Research Proceedings **6** (2018) 95-100   doi: http://dx.doi.org/10.21741/9781945291890-16

stress to the depth of 2 mm is very low so in our opinion the power-series method can be used for the evaluation although the requirement of the stress uniformity given in the previous versions [3] (e.g. ASTM E837-08) is not met. By the use of the power-series method in drilling to the depth of 4 mm we obtain a rough estimate of the average value of the residual stresses for the mean depth of 2 mm. This value does not show any significant dispersion and it can be used for the demonstration of the requirement of the standard for the uniformity [1].

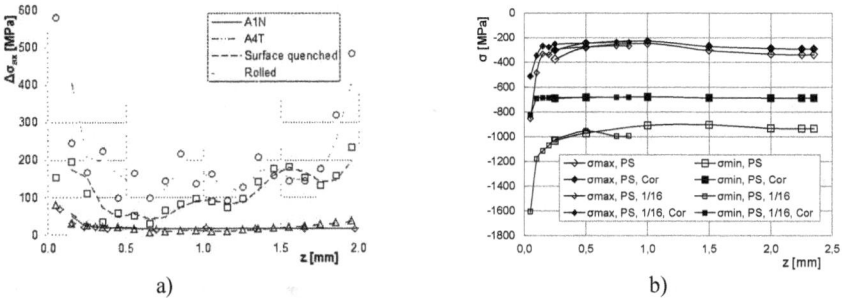

a)   b)

Fig. 5 The maximal deviation of the axial component of the residual stresses measured in the axle centre for two standard materials of the axles and for the hardened and roll-burnished axles (values non-corrected to the elastic-plastic stress state). An example of such correction of the residual stresses approximating to the yield point (b) represents shifting the distributions of the mean stresses towards the lower values (power-series method).

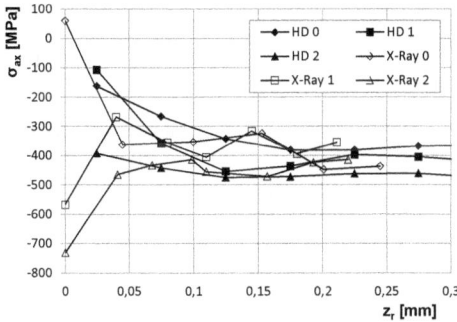

Fig. 6 Comparison of the surface stresses measured by the hole-drilling method (HD) and the X-Ray method on the original (0) and 2 new surfaces (1, 2) after removing 2×1 mm layers.

Another way how to avoid the high uncertainty of the determination the residual stress in the depth of 2 mm is the measurement on the removed layers. The investigation was carried out on the surface-hardened axle A4T on the original not machined surface and after that twice on the new surface where layers of the material of the thickness of 1 mm were removed successively by lathe-turning. The measurement was also carried out by the X-Ray method in several gradually etched depths from these surfaces. At Fig. 6 there is a comparison of the subsurface residual stresses using the 1/16 inch rosette and the X-Ray method. Three measurements on three layers overlapping one another are presented. The residual stress was induced by the lathe-turning in the surface layer but the results are steady for all the layers from the depth of approx. 0.1 mm.

Summary

The hole-drilling strain-gauge method is an acceptable method to demonstrate the level of the residual stresses in the railway axles required by the standards EN 13 260+A1 and EN 13261 under certain circumstances.

i) The evaluation of the residual stresses for reducing the measurement uncertainty requires carrying out the calculation corrections which are not contained in the standard ASTM E837-13. Therefore the new version of the standard EN 13 260 + A1 being prepared should contain that the measurement of residuals stress shall be carried out in compliance with the methodology of ASTM E837-13 instead of directly using this standard in order to avoid problems with this formulation during the accreditation of the measurement method.

ii) Then, the demonstration of the requirements of the EN 13 260 + A1 standard can be implemented without any problems for the axles made of the standard materials A1N and A4T with the level of the residual stresses up to 100 MPa.

iii) In order to evaluate the residual stresses of the hardened or roll-burnished railway axles in the depth of 2 mm below the surface we recommend using the method of layers removing or the power-series method with the correction to the elastic-plastic stress state as the distribution of state of stress in the subsurface layer is roughly homogeneous. It is problematic to use the integral method for the reason of a high uncertainty comparable with the requirement of the standard for the inspection of the railway axles.

iv) The measurement of the residual stresses in the depth of 0.1 mm by the hole-drilling method is also burdened with a big error. The demonstration of the level lower than the demanded tensile 100 MPa according EN 13 260 + A1 standard usually is without any problems if all the necessary corrections are carried out as the residual stresses in the existing axles are nearly always compressive.

The article has originated in the framework of the institutional support of the long-term conceptual development of the research organization.

References

[1] EN 13 261+A1 Railway applications – Wheelsets and bogies – Axles – Product requirements (2010).

[2] EN 13 260+A1 Railway applications – Wheelsets and bogies – Wheelsets – Product requirements (2010).

[3] ASTM E837-13a Determining Residual Stresses by the Hole-Drilling Strain-Gage Method. American Society for Testing and Materials, West Conshohocken, PA (2013).

[4] M. Scafidi, E. Valentini, B.Zuccarello, Error and Uncertainty Analysis of the Residual Stresses Computed by Using the Hole Drilling Method, Strain 47 (2011) 301-312. https://doi.org/10.1111/j.1475-1305.2009.00688.x

[5] M. Scafidi, E. Valentini, B.Zuccarello, Effect of the Hole-Bottom Fillet Radius on the Residual Stress Analysis by the Hole Drilling Method, JCPDS-International Centre for Diffraction Data (2009) 263-270.

[6] D. Halabuk, T. Návrat, The Effect of Third Principal Stress in the Measurement of Residual Stresses by Hole-drilling Method, DMEE 2018, in print.

Residual Stresses 2018 – ECRS-10   Materials Research Forum LLC
Materials Research Proceedings 6 (2018) 101-106   doi: http://dx.doi.org/10.21741/9781945291890-17

Residual Stress Measurement of the Engineering Plastics by the Hole-Drilling Strain-Gage Method

Ami Kohri[1,a,*], Takao Mikami[2,b] and Yuhei Suzuki[1,c]

[1] IHI Inspection & Instrumentation Co., Ltd., 6-17, Fukuura, 2-chome, Kanazawa-ku, Yokohama-city, Kanagawa 236-0004, Japan

[2] IHI Inspection & Instrumentation Co., Ltd., 25-3, Minami-Ohi 6-chome, Shinagawa-ku, Tokyo 140-0013, Japan

[a]a_kohri@iic.ihi.co.jp, [b]t_mikami@iic.ihi.co.jp, [c]y_suzuki@iic.ihi.co.jp

Keywords: ASTM E837, Hole-Drilling, Plastics, Orthotropic Material, Injection Molding

Abstract. The hole-drilling method is regulated as the ASTM E837 standard and has been widely used for the residual stress measurement of metallic materials. We applied this method to orthotropic plastic specimens. Known stress distributions with depth were given by the four points bending tests and the calculated stresses from the measured relieved strains were compared with the given values analyzed by the beam theory. Consequently, the calculated uniform stresses based on both orthotropic and isotropic approaches were equivalent to the beam theory analyses. In addition, the calculated non-uniform stresses with depth in accordance with ASTM E837 standard agreed well with the beam theory values. It was confirmed that the residual stress measurement by the hole-drilling method is effective for the orthotropic plastic materials.

Introduction

In recent years, high performance plastic materials with superior mechanical properties and high thermal stability have been developed. They are called 'engineering plastics' and have been applied to industrial products. Particularly, in the field of automotive and aerospace industries glass-fiber reinforced plastics have been actively applied. Most of them have anisotropic elastic properties due to their molding processes. Residual stresses are present in these plastics without exception and are to be measured from the point of view of the quality control.

There are many methods for measuring residual stresses. The hole–drilling method is considered to be the most suitable one for plastic materials. This method is standardized as the ASTM Standard E837 [1] and has been widely used for the residual stress measurement of elastic materials. However this method is limited to isotropic materials. Schajer *et al.* [2] and Pagliaro *et al.* [3] have generalized the computational procedures for the hole-drilling method to extend the use of the method to orthotropic materials.

In this study, known bending stresses were given by four point bending tests on the rectangular plate specimens made of glass-fiber reinforced plastics and the measured stresses by the hole-drilling method were compared with the analyzed values based on the beam theory.

Experiment

Specimen. The material of the specimen is Polyamide 6 (PA6) added with 30% of glass fiber. After manufacturing a plate (150mm×150mm×4mm) by injection molding, it was cut into rectangular specimens (110mm×22mm×4mm) with longitudinal axis aligned with material direction (MD).

Four point bending test. The residual stresses were simulated by applying a bending load. For this purpose, the rectangular specimen was loaded by using a four point testing apparatus with a

Residual Stresses 2018 – ECRS-10   Materials Research Forum LLC
Materials Research Proceedings 6 (2018) 101-106   doi: http://dx.doi.org/10.21741/9781945291890-17

span of 80mm and a loading point distance of 30mm as shown in Fig.1. We conducted the tests for two specimens and maximum bending stresses of 15MPa and 20MPa were given on the surfaces of them respectively. TML's FRS2-23 rosette strain gages with gages 1, 2, 3 oriented at $0°/135°/270°$ respect to x axis (see Fig. 2) were attached to tensile side surface of each bending specimen. The applied bending stress was monitored by two strain gages aligned with longitudinal direction of each specimen during the test (see Fig. 2).

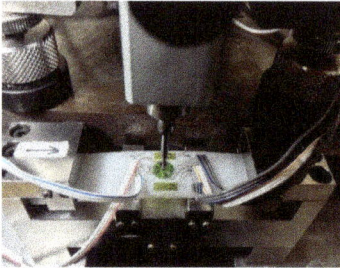

Fig. 1. General view of hole-drilling during four point bending test

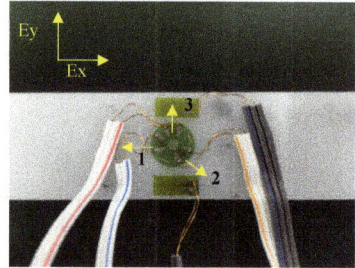

Fig. 2. Rosette strain gage after drilling

Orthotropic elastic properties. The elastic properties of the specimen were measured by using a tensile testing machine. The measured elastic properties are sown in Table. 1.

Where,

E_x, E_y : Elastic moduli along x and y (elastic symmetry) axes

G_{xy} : x-y shear modulus

v_{xy}, v_{yx} : x-y Poisson's ratios

Table 1. Elastic properties

E_x	7.74 GPa
E_y	4.83 GPa
G_{xy}	2.26 GPa
v_{xy}	0.365
v_{yx}	0.287

Hole-drilling measurement. Using the MTS 3000–RESTAN system, a blind hole with a diameter of about 1.8mm was drilled at the rosette center of each specimen in each loading condition [see Fig.1 and Fig.2]. The hole was drilled with an increment of 0.05mm in the depth direction and the final depth was 1.0mm (i.e. 0.05mm×20steps). At each step, relieved strains were measured and recorded automatically. The measured cumulative relieved strains for the two specimens are shown in Fig.3 and Fig.4. Comparing the relieved strains measured on gage 1(i.e. x direction) in these Figures are almost proportional to the given stress ratio of 1.33 (20MPa/15MPa). The final cumulative relieved strains at the depth of 1mm were used in the orthotropic analysis described in the next section. On the other hand, all the relieved strains were used in the isotropic analysis in accordance with the ASTM E837 standard.

Materials Research Forum LLC

doi: http://dx.doi.org/10.21741/9781945291890-17

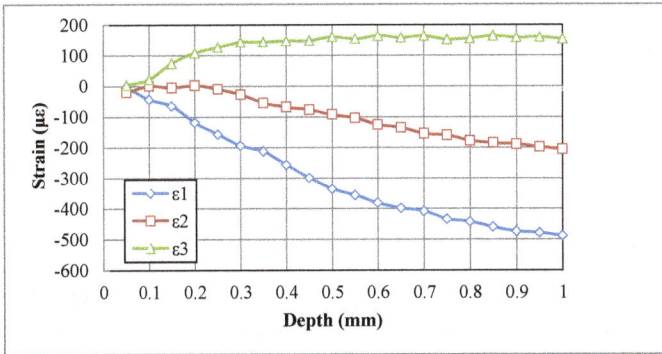

Fig. 3. Measured relieved strains for the applied stress of 15MPa

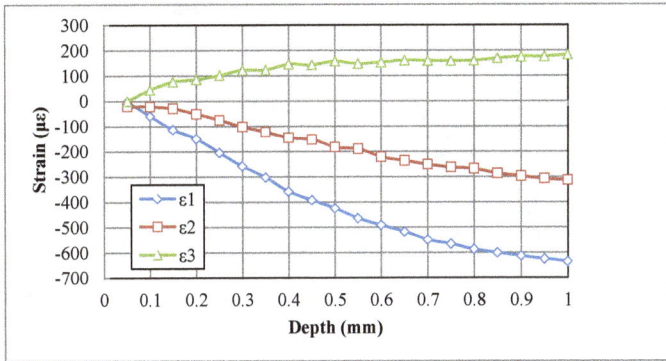

Fig. 4. Measured relieved strains for the applied stress of 20MPa

Orthotropic hole-drilling analysis. We applied the Eq. (1) suggested by Schajer et al [2] for the anisotropic analysis.

$$\frac{1}{\sqrt{E_x E_y}} = \begin{bmatrix} C_{11} & C_{12} & C_{13} \\ C_{21} & C_{22} & C_{23} \\ C_{31} & C_{32} & C_{33} \end{bmatrix} \begin{bmatrix} \sigma_x \\ \tau_{xy} \\ \sigma_y \end{bmatrix} = \begin{bmatrix} \varepsilon_1 \\ \varepsilon_2 \\ \varepsilon_3 \end{bmatrix} \qquad Eq.(1)$$

Where,

C_{**}: orthotropic strain relief compliances

σ_x, σ_y: x-y Cartesian normal stresses

τ_{xy} : x-y Cartesian shear stress

Residual Stresses 2018 – ECRS-10   Materials Research Forum LLC
Materials Research Proceedings 6 (2018) 101-106   doi: http://dx.doi.org/10.21741/9781945291890-17

$\varepsilon_1, \varepsilon_2, \varepsilon_3$: measured relieved strains by rosette strain gage

The values of the elastic compliances in Eq. (1) depend on the orthotropic elastic properties of the specimen, the hole diameter and the rosette strain gage geometry.

Considering the elastic properties in Table 1, the values of the elastic compliances were calculated by using a Table given in [2] and the results are shown in Table 2.

Table 2. The calculated elastic compliances for this study from [2]

C_{11}	C_{13}	C_{21}	C_{22}	C_{23}	C_{31}	C_{33}
-0.284	0.108	-0.094	0.443	-0.118	0.120	-0.369

Isotropic hole-drilling analysis. We applied an average value of elastic moduli along x and y axes as an isotropic elastic modulus in order to analyze the stress by using the ASTM E837-13 standard.

Results

The results of the orthotropic and isotropic analyses for uniform stresses with depth are shown in Table 3. In Table 3, 'Applied average stress' means the averaged theoretical bending stress from the surface to the depth of 1mm as shown in Fig. 5.

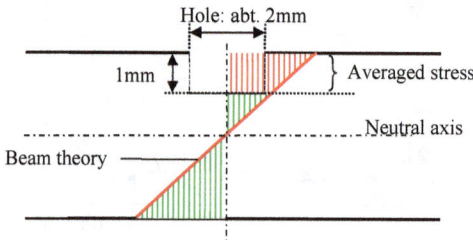

Fig. 5. Stress distribution in the bent specimen and the measured region by hole-drilling method

Table 3. Comparison between applied stresses and calculated stresses by orthotropic and isotropic analyses

Applied bending stress : σ_{max} (MPa)	Applied averaged stress : σ_{ave} (MPa)	Orthotropic analysis : σ_x (MPa)	Isotropic analysis : σ_x (MPa)
15.0	11.2	10.9	10.3
20.0	15.0	14.4	13.7

As shown in Table 3, the uniform stresses calculated by orthotropic and isotropic analyses based on the measured relieved strains agreed well with the applied averaged stresses (i.e. beam theory values).

Furthermore, we calculated a non-uniform stress distribution with depth in accordance with the ASTM E837 standard for the15MPa load case. The result is shown in Fig.6. This result agreed well with the theoretical bending stress distribution based on the beam theory. The deviation

Residual Stresses 2018 – ECRS-10   Materials Research Forum LLC
Materials Research Proceedings **6** (2018) 101-106   doi: http://dx.doi.org/10.21741/9781945291890-17

from the beam theory is considered to be the influence of initial residual stresses present in the specimen. The initial residual stress distribution with depth is considered to be different between specimens. We measured therefore initial residual stress distributions for another three specimens. The measured initial residual stress distributions for σ_x component are shown in Fig. 7 and it confirms that the initial residual stress distributions are significantly different each other depending on the specimens.

In this study, we applied the plastic specimens with the elastic modulus ratio: E_y/E_x of 0.62 and confirmed that the effect of the anisotropy on the stress analysis results is not so significant. The elastic modulus ratio: E_y/E_x of commercially available glass-fiber reinforced plastics is from 0.5 to 0.7. Consequently, it can be stated that in the residual stress measurement with the hole-drilling strain gage method for the glass-fiber reinforced plastics, we can obtain the residual stresses with sufficient accuracy for both "uniform" and "non-uniform" stress cases by applying the averaged elastic properties in the analyses based on ASTM E837 standard.

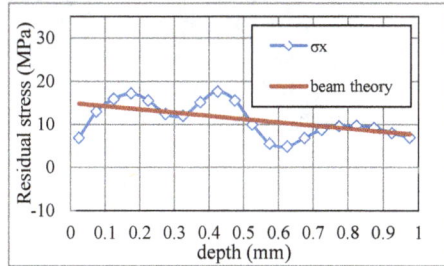

*Fig. 6. Comparison between the measured
non-uniform stress and the beam theory value
for the applied bending stress of 15MPa*

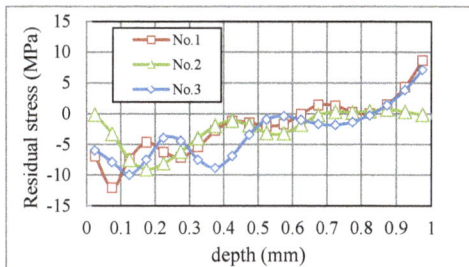

*Fig. 7. Measured initial residual stress distributions
in another three specimens*

Residual Stresses 2018 – ECRS-10   Materials Research Forum LLC
Materials Research Proceedings **6** (2018) 101-106   doi: http://dx.doi.org/10.21741/9781945291890-17

Summary

Known bending stresses were given by four point bending tests on the rectangular plate specimens made of glass-fiber reinforced plastics with the elastic modulus ratio: E_y/E_x of 0.62 and the measured stresses by the hole-drilling method were compared with the analyzed values based on the beam theory. As a result, the uniform stresses calculated by orthotropic and isotropic analyses based on the measured relieved strains agreed well with the applied averaged stresses (i.e. beam theory values). In addition, non-uniform stress distribution with depth was calculated in accordance with the ASTM E837 standard and the calculated results agreed well with the theoretical bending stress distribution based on the beam theory for both "uniform" and "non-uniform" stress cases.

References

[1] ASTM E 837-13a, Standard Test Method for Determining Residual Stresses by the Hole-Drilling Strain-Gauge Method 2013.

[2] G.S. Schajer and L. Yang, Residual-stress measurement in orthotropic materials using the hole-drilling method, Exp. Mech., 34, 1994. https://doi.org/10.1007/BF02325147

[3] P. Pagliaro, B. Zuccarello, Residual Stress Analysis of Orthotropic Materials by the Through-hole Drilling Method, Exp. Mech. 47, 2007. https://doi.org/10.1007/s11340-006-9019-3

Diffraction Methods

Residual Stresses 2018 – ECRS-10
Materials Research Proceedings 6 (2018) 109-114

Materials Research Forum LLC
doi: http://dx.doi.org/10.21741/9781945291890-18

Micromagnetic Analysis of Residual Stress Distribution in 42CrMo4 Steel after Thermal and Mechanical Surface Treatment

Ilya Bobrov[1,2,a,*], Jérémy Epp[1,3,b] and Hans-Werner Zoch[1,2,3,c]

[1]Universität Bremen, Badgasteiner Straße 1-3, 28359, Bremen, Germany

[2]Leibniz-Institute for Materials Engineering – IWT, Badgasteiner Str. 3, 28359 Bremen, Germany

[3]MAPEX Center for Materials and Processing, University of Bremen, Germany

[a]bobrov@iwt-bremen.de, [b]epp@iwt-bremen.de, [c]zoch@iwt-bremen.de

Keywords: Residual Stresses, Micromagnetic, X-Ray Diffraction, Barkhausen Noise

Abstract. In this study, residual stresses and micromagnetic parameters were analyzed with a Barkhausen-Noise-Eddy-Current-Microscope which allows the analysis of local micromagnetic properties with high resolution down to 20 µm. 42CrMo4 steel samples with varying initial microstructure and subjected to different thermal and mechanical surface treatments were investigated. In particular, shot peening and induction hardening were used to generate different distributions of microstructural and residual stress modifications. Calibration strategies were developed using standard methods as X-ray diffraction measurements and metallographic examinations. The results show that a quantitative evaluation of residual stress distribution can be possible, even in regions with high gradients when proper measurement and calibration strategy is used. With this method, large areas with thousands of measurement positions can be analyzed very fast and thus open new possibilities in the investigation of local residual stress distribution of components.

Introduction

Thermal and mechanical surface treatments like shot peening, inductive hardening and their combinations induce local changes in the microstructural features such hardness or residual stress state of the material, while the level of modifications and the size of the affected region can be varied by the process parameters [1]. Induced microstructural changes can be detected and evaluated with well-known methods like X-Ray diffraction or hardness measurements [2]. These methods can provide information about the material state, which can be used to evaluate mechanical properties of the materials as well. On the other hand, fast non-destructive analysis like micromagnetic methods have high potential for large area investigation or high-throughput material screening since they are very sensitive to the microstructure and mechanical properties of the material [3]. However the data obtained with these methods cannot be directly used to describe material properties, but proper calibration is required. This calibration function can be obtained after suitable calibration and can be used to evaluate the mechanical properties [4]. The advantage of such methods is the non-destructivity and faster data collection as for the other methods.

In this study micromagnetic measurement were performed with Barkhausen Noise and Eddy Current Microscope based on the 3MA (Micromagnetic Multiparameter Microstructure and Stress Analysis) method from Fraunhofer Institut for Non-Destructive Testing, Saarbrücken, Germany. This method is described in details in [5].

In this equipment 3 different testing principles are realized and can be used to obtain data from small areas with high spatial resolution (~20 µm) [6]: Barkhausen Noise (BN) analysis,

Residual Stresses 2018 – ECRS-10   Materials Research Forum LLC
Materials Research Proceedings 6 (2018) 109-114   doi: http://dx.doi.org/10.21741/9781945291890-18

incremental permeability analysis and Eddy Current (EC) method. The application of these methods with several testing frequencies provides more than 30 micromagnetic parameters, which correlate to the properties of the material structure and layered systems. The combination of the micromagnetic characteristics in a multiparameter approach enables the separation of overlapped influences, for example stress and layer thickness [7]. The 2D-maps of selected micromagnetic parameters can allow to gain information on material homogeneity and local property variations. Furthermore, the comparison and combination of the characteristics of several samples allows for a qualitative estimation of the differences in material properties [8].

In the present study 42CrMo4 steel samples were subjected to different surface treatments to induce varying microstructural and mechanical property distributions. Micromagnetic measurements were performed using the BEMI system, while X-ray diffraction analyses and hardness measurements were used to calibrate and interpret the micromagnetic parameters. The aim of the investigations was to achieve quantitative evaluation of local properties and to predict hardness and residual stress distributions in the different treatment conditions. Single micromagnetic parameters used in the function were evaluated individually to determine their sensitivity regarding the different target properties.

Experimental methods

The experiments were performed on samples from 42CrMo4 steel (AISI 4140H). The chemical composition of this steel contains 0.43% C, 0.26% Si, 0.74% Mn, 0.01% P, <0.001% S, 1.09% Cr and 0.25% Mo. For this study two initial material states with different hardness were prepared. The ferritic-pearlitic (FP) material state was achieved by soaking for 0.5 h at 840 °C and subsequent air cooling while a quenched and tempered (QT) state was achieved by quenching in 60 °C oil after prior austenitizing with additional tempering at 450 °C for 1 hour.

X-ray diffraction was performed with MZ IV diffractometer of GE Inspection Technology, Ahrensburg, Germany with Cr-$K_{\alpha 1,\,2}$ radiation, produced by long fine focus tube. The primary beam size was 200 µm in diameter and was defined using a focusing mini-lens from IFG, Berlin, Germany. Diffracted signal was detected by position sensitive detector "Miostar 2" (Photron-X), equipped with vanadium filter. For residual stress, the {211} peak of α-Fe was measured at each position along 13 χ-angles between -45° and +45° and the residual stresses were calculated by standard $\sin^2\psi$-method using X-ray elastic constants ½ S_2 = 5.81 10^{-5} MPa-1 [9].

The basic part of the BEMI (Figure 1, A, B) is the 3-axes positioning system with a holder unit for the inductive probes and the reference Hall probe. The test specimen is fixed on an electromagnetic yoke with an additional Hall probe below the specimen. The magnetization unit consists of a function generator and an amplifier. This gives the ability to magnetize with a field strength of 0-125 A/cm and frequencies up to 250 Hz. The manipulation control unit is a commercial system with an accuracy of 10 µm in x, y, and z directions. The major parts of the microscope are the units of EC testing and BN measurement that can be selected by a software controlled switch. The EC unit allows simultaneous testing with four multiplexed frequencies between 10 Hz -10 MHz. The BN unit detects the induced BN in three filters ranges between 20 kHz and 10 MHz with an amplification up to 100 dB. The computer controls the system, the signal processing and stores the measured data [10].

In the present investigations, EC measurements were performed with 4 different exciting frequencies: 500 kHz; 1 MHz; 2 MHz; 3 MHz. Base magnetization frequency for the two other measurement methods was defined at 100 Hz. BN analysis was performed using 25 A/cm magnetization amplitude. Obtained signal was filtered with lowpass frequency of 100 kHz and highpass frequency of 1 MHz. The magnetization amplitude for incremental permeability was chosen by 15 A/cm and the EC loop frequency for this analysis was chosen by 1 MHz

The samples used for this study were 200×30×19 mm blocks of 42CrMo4 steels from which transversal cross section samples with 20×20×10 mm were prepared. Cross section surface was

Residual Stresses 2018 – ECRS-10   Materials Research Forum LLC
Materials Research Proceedings 6 (2018) 109-114   doi: http://dx.doi.org/10.21741/9781945291890-18

ground and polished. The treatment combination for the samples are: quenched and tempered with additional shot peening (QT+SP); quenched and tempered with additional induction hardening with surface hardening depth (SHD) of about 1 mm (QT+IH); previous combination with additional shot peening (QT+IH+SP); induction hardened sample with SHD= 1 mm, with ferritic-pearlitic initial microstructure (FP+IH). All measurements were performed in the area from the treated surface towards the sample center as shown in Figure 1C. Residual stress measurements were performed only for transversal stress component.

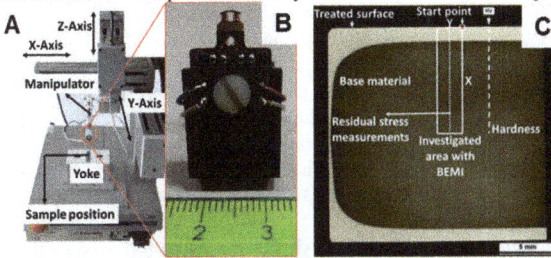

Fig 1. A) BEMI B) BEMI-Sensor C) Investigated regions on the cross section of the sample

The micromagnetic measurements were performed on 8×2 mm area with 0.02 mm step in each direction (Figure 1, C). From the obtained 2D-maps, average of the 100 data points was calculated over the measured 2 mm in y-direction for each x-axis step, giving a mean parameter evolution from the surface to the depth.

Results and discussion

The results of hardness measurements, residual stress and two micromagnetic parameters for four different samples are shown in Figure 2.

After QT+SP, the hardness is almost constant around 420 HV1 over the entire region, while the residual stress distribution exhibits increasing compression stresses towards the surface and slight compression at distances larger than 0.6 mm under the surface. In induction hardened state, the sample QT+IH has almost constant high hardness in the first mm, followed by a decrease to minimum values at 1.3 mm depth slightly below the original hardness (about 80 HV1 lower) which is then reached at 2 mm depth. This decrease is due to tempering effects below the hardened layer. Additional shot peening of sample QT+IH+SP leads to the same overall hardness distribution. Slight difference (~30HV) between QT+IH and QT+IH+SP samples in the first measured depth can be attributed to the shot peening process. In the case of sample FP+IH similar high hardness is achieved in the first 0.8 mm, followed by a decrease and almost constant value of 200 HV1 below 1.3 mm. On the other hand, it can be observed that the residual stress distributions of induction hardened samples with high compression in the hardened layer still present strong evolutions up to a depth of ca. 3 mm. In particular, the QT+IH samples exhibit high tensile peaks. This different behavior of the properties can be used to separate both effects and calculate calibration functions on the base of micromagnetic parameters.

Distributions of maximum BN amplitude (M_{max}) and of the coercivity value determined from Barkhausen peaks (H_{cm}) are also shown in Figure 2 (C, D) over the distance from the treated surfaces for the 4 samples. Both micromagnetic parameters exhibit specific evolutions as a function of the distance from the surface, what is a consequence of the residual stress and hardness gradients in this region. For the QT+SP condition, only the surface near zone is affected by residual stress variations, while at depth higher than 0.6 mm, more or less constant properties were detected. Nevertheless, the micromagnetic parameter exhibit linear evolution over the complete investigated depth which was not expected. Additional experiments with fully homogeneous samples showed, that this effect is the consequence of a magnetic field inhomogeneity within the sample. Strategies to eliminate or at least to compensate this

Residual Stresses 2018 – ECRS-10   Materials Research Forum LLC
Materials Research Proceedings **6** (2018) 109-114   doi: http://dx.doi.org/10.21741/9781945291890-18

instrumental influence will have to be developed in further investigations in order to increase the reliability of these analyses. The three samples in IH condition exhibit comparable evolution of M_{max} with low values at the surface, followed by increasing tendencies. The two QT+IH samples have maximum values around 2 mm below the surface which decrease again towards the core, while the FP+IH sample shows only a continuous increase over the depth. The distribution of M_{max} for QT+IH+SP sample has a local maximum on the 0-0.6mm distance from the surface. Finally, it can be observed that the M_{max} curves converge to comparable values at highest depth for all samples. These distributions are very similar to the residual stress curves for each of the samples. On the other hand, H_{cm} values exhibit different characteristics. The H_{cm} values distribution for inductive hardened samples with QT initial state can be divided into 3 regions. First constant or only slightly increasing values from the surface to a depth around 1.3 mm, where it then starts to decrease. This correlates well with the SHD of the samples. Then H_{cm} is decreasing continuously up to the 2 mm and stabilizes at constant value below 3 mm. This further evolution is not as sharp as the hardness profile what indicates that both hardness and residual stresses influence this parameter. In the case of QT+IH+SP, a slight increase of H_{cm} occurs close to the treated surface, what is a consequence of the shot peening. The sample with the FP initial state has has a maximum of H_{cm} values by the depth of 1.3 mm followed by a slow continuous decrease and ends at much lower value than the QT initial microstructure. This behavior also reflects the combination of hardness and residual stress influences for the FP initial microstructural state.

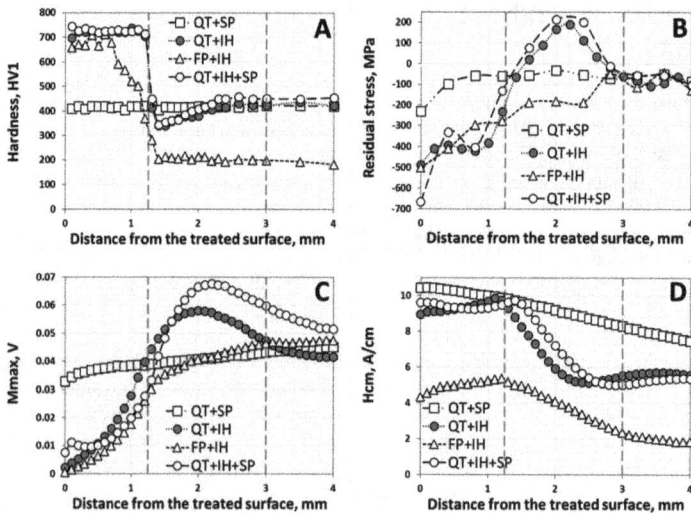

Fig. 2. Average evolution of properties for 4 different surface treatment combinations (QT= Quenched and Tempered; IH=Induction Hardening; FP= Ferrite-Pearlite; SP= Shot Peened): A) Hardness; B) Residual stresses; C) Maximum BN amplitude (M_{max}); D) Coercivity value from BN analysis at $M(H)=M_{max}$

The results were then used to determine suitable calibrations. Preliminary studies showed small difference in r^2-factors using more than 6 coefficients. Due this fact the calibration functions were obtained using regression analysis with 5 coefficients allowed. To achieve that, the reference points were chosen close to real X-Ray spot position with the averaging radius of 200 μm. Calibration was performed using all reference points for all investigated samples at the same time with the tool implemented in the 3MA software. Calibration functions were determined to analyze the hardness and residual stress distribution over the investigated area.

The following calibration functions were obtained:

$$HV = 7.43 \times 10^2 - 3.24 \times 10^4 \times M_{mean} + 7.13 \times 10^4 \times M_{max}^2 - 1.99 \times 10^3 \times \sqrt{|M_{mean}|} - 9.69 \times 10^2 \times \sqrt{|M_r|}. \tag{1}$$

$$RS = -5.04 \times 10^2 + 1.26 \times 10^5 \times M_{max}^2 + 6.96 \times 10^1 \times \sqrt{|H_{cm}|} + 1.77 \times 10^{-1} \times DH50m^2 - 6.49 \times 10^3 \times Im_4^2. \tag{2}$$

Here, HV is the hardness, RS is the transversal residual stress value, M_{mean} is the time averaged value of M(H)-curve from BN analysis over 1 period, M_{max} is the maximum amplitude of BN, M_r is the M(H) value by H=0, H_{cm} is coercive field strength from BN analysis measured from M(H) curve by M=M_{max}, DH50m is the BN peak broadening by M(H) = 0,5 $\times M_{max}$ and Im_4 is the imaginary part of the impedance for highest frequency of the EC test.

These calibration functions were then used to calculate hardness and residual stress distributions from the micromagnetic investigations. The results are shown in Figure 3.

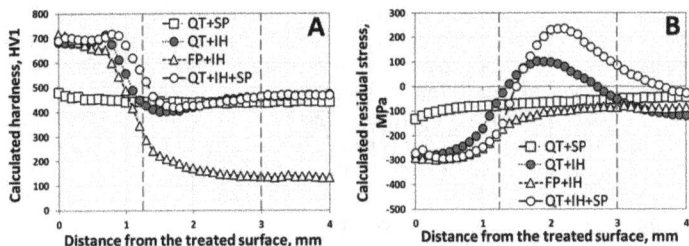

Fig. 3. Average evolution of calculated properties for the 4 different surface treatment combinations (QT= Quenched and Tempered; IH=Induction Hardening; FP= Ferrite-Pearlite; SP= Shot Peened): A) Hardness; B) Residual stresses;

Calculated hardness distributions show good agreement with the measured hardness curves. The deviations from the measured hardness can be divided into 2 groups. First are the small deviations: ca. ~50HV difference for the QT+SP sample and also for the hardness close to the surface and of the base material of the IH samples. Larger deviations reaching 100HV are resulting for base material hardness of FP sample. In this case, the calculated depth evolution is not as sharp as the measured values and a more progressive change is calculated over the depth. This can be explained by the fact that there is no micromagnetic values, which have sharp evolutions as the hardness curve. Indeed, hardness is a complex property of several microstructural features, and therefore several contributions can be expected on the micromagnetic parameters. By increasing the number of terms in the calibration function the reliability of calculated hardness can be further increased. Calculated residual stress distribution exhibit larger deviations from the measured curves. In particular, the surface near high compression stresses cannot be predicted reliably, while tensile peaks in QT+IH samples and low compression towards the core are close to the measured values for all samples. This can be explained by the saturation behavior of several micromagnetic parameters in compression range, while high sensitivity is achieved in tensile regime [11].

From the obtained results, it could be shown that measurements and calibration of micromagnetic scans using BEMI is capable to achieve quantitative analyses of residual stresses and hardness distributions in surface treated samples, but optimization of the procedure is still required to improve the reliability of the predicted values.

Conclusions

In the present study, steel grade 42CrMo4 was produced in two different heat treatment conditions and submitted to thermal and mechanical surface hardening treatments. Various distributions of residual stresses and hardness were generated over the depth. By using a microscale micromagnetic measurements, evolution of micromagnetic parameters were determined in complete areas with high spatial resolution, and compared to the material properties measured by standard methods. As already known for macroscopic micromagnetic analysis the parameters can have different sensibility to the different microstructural features and properties.

It was shown, that BN analysis is very sensitive to hardness and residual stresses. Calibration function obtained for the hardness distribution contains only the values from BN analysis, when the calibration function is restricted to contain only 5 micromagnetic values. M_{max} values distributions are more representative for residual stress distribution but the calculation requires more parameter. Further investigations are necessary to obtain reliable calibration function using this results. On the other hand there H_{cm} values from the same analysis method are influenced by both hardness and residual stresses. For improvement of calculation reliability, more parameters should be included into the calibration function.

From the obtained results, it could be shown that the calibration of the system can allow the quantitative analyses of residual stresses and hardness distributions in surface treated samples with high spatial resolution, but optimization of the procedure is still required to improve the reliability of the predicted values. In particular, the effect of non-homogeneous magnetization should be taken into account. This factor is particularly relevant for larger samples dimensions. For the elimination of this effect a correction/compensation procedure is required, which is currently under development. Finally, multiple-objective optimization methods might be applied to take the single contributions and their overlapping into account.

Acknowledgments

Financial support of subproject D01 "Qualification of material conditions with mechanical and physical measuring methods" within the Collaborative Research Center SFB 1232 "Farbige Zustaende" by the German Research Foundation (DFG) is gratefully acknowledged.

References

[1] Schulze, Volker. Modern mechanical surface treatment: states, stability, effects. John Wiley & Sons, 2006.

[2] J. Epp, XRD methods for materials Characterization in Material Characterization Using Nondestructive Evaluation (NDE) Methods, Woodhead publishing pp. 81–124 (2016) https://doi.org/10.1016/B978-0-08-100040-3.00004-3

[3] Blitz, Jack. Electrical and magnetic methods of non-destructive testing. Vol. 3. Springer Science & Business Media, 2012.

[4] J. Epp, T. Hirsch: Residual Stress State Characterization of Machined Components by X-ray Diffraction and Multiparameter Micromagnetic Methods, Experimental Mechanics 50 (2010) 1, 195-204

[5] I. Altpeter, et al., Electromagnetic and Micro-Magnetic Non-Destructive Characterization (NDC) for Material Mechanical Property Determination and Prediction in Steel Industry and in Lifetime Extension Strategies of NPP Steel Components, Inverse Problems, 18 (2002) 1907 -1921. https://doi.org/10.1088/0266-5611/18/6/328

[6] K. Szielasko, et al., Ortsauflösende Charakterisierung ferro- und ferrimagnetischer Schichten für magneto-resistive und magnetooptische Sensoren. GZfP-Jahrestagung 2014 - Di.1.B.4

[7] M. Abuhamad, I. Altpeter, G. Dobmann, M. Kopp, Non-destructive characterization of cast iron gradient combustion engine cylinder crankcase by electromagnetic techniques (in German), Proceedings of the DGZfP-Annual Assembly (2007), Fürth

[8] B.Wolter, G. Dobmann, Micromagnetic Testing for Rolled Steel, European Conference on Non-destructive Testing (9) (2006) Th. 3.7.1, 25.-29. 09. 2006, Berlin.

[9] Noyan, Ismail C., and Jerome B. Cohen. Residual stress: measurement by diffraction and interpretation. Springer, 2013.

[10] Bender, J., D. O. Thompson, and D. E. Chimenti. "Barkhausen Noise and Eddy Current Microscopy (BEMI): Microscope Configuration, Probes and Imaging Characteristics in 'Review of Progress in Quantitative Nondestructive Evaluation." (1997): 212Iff.

[11] Stewart, D. M., K. J. Stevens, and A. B. Kaiser. "Magnetic Barkhausen noise analysis of stress in steel." *Current Applied Physics* 4.2-4 (2004): 308-311.

Composites, Nano and Microstructures

Residual Stresses 2018 – ECRS-10
Materials Research Proceedings 6 (2018) 117-122

Materials Research Forum LLC
doi: http://dx.doi.org/10.21741/9781945291890-19

In Situ Mechanical Behavior of Regenerating Rat Calvaria Bones Under Tensile Load via Synchrotron Diffraction Characterization

Ameni Zaouali[1,*], Baptiste Girault[1], David Gloaguen[1], Fabienne Jordana[2], Marie-José Moya[1], Pierre-Antoine Dubos[1], Valerie Geoffroy[2], Matthias Schwartzkopf[3], Tim Snow[4], Himadri Gupta[5], Olga Shebanova[4], Konrad Schneider[6], Baobao Chang[6]

[1] Institut de Recherche en Génie Civil et Mécanique, GeM (UMR CNRS 6183), Université de Nantes, 58, rue Michel Ange - BP 420, 44606 Saint-Nazaire Cedex, France

[2] Regenerative Medicine and Skeleton, RMeS (UMR_S 1229), Université de Nantes, Faculté d'Odontologie, 1, place Alexis Ricordeau, 44000 Nantes, France

[3] Deutsches Elektronen-Synchrotron (DESY), Notkestr. 85, 22607 Hamburg, Germany

[4] Diamond Light Source Ltd, Diamond House, Harwell Science & Innovation Campus, Didcot, Oxfordshire OX11 0DE, United Kingdom

[5] Institute of Bioengineering, Queen Mary University of London, London, E1 4NS, United Kingdom

[6] Leibniz-Institut für Polymerforschung Dresden e. V, Dresden, 01069, Germany

* ameni.zaouali@etu.univ-nantes.fr

Keywords: Bone Regeneration, Mechanical Properties, Small Angle X-Ray Scattering, Wide Angle X-Ray Scattering, Mineral Crystals

Abstract . The major challenge of Research in bone surgery is to develop strategies to repair large bone defects. Bone grafting technique is the gold standard to fill and heal these kind of defects. This work address the evolution of the mechanical properties as regard to bone regeneration microstructure. Managing such a time dependent (different regeneration step) and spatially resolved (strain field across natural/reconstructed bone interface) process is achieved through a quantitative analysis of the mechanical strain distribution supported by the mineral part of bone architecture (hydroxyapatite - Hap) and crystal microstructural features (size distribution, spatial pattern). SAXS/WAXS (Small- and Wide- Angle X-ray Scattering) experiments will highlight strain distribution respectively in the reconstructed bone's collagen fibrils and minerals, through mechanical state mapping over a surgically created defect under *in situ* tensile testing on samples harvested at different regeneration stages.

Introduction

Today regenerative medicine is moving towards the development of less and less invasive surgical techniques with the objective of reducing morbidity and the duration of hospitalization. Adult stem cells are a very promising avenue in regenerative medicine. Unfortunately, direct injection into the body suffers from major limitations, in particular for osteoarticular and cardiovascular regeneration. It has been shown that after direct injection most of the cells escapes from the injection site or are not displaying the expected physiological functionality. This quest for reduced-invasive surgery has motivated the development of injectable matrices for bone and cartilage tissue engineering. Once implanted, these injectable matrices must be able to settle down, acquire the desired form, and enable oxygen, nutriment and cell diffusion. Indeed, different types of implants (notably biocomposites) have been developed and adapted in bone tissue engineering. The evaluation of the clinical success of the latter is translated into the quantification of bone formation [1]. However, the microstructural feature evolution of the newly formed bone and its related mechanical behavior after implantation are still poorly understood. This work aims to study the biomechanical properties of healing bone at different stages of regeneration in order to offer tools for implant material optimization in bone engineering and their clinical success evaluation. The major challenge of this work is therefore to correlate time- and space- resolved mechanical behavior of bone to the related microbiological processes and regeneration kinetics. Proposed experiments intend to particularly highlight the distribution of mechanical stresses induced by the reconstruction process and mainly supported by the mineral part of bone architecture (hydroxyapatite - Hap) through mechanical state mapping thanks to X-ray in various regeneration levels until complete reconstruction thanks to samples harvested at different regeneration stages.. This study is achieved through *in situ* tensile testing under SAXS/WAXS synchrotron μ-beam. The size and orientation of bone mineral particles as well as the space distribution of particle agglomerates in rat calvaria defects, were investigated at different healing stages. The resulting two dimensional maps of crystal strain, mean thickness and degree of orientation, revealed the strong correlation between the bone mechanical resistance and the crystal organization.

Materials and methods
Samples. Eighteen white laboratory rats have been used for bone-regeneration tests (11 animals as implantation recipient and 7 rats as donors). Two bilateral defects of 3 mm diameter were created by surgical way in rat calvaria under general anesthesia. Defects were then treated with a mixture of compact bone powder (femoral head) and blood. Rats were then sacrificed by an overdose of CO_2 at 2, 4, 6 and 8 weeks after surgery. The calvaria parallelepiped-shaped specimens ($10 \times 5 \times 0.6$ mm^3) were harvested from the animals and immersed in a 70 % alcohol solution in order to preserve its mechanical attributes. Mechanical investigations were carried out on samples harvested at the different regeneration stages, emblematic of the osseous reconstruction level. Dogbone sample shaped were managed by embedding 2 mm of the end of their longest dimension (10 mm) in a non-invasive, low polymerization temperature, high stiffness resin.

Small and Wide-angle X-ray scattering. X-ray scattering techniques are particularly appropriate for bone ultrastructure analyses [2]. Indeed, they enable to obtain representative and quantitative information on the nanoscale structure as well as on the shape and organization of the organic and inorganic components. At sub-nanometric (crystal) and nanometric (collagen fiber) levels, Wide and Small Angle X-ray Scattering (respectively WAXS and SAXS) have therefore been used to determine microstructural features over different scales. WAXS patterns enable investigations of crystalline structure, crystallographic orientation, mineral particle distribution and averaged crystal size whereas SAXS analysis enable information onto crystal particles aggregates (clusters) set along collagen fibrils such as their size, orientation distribution or mean distances. Using interreticular distances (WAXS) or cluster distances (SAXS) as strain gages, such techniques enable resolving deformation undergone by either Hap or collagen fibrils, respectively. This leads to a two-scale mechanical behavior characterization (Collagen fibers: > 200 nm; crystals: < 10 nm) when a mechanical load is applied. SAXS and WAXS measurements were conducted at the P03 (MiNaXS) beamline at Petra III Light Source (Hamburg, Germany). Ten samples (3 rat specimens × 3 regeneration stages + reference sample) were studied through *in situ* tensile testing in the elastic regime in a step-by step mode (initial mechanical stress state + 4 incremental loads) up to 50 N. For each loading step, SAXS and WAXS measurements were achieved in transmission mode thanks to two Pilatus detectors with scanning steps of 60 μm in a continuous line-scan mode along (longitudinal strain) and perpendicular (transverse strain) to the tensile axis with 16×22 μm² spot size and 15 keV beam energy (1.0 Å wavelength). The drawn map ensure an investigation of the strain distribution in both collagen fibrils (SAXS) and Hap nanocrystallites (WAXS) across the interface between natural and regenerated bone over a 2×2 mm² area of acquisition (a fourth of the area of interest, symmetrically representative of the entire circular defect).

Data analysis
WAXS data evaluation. Experimental data were processed with DAWN software (Version 2.8, DIAMOND, UK) [3]. In the first place, the 2D WAXS patterns were converted to 1D diffraction patterns (intensity vs. diffraction angle) using azimuthal integration. The position of obtained diffraction peak gives information on the lattice spacing. In fact, Bragg's law gives a relationship between crystallite structure and parameters as determined from x-ray scattering:

$$n \lambda = 2d(hkl) \sin\theta. \tag{1}$$

Here, $d(hkl)$ is the interreticular distance for a $\{hkl\}$ plane family, θ is the Bragg's angle, λ is the X-ray wavelength and n is an integer given by the reflection order. The stress state evolution is reflected by an interreticular distance variation of the crystallographic planes. The mean true (rational) strain $\langle \varepsilon(hkl) \rangle_{V_d}$ in the probed diffracting volume V_d, for a given measurement direction is determined from the variation between the unloaded state $d_0(hkl)$ and the loaded state, $d(hkl)$. This deformation is connected through the Bragg law to the diffraction angles θ_0 for the reference state and θ for the stressed state by [4]:

$$\langle \varepsilon(hkl) \rangle_{V_d} = \ln\left(\frac{d(hkl)}{d_0(hkl)}\right) = \ln\left(\frac{\sin\theta_0(hkl)}{\sin\theta(hkl)}\right). \tag{2}$$

The WAXS data interpretation was limited to the {002} reflection. The lattice strains for this peak were calculated on both longitudinal (i.e., loading direction) and transverse directions. Bone tissue can be describe as a crystal dispersion material with an organic matrix, where polycrystal grains are the monocrystalline Hap platelets. In this respect, 2D WAXS diffraction patterns form concentric rings and a preferential crystallographic orientation of the crystals comes out as a strengthening of intensity in the diffraction rings over particular directions. The ensuing azimuthal intensity dependence directly relates to the predominant orientation of the mineral plates. Herman's factor [5] was used to describe and quantify crystals orientation in bone matrix by analyzing WAXS patterns. This factor, noted $F_{x/r}$, quantifies the relative orientation of a given crystallographic orientation x (in our case <001>: the Hap have a hexagonal structure and the <001> direction corresponds to the c-axis oriented along the collagen fibrils [6]) with respect to a fixed reference direction r (for example, the tensile axis). The mathematical expression of Herman's factor is given by the following relation:

$$F_{x/r} = \frac{3\langle\cos^2(a)\rangle - 1}{2}. \tag{3}$$

$$\text{where } \langle \cos^2(a) \rangle = \frac{\int_{\pi/2} I \cos 2\varphi \sin \varphi \, d\varphi}{\int_{\pi/2} I \sin \varphi \, d\varphi}. \tag{4}$$

a is the angle defined by the two directions x and r.
I being the intensity distribution of {hkl} diffraction peak in a given azimuthal angle φ.
The value of the Hermans factor is between -1/2 and 1.
$F_{x/r} = -1/2$: the orientation of the crystals is perpendicular to the reference orientation,
$F_{x/r} = 0$: an isotropic distribution of the orientation of crystals in space (uniform intensity on the ring),
$F_{x/r} = 1$: the orientation of the crystals is parallel to the reference orientation.

SAXS data evaluation. In a SAXS experiment, the scattering of X-rays is used to obtain information about collagen fibril strain distribution and mineral crystals size. The SAXS signal intensity is based on differences in electron density on the nanostructure of bone. Therefore,

Residual Stresses 2018 – ECRS-10   Materials Research Forum LLC
Materials Research Proceedings **6** (2018) 117-122   doi: http://dx.doi.org/10.21741/9781945291890-19

peaks obtained in SAXS indicate the periodicity of a specific phase in bone. Indeed, the variation of electron density is mainly due to those occurring between the mineral phase and the organic matrix (collagen fiber). As a two-phase material, the information on the average thickness of the mineral particles in the bone can be derived from the decay of the diffusion signal as regard to the total intensity diffused through the parameter T, thanks to Porod's law [5,7].

Here the diffusion vector $\vec{q}$ is defined as the difference between the incident and the scattered beam vector.

$$|\vec{q}| = q = \frac{4\pi}{\lambda}\sin\theta. \tag{5}$$

And according to Porod's law, the intensity of diffusion at large q values decreases with q^4.

$$\lim_{q\to\infty} I(\vec{q}) = B + \frac{P}{q^4}. \tag{6}$$

B is a background resulting from an incoherent and inelastic scattering and P is the Porod constant given by:

$$P = \lim_{q\to\infty} q^4 I(q). \tag{7}$$

Finally, T parameter is given by the following expression:

$$T = \frac{4}{\pi P}\int_0^\infty q^2 I(\vec{q})\, dq = \left(\frac{4\,\Phi(1-\Phi)}{S}\right). \tag{8}$$

Φ is the volume fraction of the mineralized crystals

S is the mineralized particle surface per volume unit of tissue

Assuming that the mineral phase has the shape of a uniform parallelepiped of lengths a, b and c, the expression of T is therefore:

$$T = 2(1-\Phi)\left(\frac{abc}{ab+bc+ac}\right). \tag{9}$$

The inorganic (mineral) matrix is composed of 85 % of calcium phosphate crystallized in the form of thin crystals (platelets or needles) of hydroxyapatite: $Ca_{10}(PO_4)_6(OH)_2$. These apatite crystals are generally in the form of small platelets (20 to 50 nm in length, 12 to 20 nm width and 2 to 5 nm thick). It is therefore possible to simplify the T parameter expression by assuming that $a \ll b, c$. the equation can be modified to:

$$T = 2a(1-\Phi). \tag{10}$$

For bone, we assume that Φ equals 0.5 [8]. Indeed, the extracellular matrix is composed by 60 to 70 % of an inorganic phase, composed essentially of hydroxyapatite crystals. Thus, T parameter gives directly the thickness of mineral particles.

Results

The map of longitudinal deformations (along loading direction) for the {002} plane family in mineral particles around the defect (marked out by dotted lines) implanted by bone graft for a load of 20 N and the related sample micro-Computed Tomograph (μ-CT), are shown in Fig. 1, for 6 weeks of regeneration. While the micro-tomography scans clearly show a complete filling of the defect after 6 weeks, the strain fields indicate that the mechanical function of the bone regarding the implant is not yet recovered at this stage of regeneration. Although the mechanical load supported by the defect is already greater at 6 weeks as compared to those at 2 and 4 weeks (not presented here) after implantation. Thus, at 6 weeks post-implantation, the regenerated bone process is not yet completed, preventing it from recovering an elastic modulus similar to that of the natural bone.

Fig. 2(a) shows the evolution of Hap plate thickness at 6 weeks post-implantation through the T parameter evaluation, as well as the distribution of their orientations thanks to Herman's orientation factor. As shown in Fig. 2(a), crystal thickness results after 6 weeks show a homogenized morphology (close thicknesses) between the crystals of the new matrix and those of the natural parts, contrary to shorter regeneration times where we notice a strong

Residual Stresses 2018 – ECRS-10   Materials Research Forum LLC
Materials Research Proceedings **6** (2018) 117-122   doi: http://dx.doi.org/10.21741/9781945291890-19

heterogeneity between the two parts [7]. The hatched white dots in Fig 2.(a) correspond to unproductive raw data.

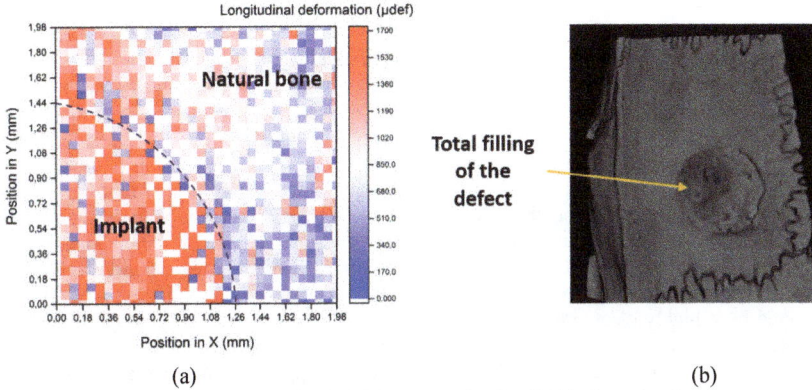

(a)   (b)

Figure 1: (a) Mapping of longitudinal deformations for the {002} plane family in Hap particles around the defect implanted by bone graft for a load of 20 N, (b) associated μ-CT at 6-week post-implantation. (lattice strains are given as micro-strain (με, units of 10^{-6})).

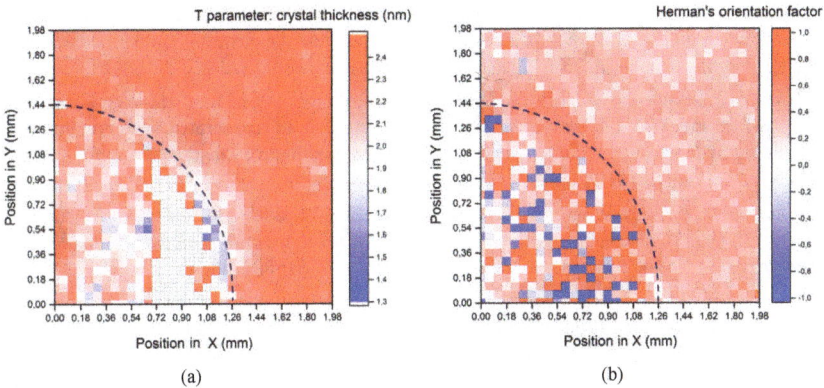

(a)   (b)

Figure 2: Mapping of (a) T parameter: crystal thickness of Hap particles at 6-week post-implantation; (b) Herman's orientation factor.

Fig 2(b) presents the maps of the Herman's orientation factor at 6-week post implantation by bone graft. The synthesized tissue (organic and mineral) in the new matrix is mostly without particular orientation for this regeneration time except at the interface, where a perfect homogenization with the natural bone is highlighted. We have seen previously (Fig. 2(a)) that the crystals of the new matrix have a thickness close to that of the natural bone after 6 weeks of regeneration. On the other hand, Fig. 2(b), clearly shows that the reorganization of Hap crystals is not yet complete at this stage. This analysis is in agreement with the results from Nakano and al. [6] showing perfect homogenization between regenerated bone and natural bone at a duration of 12 weeks for a cranial bone. This suggests that, initially, bone regeneration consists in restoring the microarchitecture of the structural elements (translated here by the thickness of the

Hap crystals). In a second stage, a transformation phenomenon is initiated leading to a reorganization of the collagen fibers and mineral crystals so that they line up with those present in the natural bone.

Discussion

Bone quality is significantly altered between the natural bone and the newly formed one at this stage of regeneration (6 weeks) due to the random distribution of mineralization in the newly formed bone. Indeed, the newly formed bone has a lower mineral content, which results in a lower modulus of elasticity as shown by the high level of deformation observed in the defect compared to the natural bone at 6 weeks of regeneration (Fig. 1(a)). The analysis of the evolution of the mechanical and microstructural states during regeneration shows a correlation between the bone mechanical resistance and the microarchitecture of the crystals as well as their spatial distribution. The correlation between the size of the crystals, their organization in the newly synthesized matrix and their mechanical behavior at different stages of regeneration showed that this crystal arrangement maximizes tensile strength. It also showed that defects were initially recovered by a highly disordered tissue to be subsequently reproduced and reorganized during regeneration.

As a conclusion, osseous reconstruction takes on complex process involving a progressive set up of the bone architecture given by degree of alignment and mineral particle size, leading eventually to the microscopic and macroscopic mechanical behavior of a regenerated bone.

References

[1] Zhang, J., Wang, H., Shi, J., Wang, Y., Lai, K., Yang, X., & Yang, G. (2016). Combination of simvastatin, calcium silicate/gypsum, and gelatin and bone regeneration in rabbit calvarial defects. *Scientific reports*, *6*, 23422. https://doi.org/10.1038/srep23422

[2] Rossi, A. L., Barreto, I. C., Maciel, W. Q., Rosa, F. P., Rocha-Leão, M. H., Werckmann, J., Rossi, AM., Borojevic, R.& Farina, M. (2012). Ultrastructure of regenerated bone mineral surrounding hydroxyapatite–alginate composite and sintered hydroxyapatite. *Bone*, *50*(1), 301-310. https://doi.org/10.1016/j.bone.2011.10.022

[3] Basham, M., Filik, J., Wharmby, M. T., Chang, P. C., El Kassaby, B., Gerring, M., & Sneddon, D. (2015). Data analysis workbench (DAWN). *Journal of synchrotron radiation*, *22*(3), 853-858. https://doi.org/10.1107/S1600577515002283

[4] François, M., Ferreira, C., Reference specimens for x-ray stress analysis: The French experience. Metrologia 41 (2004) 33–40. https://doi.org/10.1088/0026-1394/41/1/005

[5] Hermans, P. H., & Weidinger, A. (1948). Quantitative x-ray investigations on the crystallinity of cellulose fibers. A background analysis. Journal of Applied Physics, 19(5), 491-506. https://doi.org/10.1063/1.1698162

[6] Nakano, T., Kaibara, K., Ishimoto, T., Tabata, Y., & Umakoshi, Y. (2012). Biological apatite (BAp) crystallographic orientation and texture as a new index for assessing the microstructure and function of bone regenerated by tissue engineering. Bone, 51(4), 741-747. https://doi.org/10.1016/j.bone.2012.07.003

[7] Fratzl, P., Schreiber, S., & Boyde, A. (1996). Characterization of bone mineral crystals in horse radius by small-angle X-ray scattering. Calcified tissue international, 58(5), 341-346. https://doi.org/10.1007/BF02509383

[8] Rinnerthaler, S., Roschger, P., Jakob, H. et et al., Scanning small angle X-ray scattering analysis of human bone sections. Calcified Tissue International, 1999. 64(5): p. 422-429. https://doi.org/10.1007/PL00005824

Films, Coatings and Oxides

Residual Stresses 2018 – ECRS-10   Materials Research Forum LLC
Materials Research Proceedings 6 (2018) 125-130   doi: http://dx.doi.org/10.21741/9781945291890-20

Residual Stress Analysis in the Oxide Scales Formed on 316L Stainless Steel at 700 °C under Humid Air

LI Linwei[1,a], JI Vincent[1,b,*]

[1]ICMMO/SP2M UMR CNRS 8182, Université Paris-Sud, Bat. 410, 91405 Orsay Cédex, France

[a]linwei.li2@u-psud.fr, [b]vincent.ji@u-psud.fr

Keywords: Residual Stresses, High Temperature Oxidation, Water Vapor, X-Ray Diffraction

Abstract. The effects of water vapor on residual stresses in the oxide scales formed on 316L austenitic stainless steel are investigated. Samples were oxidized in thermogravimetric analyzer at 700°C for 6 hours - 96 hours with different amounts of water vapor (air, air+0.5%H_2O, air+4.0%H_2O). Grazing incidence X-ray diffraction (GIXRD) at different incident angles was used to study the phases and residual stresses in the oxide scales. The results demonstrate the formation of an inner chromia (Cr_2O_3) or chromium and iron solid solution ($Fe_xCr_{2-x}O_3$) layer and an outer hematite (Fe_2O_3), iron and nickel metallic compound ($FeNi_3$) and spinel layer. With the presence of water vapor, few wüstite (FeO) was also detected near the substrate. The residual stresses in the oxide scales are compressive, while the ones in the substrate are mostly tensile. Water vapor influenced not only the composition ratio of oxide scales and the residual stress levels but also the approach of oxide film damage.

Introduction

For austenitic stainless steels, which are widely used as constructional materials in power generation and petrochemical industries, a dense and continuous Cr-rich oxide film usually plays a protective role in the oxidation corrosion resistance at elevated temperature by acting as a barrier to prevent oxygen anion and metal cation inter-diffusion and reaction. However, these materials are meeting new challenges: higher working temperatures and more water vapor, among which the former is for gaining higher reaction efficiency and the latter is due to the more usage of biomass energy for environment cleanness [1]. For years many works concerning the high temperature oxidation mechanism of austenitic stainless steel have been reported [2-4], which intend that oxidation behaviors could be affected by various factors, such as the element composition in the substrate, the treatment of the substrate surface, the reactive gas, humidity and temperature. In addition, residual stresses which directly reflect the adhesion between the film and substrate are also taken attention [5]. The purpose of this work is to investigate the effects of water vapor on residual stress evolution in the Cr-containing oxide scale formed on 316L austenitic stainless steel at 700°C.

Experimental

A 316L stainless steel plant (the chemical composition is given in Table 1) was cut into a dimension 10×10×1mm with a 1mm in diameter hole drilled near the edge for being hung in the Thermal-gravimetric (TG) analyzer (SETARAM 92-16.18). Before oxidation, each side of samples were polished with SiC papers from P240 grit to P4000 grit to get a uniform surface roughness around 0.2μm, later immersed in the mixture solution of ethanol and acetone and kept in the ultrasonic cleaner for 10 minutes, then dried with pressed gas. TG experiments were carried out at 700°C from 6 hours to 96 hours in the air with different amounts of water vapor (dry air, air+0.5%H_2O and air+4.0% H_2O).

After oxidation, the oxide scales were identified by GIXRD at incident angles of 0.5° and 2° respectively, using PANalytical X'Pert PRO MRD with copper radiation source (λ=0.154nm). Besides, Raman spectroscopy (λ=472.99nm) was also used to identify the chemical compositions of oxide sales as a reference. The microstructures of cross-section were studied by field emission gun scanning electron spectroscopy (FEG-SEM) equipped with an energy dispersive X-ray spectroscopy (EDX), using ZEISS SUPRA 55VP.

The residual stresses in both oxide scales and substrates were characterized by GIXRD, followed European standard NF EN 15305 (published in April 2009). The strongest peak of each phase was chosen (the {104} peaks for M2O3 and {111} peaks for substrate) in order to obtain sufficient intensities as well as positions of each peak at 13 distinct Psi angles (from -30° to +30°) were determined by using a Gauss function fitting.

Table 1 Chemical composition of 316L stainless steel (by weight %)

Fe	Cr	Ni	Mo	Mn	C	Si	P	S	N
Bal.	16-18	10-14	2-3	<2	<0.03	<0.75	<0.045	<0.03	<0.1

Oxidation product characterization

The 0.5° peak patterns of 316L stainless steel oxidized for 6 hours with different amounts of water vapor are presented in Fig. 1a, implying the generation of a corundum oxide M_2O_3 (M refers Cr, Fe or their solid solution) and an intermetallic compound ($FeNi_3$) during the initial oxidation. From Fig. 1b, the enlarged picture of Fig. 1a, it can been seen that the {104} peaks of Cr_2O_3 and Fe_2O_3 are overlapped under the condition of water vapor participation, while a more narrow peak without water vapor clears Cr and Fe solid solution phase ($Cr_xFe_{2-x}O_3$), because water vapor could accelerate the oxidation, promoting the stratification between Cr and Fe oxides from the beginning of reaction [6]. Fig. 1c shows the peak patterns obtained at different incident angles of 0.5° and 2° respectively after oxidation for 72 hours, suggesting the formation of few spinel structure oxides AB_2O_4 (A refers Mn or Fe, B refers Fe or Cr). Besides, wüstite (FeO) was detected near the substrate after oxidation in humid air, while spinel oxides preferred forming on the surface in dry air. In addition, the weight percent of M_2O_3 in the oxide scales were calculated by the software HighScore Plus and listed in Fig. 2.

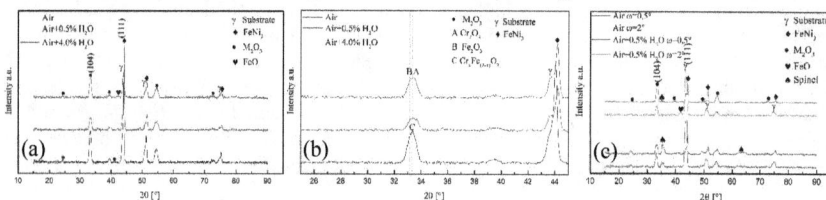

Fig. 1 GIXRD patterns of 316L stainless steel oxidized at 700°C with different amounts of water vapor (a) 6 hours, incident angle of 0.5° (b) enlarged figure of Fig. 1a (c) 72 hours, incident angles of 0.5° and 2°

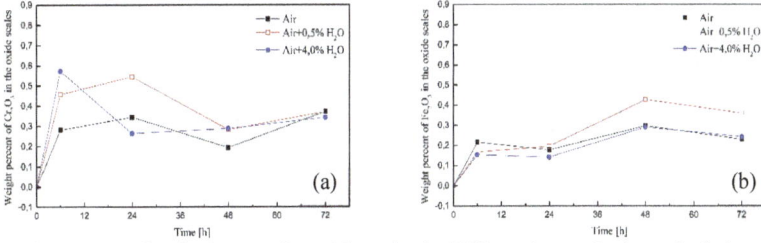

*Fig. 2 Weight percent of each phase in the oxide scales by XRD semi-quantitative calculation (a)
Cr_2O_3 (b) Fe_2O_3*

Raman spectra depicts the evolution of oxidation products versus oxidation time under different moisture conditions, confirming the existence of Cr and Fe corundum oxide (Cr_2O_3, Fe_2O_3 or $Cr_xFe_{2-x}O_3$), demonstrated in Fig 3a-c. That a phenomenon matched with the XRD results can also be found. Cr-rich oxides grew within the first 24 hours but became replaced by Fe-rich oxides later, which means the protective Cr-rich oxide layer would lose their oxidation resistance properties gradually, and then lead to crack growth or delamination of oxide films [7]. Fig 4a gives the microstructure and element mapping of oxide scale after 200-hour oxidation in the dry air, revealing that the two-layer oxide scale is composed by the inner Cr-rich oxides and the outer Fe-rich oxides. Delamination happened along the interface between the two oxide layers and cracks generated through the outer layer. Fig. 4b describes the microstructure of oxide layer after 96-hour oxidation with 4.0% water vapor, in which wavy oxide layer formed and plenty of vacancies concentrated at the interface between the metal and oxide layer (O/M interface).

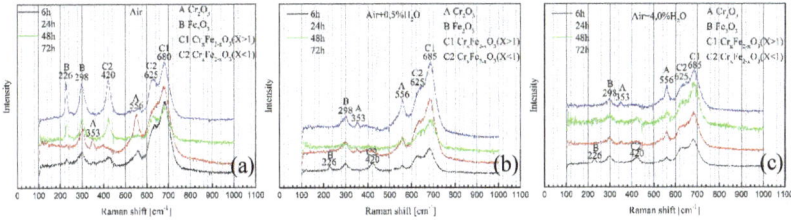

*Fig. 3 Raman spectra of oxide scales formed at 700°C with different amounts of water vapor (a)
air (b) air+0.5%H_2O (c) air+4.0%H_2O*

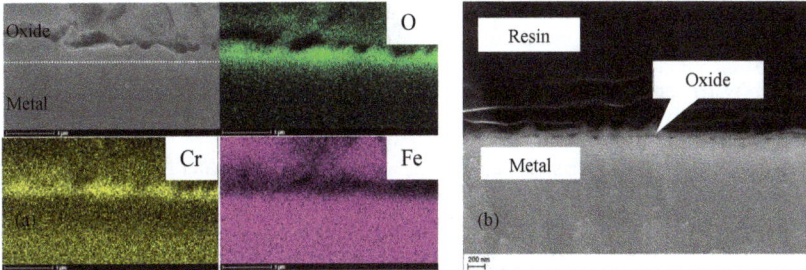

*Fig. 4 SEM images and element mapping of oxide scales formed on 316L stainless steel at 700°C
(a) 200-hour oxidation without water vapor (b) 96-hour oxidation with 4.0% water vapor*

Residual stresses

The residual stresses in the oxide scales and substrate were studied by GIXRD technique at the incident angles of 0.5° and 2° respectively. The overlapping peaks of Cr_2O_3 and Fe_2O_3 were distinguished by profile fitting for all the patterns obtained at each Psi angle to determine the positions of each phase. The parameters needed for stress calculation were listed in Table 2 and the results are expressed in Fig. 5. It can be deduced from the fluctuating curves that the stress accumulation together with relaxation influenced the final values. The thermal stress produced during the cooling process makes contribution to the residual stresses simultaneously, which can be calculated according to Eq. 1.

$$\sigma_{thermal} = \frac{E_O(\alpha_O - \alpha_M)}{(1-v_O)}\Delta T,$$
(1)

where ΔT is the temperature difference, E is the Young's modulus, α is the thermal expansion coefficient and v is the Poisson's ratio (O and M refer the oxide and metal respectively).

Fig. 5 The evolution of the residual stresses of the substrate and oxide scales (a) residual

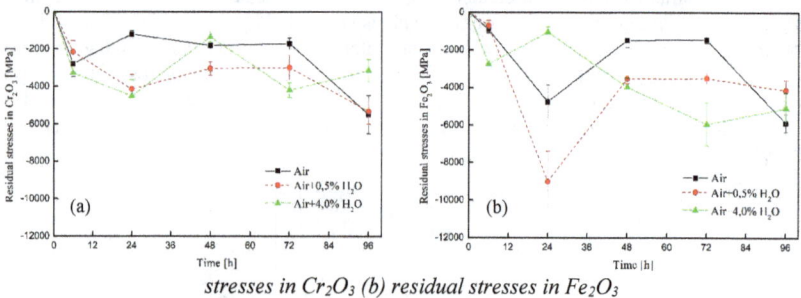

stresses in Cr_2O_3 (b) residual stresses in Fe_2O_3

During the first 6 hours, the oxide layer formed in the dry air did not have a two-layer structure because of the diffusion rate limit, which is the same in the air with 0.5% H_2O. However, the oxide scale formed in the air with 4.0% H_2O contained much more Cr_2O_3 ($\approx$57.1% in Fig. 2a), leading to stratification. So, the thermal stresses of Fe_2O_3 in those two cases are -734 MPa and +540 MPa, and of Cr_2O_3 are -2630MPa and -3410MPa respectively. Compared with the results in Fig. 5, it can be demonstrated that the residual stresses of Fe_2O_3 and Cr_2O_3 basically came from thermal stresses, except the one of Fe_2O_3 layer formed in the air with 4.0% water vapor, in which growth stress played an important role. That the residual stress of Cr_2O_3 formed in the air with 0.5% water vapor was less than the one formed in the dry air should be noticed, because the formation of FeO which has ability of deformation at high temperature could help relax partial stress [8].

Later but before 24 hours, the situations became more complicated. After oxidation for a while, the oxide scales had a two-layer distribution progressively, with a sharp rise of Cr_2O_3 ($\approx$34.4% dry air, $\approx$54.3% 0.5% H_2O). There is a common agreement that Cr_2O_3 usually forms by Cr ion outward diffusion, which rarely causes stress except at the grain boundaries and defects [8]. In the dry air, the adhesion between Cr_2O_3 and the substrate was quite good after a long oxidation, illustrated in Fig. 4a. As a result, the residual stress of Cr_2O_3 formed in the dry air declined, which was affected by the deformation of the substrate at high temperature due to their compact contact, indicating a tensile status during the oxidation. But in the humid air, vacancies preferred generating and gathering at the O/M interface, implying a bad adhesion [9]. So the residual stress of Cr_2O_3 formed in the humid air was less influenced by the substrate during the oxidation since the stress could relax through the pore formation. The residual stresses in Fe_2O_3 layer generated in the air with 4.0% H_2O shows an opposite trend compared with the other two cases, which is similar with Cr_2O_3 ratio ($\approx 26.3\%$) in that condition. Due to the rapid consumption and diffusion of Cr, plenty of vacancies appeared at O/M interface and Cr depletion in the subscale of the substrate occurred [9], which reduced Cr_2O_3 and Fe_2O_3 to increase sustainably. Considering Cr-containing oxide usually has priority to generate, the residual stress of Fe_2O_3 decreased even though the residual stress accumulated in Cr_2O_3 layer.

Between 24 hours and 48 hours, cracking appeared in those three cases, but the fissure positions were different. For the condition of the dry air, the cracking preferred forming through the outer layer, which was displayed in Fig. 4a. The formation of the spinel oxides ($FeCr_2O_4$ or Fe_3O_4) above Cr_2O_3 was like an obstacle [10], impeding the anion and cation inter-diffusion, so that Fe_2O_3 generated continuously and the stress accumulated until cracking. The fissure was also situated in Fe_2O_3 layer in the air with 0.5% H_2O, because the residual stress of Fe_2O_3 decreased suddenly. However, for the condition of 4.0% water vapor, Cr_2O_3 layer was broken, which can be demonstrated by the new Cr_2O_3 formation ($\approx 28.7\%$) and the decrease of residual stress of Cr_2O_3.

From 48 hours on, cracking or detachment continued happening, creating an unpredictable stress status. The competition of stress accumulation and relaxation were intense along with the new oxide generation, as well as fissure or delamination formation.

Table 2 Parameters for stress calculation

Phase	Young's modulus (E) [GPa]	Poisson's ratio (v)	Thermal expansion coefficient (α) [10^{-6} K^{-1}]
Cr_2O_3	280	0.29	9.6
Fe_2O_3	220	0.19	12.5
Substrate	193	0.25	19.4

Summary
In conclusion, the oxide scales were characterized in this work, confirming the existence of a two-layer structure with an inner protective Cr-rich oxide and an outer non-protective Fe-rich oxide. The evolution of residual stress in the Cr_2O_3 layer was discussed, concerning the adherence between the oxide and substrate. During the incubation period, the residual stresses of Cr_2O_3 derived from the thermal stress during the cooling process for all the three conditions. After the two-layer structure formed, the residual stress of Cr_2O_3 began to be influenced by the growth stress simultaneously, which means that the outer layer may apply a compressed stress on it while the substrate may cause a contrast effect during the high temperature oxidation. Consequently, distinct oxidation conditions brought different results to the Cr_2O_3 layer. In the dry air, the Cr_2O_3 film kept compact and continuous, in which the spallation happened in the outer layer. Similarly, with few water vapor, the generation of vacancies could relieve stress to some extent, leading to an intact Cr_2O_3 film. However, with much more water vapor, the collective effects of pores at the O/M interface and the compressed loads from either the outer layer or the substrate contributed to the fold or damage generation in the Cr_2O_3 layer. For long oxidation, the competition of stress accumulation and relief excited the cracking or spalling in the oxide scale as well as the new oxide generation. It can be predicted that the oxidation resistance would lose rapidly in the case of more water vapor participation, and disastrous corrosion would happen in the case of dry air due to the replacement of Cr-rich oxide by Fe-rich

oxide. Less water vapor which could promote Cr_2O_3 formation and extend the exhausted time of Cr (reduce vacancies) may be a better choice.

References

[1] M. P. Brady, M. Fayek, J. R. Keiser, H. M. Meyer III, K. L. More, L. Anovitz, D. J. Wesolowski and D. R. Cole, Wet oxidation of stainless steels: New insights into hydrogen ingress, Corrosion Science. 53 (2011) 1633-1638. https://doi.org/10.1016/j.corsci.2011.02.011

[2] H. Falk-Windisch, J. E. Svensson and J. Froitzheim, The effect of temperature on chromium vaporization and oxide scale growth on interconnect steels for Solid Oxide Fuel Cells, Journal of Power Sources. 287 (2015) 25-35. https://doi.org/10.1016/j.jpowsour.2015.04.040

[3] X. Peng, J. Yan, Y. Zhou and F. Wang, Effect of grain refinement on the resistance of 304 stainless steel to breakaway oxidation in wet air, Acta Materialia. 53 (2005) 5079-5088. https://doi.org/10.1016/j.actamat.2005.07.019

[4] W. Kuang, X. Wu and E.-H. Han, The oxidation behaviour of 304 stainless steel in oxygenated high temperature water, Corrosion Science. 52 (2010) 4081-4087. https://doi.org/10.1016/j.corsci.2010.09.001

[5] J. Xiao, N. Prud'homme, N. Li and V. Ji, Influence of humidity on high temperature oxidation of Inconel 600 alloy: Oxide layers and residual stress study, Applied Surface Science. 284 (2013) 446-452. https://doi.org/10.1016/j.apsusc.2013.07.117

[6] C. Tedmon, The effect of oxide volatilization on the oxidation kinetics of Cr and Fe-Cr alloys, Journal of the Electrochemical Society. 113 (1966) 766-768. https://doi.org/10.1149/1.2424115

[7] M. Halvarsson, J. E. Tang, H. Asteman, J. E. Svensson and L. G. Johansson, Microstructural investigation of the breakdown of the protective oxide scale on a 304 steel in the presence of oxygen and water vapour at 600°C, Corrosion Science. 48 (2006) 2014-2035. https://doi.org/10.1016/j.corsci.2005.08.012

[8] T. Mitchell, D. Voss and E. Butler, The observation of stress effects during the high temperature oxidation of iron, Journal of Materials Science. 17 (1982) 1825-1833. https://doi.org/10.1007/BF00540812

[9] X. Wei, X. Peng, X. Wang and Z. Dong, Development of growth and thermal stresses in NiO scale on nanocrystalline Ni without and with dispersion of CeO2 nanoparticles, Corrosion Science. 118 (2017) 60-68. https://doi.org/10.1016/j.corsci.2017.01.014

[10] C. Fujii and R. Meussner, Oxide Structures Produced on Iron-Chromium Alloys by a Dissociative Mechanism, Journal of the Electrochemical Society. 110 (1963) 1195-1204. https://doi.org/10.1149/1.2425624

Residual Stresses 2018 – ECRS-10
Materials Research Proceedings 6 (2018) 131-136

Materials Research Forum LLC
doi: http://dx.doi.org/10.21741/9781945291890-21

Evaluation of Residual Stresses in PVD Coatings by means of Tubular Substrate Length Variation

Harri Lille[1,a,*], Alexander Ryabchikov[1,b], Jakub Kõo[1,c], Eron Adoberg[2,d],
Valdek Mikli[2,e], Jakob Kübarsepp[2,f], Priidu Peetsalu[2,g]

[1]Institute of Forestry and Rural Engineering, Estonian University Life of Sciences, Kreutzwaldi 5, 51014 Tartu, Estonia

[2]Department of Mechanical and Industrial Engineering, Tallinn University of Technology, Ehitajate tee 5, 19086 Tallinn, Estonia

[a]harri.lille@emu.ee, [b]alexander.ryabchikov@emu.ee, [c]jakub.koo@emu.ee,
[d]eron.adoberg@ttu.ee, [e]valdek.mikli@ttu.ee, [f]jakob.kubarsepp@ttu.ee, [g]priidu.peetsalu@ttu.ee

Keywords: TiAlN Hard PVD Coating, Residual Stresses, Length Variation

Abstract. The aim of the study was to determine macroscopic residual stresses in PVD coatings. The device for measurement of the length of the substrate was improved, where a change in tube length was reduced to the deflection of the middle cross-section of the elastic element whose deformation was measured by four strain gauges. The formulas for calculation of residual stresses are presented. For comparison a unilateral coating was deposited on a vertically fixed plate using the conventional curvature method. As an application, residual stresses in hard PVD TiAlN coatings were investigated. The microstructure and thickness of the studied coatings were investigated by means of scanning electron microscopy (SEM) in Zeiss EVO MA-15. The mean values of compressive residual stresses determined by both methods, for the studied coatings, were very high (3.1-6.5 GPa), irrespective of coating thickness, and practically equal with the measurement uncertainty of the method. The developed tube length variation method is reliable and applicable for determination of residual stresses in PVD coatings.

Introduction

Physical Vapour Deposition (PVD) coatings are used inter alia for blanking, punching and cutting applications and can be deposited both on plain and more complex surfaces [1, 2]. It is wellknown that residual stresses arising in coatings during the deposition process have an important effect on the service life of the coating through influencing its mechanical and tribological properties and adhesion.

The aim of the study was to determine macroscopic residual stresses in coatings vapoured on a vertically fixed cylindrical surface, using the deformation method, through measurement of the longitudinal length variation of the thin-walled tube, as well as to validate the results obtained with the conventional curvature method using the plate as the substrate. One batch of vertically fixed plates was prepared by depositing a unilateral coating on the front surface and the other batch of plates, by depositing it on the back surface. Thus a considerable amount of the vapoured target material was deposited on the fixing device as well [3]. On the other hand, using the tubular substrate, most of the coating was deposited on the outer surface of the tube (a small part of the coating was deposited on the nozzle) it is possible to estimate the values of residual stresses in coatings on cylindrical surfaces (e.g. cutting tools [4]).

The measuring device for determination of the longitudinal length change of the substrate was improved (Fig. 3), where tube length variation was reduced to the deflection of the middle cross-section of the elastic element whose deformation was measured by four strain gauges [5]. As an example of application, residual stresses were measured in hard PVD TiAlN coatings which are

Residual Stresses 2018 – ECRS-10   Materials Research Forum LLC
Materials Research Proceedings **6** (2018) 131-136   doi: http://dx.doi.org/10.21741/9781945291890-21

most widely used for cutting tools [4]. Also the microstructure and thickness of the studied coating were investigated by means of scanning electron microscopy.

Evaluation of Residual Stresses in the Coating

Plates (Fig. 1b) are only deposited from one side and should be placed gripped with a claw in the fixing device made of carbon steel [3]. Depending on the deflection of the plate, modified Stoney's formula will account for biaxial stresses [6]. In order to prevent deposition of the coating on the cross-section of the tube ends, they were closed by the nozzle (Fig. 1a); at the same time, the tube was vertically fixed, by the lower nozzle, to the rotary table of the chamber and was simultaneously rotated around its axis.

a)

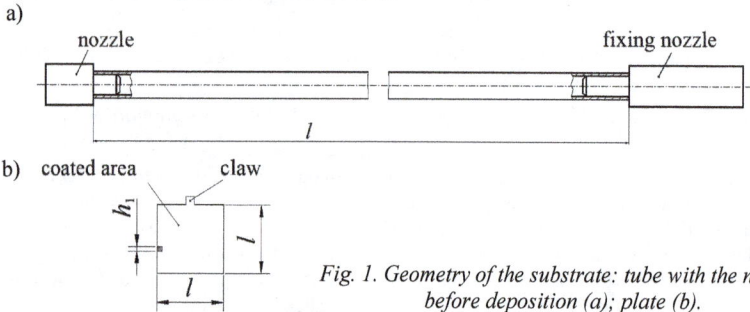

b)

Fig. 1. Geometry of the substrate: tube with the nozzles before deposition (a); plate (b).

The mean values of residual stresses in the coatings were calculated from the length change of the tubular substrate [7]. As the coating was relatively thin it was assumed that residual stresses are distributed uniformly throughout coating thickness (Fig. 2).

a)   b)

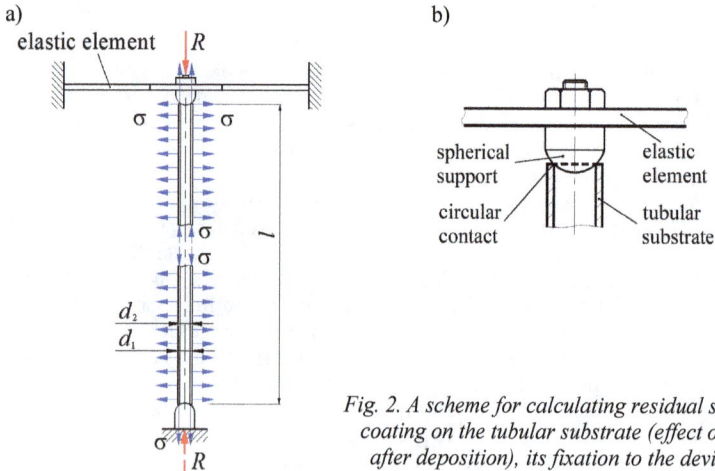

Fig. 2. A scheme for calculating residual stresses in the coating on the tubular substrate (effect on substrate after deposition), its fixation to the device (a) and circular contact of the tube (b).

Residual Stresses 2018 – ECRS-10   Materials Research Forum LLC
Materials Research Proceedings 6 (2018) 131-136   doi: http://dx.doi.org/10.21741/9781945291890-21

Hence the following expression can be used:

$$\sigma = \frac{E_1}{2(1-\mu_1)r_1}\left(\frac{r_1^2 - r_2^2}{l} + \frac{C}{E_1\pi}\right)\frac{\Delta l}{h_2},\tag{1}$$

where E_1 and μ_1 are the modulus of elasticity and Poisson's ratio of the substrate, r_1 is outer radius of the tube, r_2 is inner radius of the tube, h_2 is thickness of the coating, l is length of the tube, Δl is measured length variation of the tube, C is rigidity of the elastic element. Considering our measuring device and the elastic element and the tube dimensions used, the share of member $C/(E_1\pi)$ is 1.7 % of the second member in the brackets of Eq. 1; therefore it was not taking in to account and Eq. 1 is expressed as follows

$$\sigma = \frac{E_1}{2(1-\mu_1)r_1}\left(\frac{r_1^2 - r_2^2}{l}\right)\frac{\Delta l}{h_2}.\tag{2}$$

During measurement of length variation, the lower end of the tube was fixed stationary and the upper end was pressed to the elastic element, which prestressed also the substrate (Fig. 3).

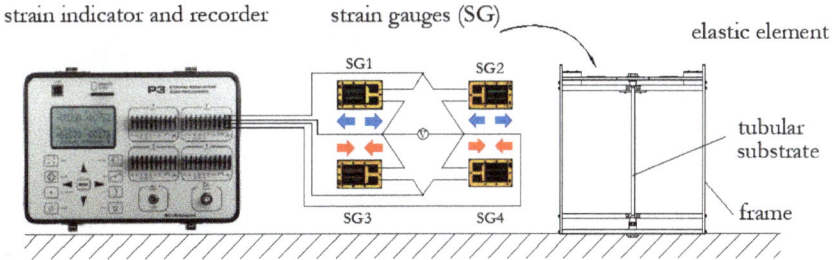

Fig. 3. A scheme of the measuring device for determination of the length of the tube and the substrate supported on the sphere.

To guarantee the centering of the tube in the measuring devise, the inner circular line of the cross section of its ends was in contact with the spherical surface of support (Fig. 2b). Consequently, the eccentric compressive load was minimized. The length of the tube was measured ten times before and ten times after deposition and mean values were used for calculating residual stresses in the coating.

Since calibration of the device is a separate task the scheme of which is presented in our paper [5], then now we can see in Fig 4a how the four strain gauges are glued in the longitudinal direction onto the free surface of the elastic element and joined to form a full bridge with four fully active arms (Fig. 4b). The elastic element was loaded manually step by step in the middle with weights of class M1, and deflection was measured with the potentiometric displacement tracer FWA050T to 50 mm/0.001 mm and recorded with the Datalogger-AHLBORN ALMEMO® 2890-9-5. At the same time, the signals from the strain gauges were measured with the VPG P3 strain indicator and recorder. The results of calibration obtained in the case of unloading (its direction corresponds to tube length measurement direction) are presented graphically in Fig. 5.

The constants should be determined so that relations are approximated in the best way by minimizing the square of error (least squares regression). The deflection of the middle cross-section of elastic element Δl, depending on of the units of the strain indicator (Fig. 5a), was found by using the program *MS Excel 2016* with the regression analysis function.

a)   b)

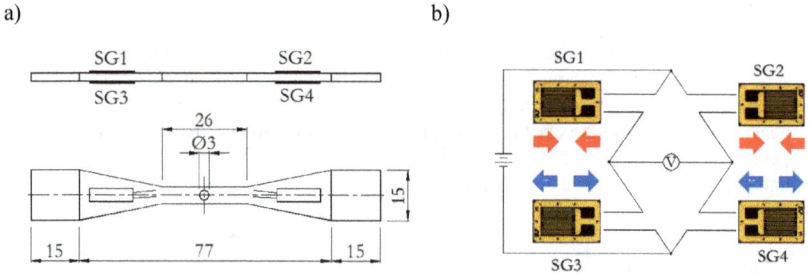

Fig. 4. The geometry of the elastic element and placement of the strain gauges (SG) (a); full bridge with four fully active arms (b).

a)   b)

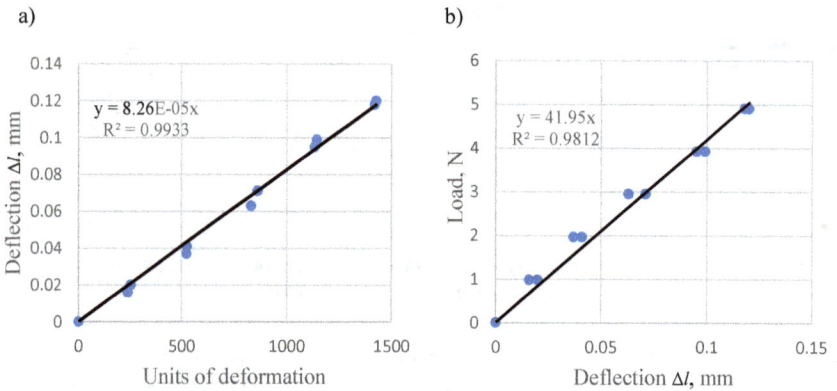

Fig. 5. Dependence of the units of the strain indicator on the deflection in the case of unloading (a); dependence of the load on the deflection of the elastic element (b).

As a result, the strain indicator constants are 8.26×10^{-5} mm per unit of the strain indicator and recorder and the rigidity of the elastic element $C = 41.95$ N/mm.

Example of application

The studied PVD coatings were produced in the Laboratory of PVD Coatings at the Tallinn University of Technology. The PVD unit Platit π-80 with two rotating cathodes, embedded in the door of the vacuum chamber, was used for deposition. The tubes were fixed in the rotary table of the vacuum chamber vertically as were the plates in the fixing device (four specimens from each sample), but so, that one batch of the plates were prepared by deposition on the front surface (directed to the edge of rotary table) and the other batch, by deposition on the back surface (directed to the centre of the rotary table). After measuring the length variation of the coated tube and the deflection of the coated plate, three pieces were cut from one tube (two from the ends and one from the middle) with a length of 10 mm and two pieces were cut from the plates of different fixing specimens with dimensions of 10×10 mm, for scanning electron microscopy (SEM).

By means of SEM in Zeiss EVO MA-15, the microstructure of the coatings (Fig. 6) was investigated and their thicknesses were measured for calculating mean values. Physical and surface properties of the substrate material, dimensions and mean values of residual stresses in the coatings are presented on Table 1.

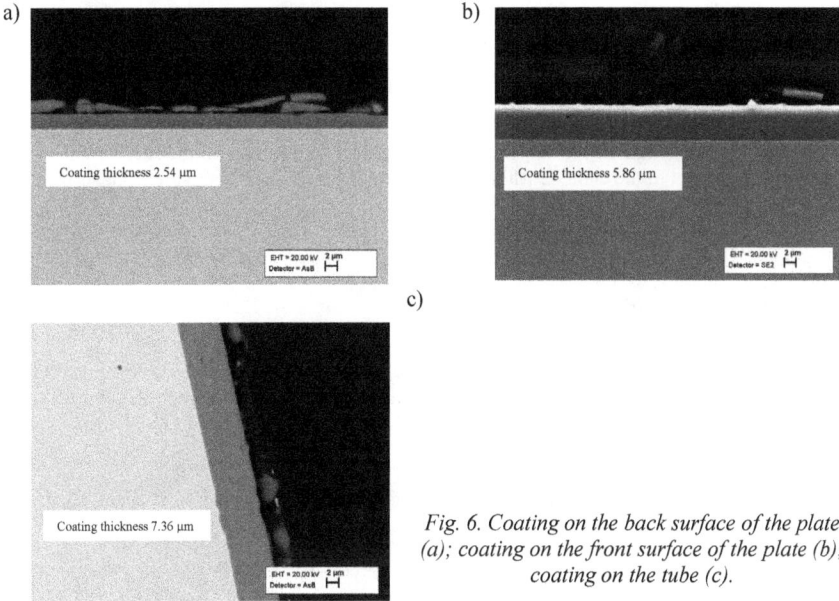

a)

b)

c)

Fig. 6. Coating on the back surface of the plate (a); coating on the front surface of the plate (b); coating on the tube (c).

Table 1. Physical and surface properties of the substrate material, dimensions and mean values of residual stresses in the coatings

Substrate material	Thickness, diameters [mm]	Usable width, length [mm]	Modulus of elasticity [GPa]	Poisson's ratio	Surface rough-ness Ra μm]	Thickness of coating, [μm]	Residual stresses, [GPa]
Steel plate	0.395* 0.395	19.75 19.70	193.0	0.28	0.024-0.029	5.86 2.54	-6.53 -4.39
Steel tube	d_1=3.0 d_2=2.5	167.5	193.0	0.25	0.024-0.029	7.36	-3.10

* The deposited surface is directed to the edge of the rotary table

We can see that the thickness of the coatings deposited on the tubular substrates and the thicknesses of the coatings deposited on the plates are significantly different. The coating deposited on the tube is thicker as some part of its coated surface is constantly forehead bombarded with target plasma atoms and ions. The coated surface of the plate is directed to the edge of rotary table and is bombarded with atoms and ions; the potential distance between the target and the substrate is minimal and the kinetic deposition parameters are larger.

Consequently, the thickness of the coating is lager and also residual stresses arising in it are higher. The coated surface of the plate is directed to the centre of on rotary table and is bombarded with atoms and ions; the potential distance between the target and the substrate is maximal and the kinetic deposition parameters are smaller (the number of atoms in the ion beam is evidently lower). Also the thickness of the coatings is lower and residual stresses arising are lower. High compressive stresses arise in the coatings which help to reduce pre-existing structural defects.

Summary

The existing device was improved to enable to measure the length of the thin walled tubular substrate before and after deposition of the coating. The obtained length variation can be used as an experimental parameter for calculating the mean values of residual stresses in coatings.

Coatings are deposited on the substrate fixed vertically in a holder on the rotary table of the chamber and parallel to the cathode while one part of the plate is fixed onto the front surface and the other, on the back surface at the time of deposition.

The calculated mean values of residual stresses in hard PVD TiAlN coatings were: in the coating on the tube, 3.10 GPa; on the front surface of the plate, 6.53 GPa and on the back surface on the plate, 4.39 GPa. The microstructure of the coating on the tube and on the plate, as well as the thickness of the coatings are presented.

Acknowledgements

This study was supported by the Estonian Ministry of Education and Research by Institutional Research Funding IUT 19-29 „Multi-scale structured ceramic-based composites for extreme applications"

References

[1] D.T. Quinto, Twenty-five years of PVD coatings at the cutting edge. Fall Bulletin. (2007) 17-22.

[2] T. Sampath Kumar, S. Balasivanandha Prabu, Geetha Manivasagam, and K. A. Padmanabhan, Comparison of TiAlN, AlCrN and AlCrN/TiAlN coatings for cutting-tool applications. Int. J. Min. Met. Mater. 21 (2014) 796-805.

[3] H. Lille, J. Kõo (et al). Comparation of Curvature and X-ray methods on Residual Stresses Measurements in Hard PVD Coatings. Materials Science Forum, 681 (2011) 455-460. Information on https://doi.org/10.4028/www.scientific.net/MSF.681.455

[4] Information on https://www.pvd-coatings.co.uk/applications/cutting-tools/

[5] H. Lille, A. Ryabchikov, J. Kõo (et al). Evaluation of Residual Stresses in PVD Coatings by means of Strip Substrate Length Variation and Curvature Method of Plate Substrate. Solid State Phenomena 267 (2017) 212-218. Information on https://doi.org/10.4028/www.scientific.net/SSP.267.212

[6] J. Kõo, J. Valgur, Residual stress measurement in coated plates using layer growing/removing methods: 100th anniversary of the publication of Stoney's paper "The tension of metallic films deposited by electrolysis". Mater. Sci. Forum. 681 (2011) 165-170.

[7] J. Kõo, A. Ryabchikov. On the determination of residual stresses in coatings from measured longitudinal deformation of a wire substrate, in: Proc. 19th Symp. on Exp. Mech. of Solids J. Stupinicki (Ed.), Warsaw Univ. of Technology, Jachranka (Poland), 2000, pp. 319-324. Information on http://hdl.handle.net/10492/3803

Cold Working and Machining

Residual Stresses 2018 – ECRS-10
Materials Research Proceedings 6 (2018) 139-144

Materials Research Forum LLC
doi: http://dx.doi.org/10.21741/9781945291890-22

Distinguishing Effect of Buffing Operation on Surface Residual Stress Distribution and Susceptibility of 304L SS and 321 SS Welds to Chloride Induced SCC

Pandu Sunil Kumar[1,a,#], Kamal Mankari[1,b,#], Swati Ghosh Acharyya[1,c,*]

[1]School of Engineering Sciences and Technology, University of Hyderabad, India

[a]sunilpandu04@gmail.com, [b]mankari.kamal.1802@gmail.com

[#]Pandu Sunil Kumar and Kamal Mankari have contributed equally to this work.

Keywords: Residual Stresses, Austenitic Stainless Steels, Stress Corrosion Cracking

Abstract. Stress corrosion cracking (SCC) of austenitic stainless steels (ASS) and its weldments in presence of chloride ions is a key concern in its successful application. AISI 304L SS in surface milled, turned, ground conditions have high tensile residual stresses on the surface which lead to early cracking in an aggressive environment. Spot welds of AISI 321 SS have shown multiple failures due to chloride induced SCC as a result of high magnitude of tensile residual stresses and improper post weld heat treatment. The present study proposes a simple surface engineering method to prevent the initiation of stress corrosion cracking in austenitic stainless steel and its welds in presence of chloride ions. 304L SS in milled, turned and ground conditions and 321 SS in spot welded condition was subjected to surface buffing operation. Surface roughness was calculated using a surface profilometer and residual stresses were determined. Residual stress distribution, and phase transformation were calculated using X-ray diffraction measurements. The detailed microstructural characterization was performed using field emission scanning electron microscopy (FESEM). 304L plates and 321 SS welds in buffed and un-buffed conditions were tested for SCC susceptibility by exposing these to boiling $MgCl_2$ as per ASTM G36. Results showed that 304L SS and 321 SS were resistant to SCC in buffed conditions as no cracking occured even after prolnged exposure to boiling $MgCl_2$. Buffing being a very simple, economic and portable operation can be easily applied on large components of AISI 304L SS after the conventional surface finishing operations and AISI 321 SS weld also can be extended to components in service in an aggressive environment.

Introduction

Austenitic stainless steels have good corrosion resistance and mechanical properties. However, these become highly susceptible to SCC when subjected to different surface finishing and welding operations [1-3]. Laser peening and shot peening are generally used for enhancing SCC resistance, but these processes induce plastic strain in the material and are not economic. Hence a simple and economic route is essential in preventing SCC. We have shown in our previous report that simple surface buffing can be used to enhance the SCC resistance of austenitic stainless steels [4]. The present study substantiates the effectiveness of buffing in preventing the initiation of Cl− induced SCC of austenitic stainless steel in presence of chloride ions. 304L SS in milled, turned and ground conditions and 321 SS in spot welded condition when subjected to surface buffing operation.

Experimental

Materials and methods

In the present study two steels namely, AISI 304L SS (C 0.03, Cr 18, Ni 8, Mn 1.6, P 0.04, Si 0.4,S 0.03, balance Fe) wt %, and AISI 321 SS welds (C 0.024, Cr 17.41, Ni 9.14, Mn 1.64, Ti 0.23, Mo 0.37, P 0.04, Si 0.4, S 0.03) wt% were used. AISI 304L SS samples were cut into a dimension of 100 mm X 25 mm X 5 mm plates and they were solution annealed in order to remove the internal stresses present in the material. AISI 304 L SS was subjected to three different surface working conditions namely a) milling b) turning and c) grinding operations at a feed rate of 0.1mm/rev to remove 0.5 mm from the surface [7]. Subsequently, the samples were buffed at 3600 rpm using and 50μm was removed from the surface. In another study, AISI 321 SS tubes having 72 mm diameter, 3 mm thickness and length of 3.6 m having spot welds on the surface were subjected to buffing and 50μm was removed. Both the samples were tested for SCC susceptibility in buffed and unbuffed condition. During buffing compressive forces are applied on the sample substrate as shown in the schematic given below (Figure 1). The buffing wheel is rotated at a set speed and is rastered on the surface of the workpiece imparting compressive stresses to the entire surface of the workpiece. No Ti_2N particles were present in the near surface region of buffed samples in each case.

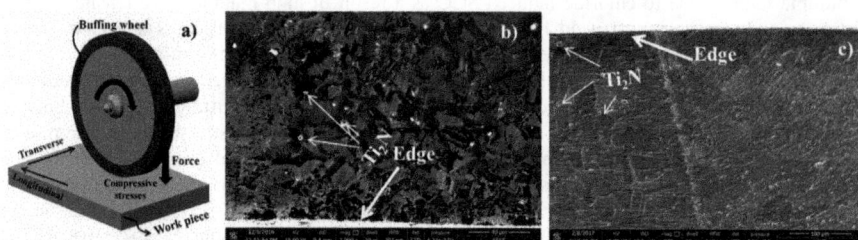

Fig 1: a) Schematic of the buffing operation b) cross sectional micrograph of 321 SS in buffed condition and c) cross sectional micrograph of 321 SS in un-buffed condition.

Surface roughness measurements

Surface roughness measurements were done on all samples using a surface profilometer (contact mode) with a scan length of 1μm at a scan speed of 0.01 mm/sec with a minimum resolution step of 1 nm. 321 SS ring samples were measured using an optical surface profilometer (non contact mode) with a scan length of 1mm at a scan speed of 47μm/sec with a resolution of 0.2 nm. The average roughness (Ra) was determined in each case.

XRD phase analysis

XRD studies were conducted on all the samples to confirm the phases present in the material using BRUKER G8 powder XRD, Cu-Kα source, λ= 1.54 Å, Bragg angle, 2θ from (30-100°), step size 0.01 and step/scan 0.5 at an accelerating voltage of 40kV and 30 A.

Residual stress measurements

Residual stresses measurements were conducted on X- stress G2R (High-resolution XRD) to find the surface residual stresses present in the material. By using Cr- Kα source, applied voltage 27 kV, a current of 70 mA, λ= 2.28, Bragg angle 147.6° was kept constant for all the samples, (hkl 311) peak was considered for diffraction with a step size of 0.1° [5]. Collimeter diameter 4 mm and exposure time of 20 s was used. The multiexposure side inclination and fixed (χ) chi method was adopted, at 0° and 90° for each measurement. 2D stress state was assumed $sin^2\chi$ technique was applied for residual stress analysis.

Residual Stresses 2018 – ECRS-10   Materials Research Forum LLC
Materials Research Proceedings 6 (2018) 139-144   doi: http://dx.doi.org/10.21741/9781945291890-22

Determination of SCC resistance

The SCC susceptibility of AISI 304L SS and AISI 321 SS in buffed and un-buffed condition was determined using ASTM G 36 [6] in surface worked and welded conditions respectively. SCC test was conducted for 3 h and 9 h in AISI 304L SS samples and 5 h and 10 h in AISI 321 SS samples. As per ASTM G 36, 600 g of magnesium chloride hexahydrate ($MgCl_2.6H_2O$) was melted and test temperature was maintained at 155± 1° C throughout the test. Care has been taken to prevent vapor losses

Results and Discussion

Surface roughness measurements:

Surface roughness values (Ra) for different surface working conditions have been tabulated in Table1. In both 304L SS and 321 SS surface roughness in buffed conditions was much less as compared to un-buffed conditions. Higher the surface roughness, greater the tendency to form localized pockets of high chloride concentration and early initiation of SCC.

Table 1: Average surface roughness values of 304L SS and 321 SS in un-buffed and buffed conditions

Material	Material conditions	Surface roughness (Ra) in μm
304L SS	Milled	2.1 ± 0.15
	Turned	4.3 ± 0.30
	Ground	0.6 ± 0.04
	Milled + Buffed	0.13 ± 0.06
	Turned + Buffed	0.11± 0.04
	Ground + Buffed	0.08 ± 0.02
321 SS	Spot welded	1.74±0.24
	Spot welded +Buffed	0.94±0.12

XRD studies

Figure 2 shows the XRD spectra of 304L SS under different surface working conditions. Solution annealed sample showed austenitic (γ) phase and other surface worked samples showed strain-induced martensite (α′) phase due to the metastable nature of austenitic stainless steel.

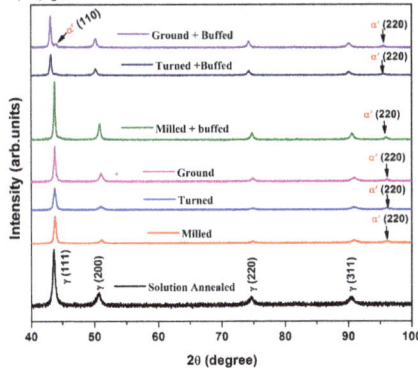

Fig. 2: Shows the XRD spectra of 304L SS in different surface working conditions.

Figure 3 shows the XRD spectra of the spot welded region of 321 SS in both buffed and un-buffed condition. Characteristic austenitic peaks were observed in both the cases together with the presence of stress induced martensite. The number of peaks for stress induced martensite was

Residual Stresses 2018 – ECRS-10   Materials Research Forum LLC
Materials Research Proceedings **6** (2018) 139-144   doi: http://dx.doi.org/10.21741/9781945291890-22

much higher for buffed condition. The Ti_2N present in the surface layers of 321SS gets removed on buffing. Optical microstructures support the observation.

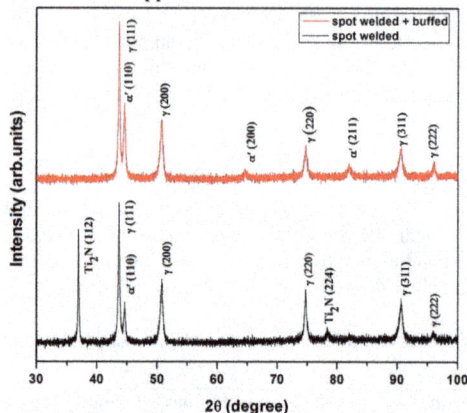

Fig. 3: Shows the XRD spectra of 321 SS in spot welded and spot welded followed by a buffed condition

Residual stress measurements

Table 2 gives the residual stresses of 304L SS and 321 SS under different conditions. The measurements were taken in longitudinal and transverse direction. The result shows that for 304L SS in milled, turned and ground conditions have high magnitude of tensile residual stresses. A similar result has been reported in earlier studies performed by some of the authors [7-8]. However, compressive residual stresses were found to be present on the surfaces in buffed condition. Similarly, the HAZ and the fusion zone of spot welds of 321 SS were found to have tensile residual stresses in un-buffed condition and compressive residual stresses in buffed condition.

Table 2: residual stress values of 304L SS and 321 SS in different conditions

304L SS conditions	0° (MPa)	90° (MPa)
Milled	740±86	639±71
Turned	397±82	69±85
Ground	192±40	15±39
Milled + Buffed	-386±21	-378±16
Turned + Buffed	-523±17	-504±26
Ground + Buffed	-481±22	-409±16
321 SS conditions	**0° (MPa)**	**90° (MPa)**
Base material	-239±23	-306±57
Spot + buffed	-351±17	-433±24

Determination of susceptibility to stress corrosion cracking (SCC)

Figure 4 and Figure 5 shows the FESEM surface micrographs of 304L SS in a) milled b) turned c) ground before and after buffing after exposure to 3 h and 9 h of SCC test respectively. The results showed that the samples in milled ground and turned condition were highly susceptible to SCC, whereas the samples in buffed condition after 3 h and 9 h of exposure. Pit initiation was observed in some cases after 9 h of exposure due to the preferential dissolution of martensite on the surface.

Residual Stresses 2018 – ECRS-10   Materials Research Forum LLC

Materials Research Proceedings **6** (2018) 139-144   doi: http://dx.doi.org/10.21741/9781945291890-22

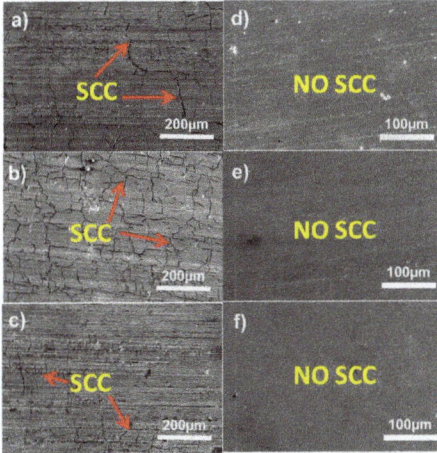

Fig. 4: Shows surface micrographs after 3 h SCC test in 304L SS a) milled b) turned c) ground d) milled and buffed e) turned and buffed f) ground and buffed.

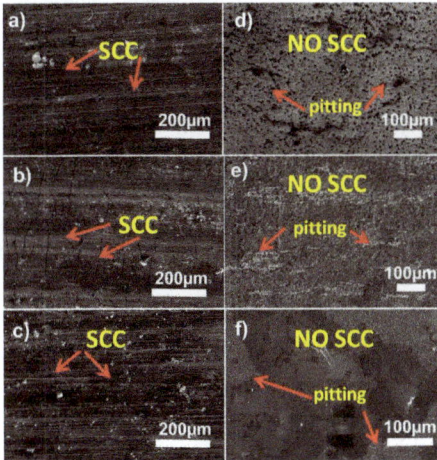

Fig. 5: Shows surface micrographs after 9 h SCC test in 304L SS a) milled b) turned c) ground d) milled and buffed e) turned and buffed f) ground and buffed conditions.

Figure 6 shows the FESEM micrographs of 321 SS spot welded region after exposure to boiling $MgCl_2$ for a) 5 h and b) 10 h. Figure 6(c-d) shows spot welded and buffed samples after 5 h and 10 h test respectively. High densities of cracks were observed in un-buffed condition whereas no cracking was observed in buffed samples.

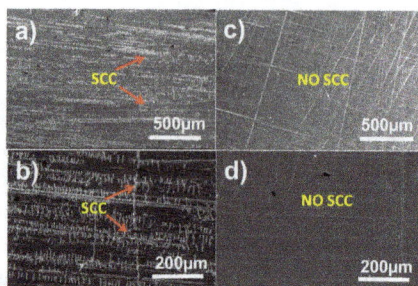

Fig. 6: Shows surface micrographs of 321 SS in spot weld condition after 5 h and 10 h SCC test respectively (a-b) before buffing and (c-d) after buffed condition.

Summary

The effect of buffing operation on the SCC susceptibility of austenitic stainless steel grade AISI 304L and 321 spot welds was established. Buffing enhances the SCC resistance of austenitic stainless steel by converting the tensile residual stresses present on the surface of machined and welded stainless steel to compressive residual stresses together with reduction in surface roughness and removal of the plastic strain near the surface formed as a result of machining.

References

[1] N. Zhou, R. Pettersson, R. Lin Peng, M. Schönning, Effect of surface grinding on chloride induced SCC of 304L, Mater. Sci. Eng. A, 658 (2016) 50-59. https://doi.org/10.1016/j.msea.2016.01.078

[2] J.Ł.S. Topolska, M. Głowacka, Failure of austenitic stainless steel tubes during steam generator operation, J. Achiev. Mater. Manuf. Eng. 55 (2012) 378–385.

[3] S. Ghosh, V. Kain, Microstructural changes in AISI 304L stainless steel due to surface machining: Effect on its susceptibility to chloride stress corrosion cracking, J. Nucl. Mater., 403 (2010) 62-67. https://doi.org/10.1016/j.jnucmat.2010.05.028

[4] Pandu Sunil Kumar, Swati Ghosh Acharyya, S.V. Ramana Rao, Komal Kapoor, Surface buffing and its effect on chloride induced SCC of 304L austenitic stainless steel, 2018 *IOP Conf. Ser.: Mater. Sci. Eng.* 314 012002, doi:10.1088/1757899X/314/1/012002.

[5] H.B. P. Withers, Residual stress. Part 1 – measurement techniques., Mater. Sci. Tech., 17 (2001) 355-365. https://doi.org/10.1179/026708301101509980

[6] ASTM G-36, Standard Practice for Evaluating Stress-Corrosion-Cracking Resistance of Metals and Alloys in a Boiling Magnesium Chloride Solution, 2006, Reapproved

[7] S. Ghosh, V. Kain, Effect of surface machining and cold working on the ambient temperature chloride stress corrosion cracking susceptibility of AISI 304L stainless steel, Mater. Sci. Eng. A 527 (3) (2010) 679–683. https://doi.org/10.1016/j.msea.2009.08.039

[8] S. Ghosh, V.P.S. Rana, V. Kain, V. Mittal, S.K. Baveja, Role of residual stresses induced by industrial fabrication on stress corrosion cracking susceptibility of austenitic stainless steel, Mater. Des. 32 (7) (2011) 3823–3831. https://doi.org/10.1016/j.matdes.2011.03.012

Residual Stresses 2018 – ECRS-10   Materials Research Forum LLC
Materials Research Proceedings 6 (2018) 145-150   doi: http://dx.doi.org/10.21741/9781945291890-23

Comparison of Different Methods of Residual Stress Determination of Cold-Rolled Austenitic-Ferritic, Austenitic and Ferritic Steels

ČAPEK Jiří[1,a,*], TROJAN Karel[1,b], NĚMEČEK Jakub[1,c], GANEV Nikolaj[1,d] and KOLAŘÍK Kamil[1,e]

[1]Department of Solid State Engineering, Faculty of Nuclear Sciences and Physical Engineering, Czech Technical University in Prague, Trojanova 13, 120 00 Prague 2, Czech Republic

[a]jiri.capek@fjfi.cvut.cz, [b]karel.trojan@fjfi.cvut.cz, [c]jakub.nemecek@fjfi.cvut.cz, [d]nikolaj.ganev@fjfi.cvut.cz, [e]kamil.kolarik@fjfi.cvut.cz

Keywords: X-Ray Diffraction, Residual Stresses, Cold-Rolling, Texture

Abstract. The aim of this contribution was to compare four methods of residual stresses determination of single-phase and dual-phase steels after cold-rolling primarily using X-ray diffraction techniques. Firstly, without taking into account the preferred orientation ($\sin^2\psi$ method), secondly from the geometry of four-point bending, thirdly without neglecting the texture (harmonic function method). And mainly, the new method, by calculating anisotropic elastic constants as a weighted average between single-crystal and X-ray elastic constants with weighting being done according to the relative intensities in the measured directions. The applicability of the new method of residual stresses determination in textured materials was proofed; however, the method needs further verification.

Introduction

The majority of practically used diffraction measurements methods and algorithms for residual stresses (RS) calculation assume the case of isotropic (non-textured) polycrystalline material. Due to the comparatively frequent existence of the preferred orientation (texture), it is more than desirable to have at disposal a method, procedure and even a computation algorithm for proper and correct RS determination. Currently, a universal method with the potential to properly evaluate RS in textured materials is, unfortunately, still missing and this issue is tackled either, in the worst scenario, by neglecting the texture (X-ray elastic constants (XEC) are used) or by choosing one of the usually proposed methods (i.e. calculation of anisotropic elastic constants (X-ray stress factors – XSF), e.g. harmonic function [1], crystallite group [2], strain pole figures [3] methods etc.).

For this purpose, a new method was developed and used for determination of residual stresses without neglecting texture. The new method is based on the Dölle model [4, 5]. However, contrary to Dölle method, this method determines the XSF (R_{ij}) as the weighted average between the single-crystal elastic constants (s_{33ij}) and the XEC (r_{ij}) where weighing is performed according to the relative intensities I in the measured directions φ, ψ, see Eq. 1.

$$R_{ij}(hkl,\ \varphi,\ \psi) = I(hkl,\ \varphi,\ \psi)\ r_{ij}(hkl) + (1 - I(hkl,\ \varphi,\ \psi))\ s_{33ij}, \tag{1}$$

where hkl are the Miller indices of the analysed planes. For texture limits, i.e. $I = 0$ and $I = 1$, the applicability and correctness of the method is automatically proofed. The general method uses function $f(I)$, which depends on the texture. To simplify and according to previous experiments, the function $f(I)$ could be approximated by a quadratic function (there is linear function in Eq. 1). Instead of other methods, this method allows to determine RS in materials with a very weak,

Residual Stresses 2018 – ECRS-10   Materials Research Forum LLC
Materials Research Proceedings 6 (2018) 145-150   doi: http://dx.doi.org/10.21741/9781945291890-23

strong and moderate texture too. The main disadvantage of this method results from the accuracy of orientation distribution function (ODF) calculation, it is the same case as for the harmonic function method.

Experiment

The tested samples of plate shape were made of AISI 420 (ferritic), AISI 304 (austenitic) and AISI 318LN (austenitic-ferritic or duplex) type of stainless steel. The samples were cold-rolled to a thickness of 1.5 mm with 0, 10, 20, 30, 40, and 50% reduction in thickness (deformation). Before deformation, the austenitic and ferritic with duplex samples were annealed in air laboratory furnace for 4 hours at 840°C and 7 hours at 650°C, respectively, in order to reduce residual stresses. According to the type of steel and thickness reduction, the tested samples were marked as F0–F50, A0–A50, D0–D50, and D_F0–D_F50 and D_A0–D_A50 for particular phases of duplex steel.

The Co$K\alpha$ radiation and the *X'Pert PRO MPD* diffractometer were used to texture and phase analyses. Texture analysis was performed on the basis of the ODF calculated from experimental pole figures (PF), which were obtained from three diffraction lines $\{110\}$, $\{200\}$, $\{211\}$ of ferrite phase (α-Fe) and $\{111\}$, $\{200\}$, $\{220\}$ of austenite phase (γ-Fe) using *MTEX* software [6]. Due to overlapping of diffraction lines $\{110\}_\alpha$ and $\{111\}_\gamma$ in duplex steel, $\{220\}_\alpha$ and $\{311\}_\gamma$ were analysed instead of them. The *ResMat* software was used to determine the harmonic coefficients, which are necessary for XSF calculation using harmonic function method.

The X'Pert PRO MPD diffractometer with manganese and chromium radiation was used to measure lattice deformations in austenite and ferrite phases, respectively, in the rolling direction. The diffraction angles $2\theta^{hkl}$ were determined from the peaks of the diffraction lines $K\alpha_1$ of planes $\{311\}$ of the austenite and $\{211\}$ of the ferrite phase. The diffraction lines $K\alpha_1$ were fitted by the *Pearson VII* function and the *Rachinger's method* was used for separation of the diffraction lines $K\alpha_1$ and $K\alpha_2$. The XEC were obtained in accordance with the *Hill model* [7] from single-crystal elastic constants, see Tab. 1. The *Winholtz & Cohen method* [2] was used for calculation of RS from lattice deformations.

Tab. 1 Single-crystal elastic constants of ferrite and austenite [8]

phase	C_{11}, GPa	C_{12}, GPa	C_{44}, GPa
ferrite	231	134	116
austenite	198	125	122

Results and discussions

The phase compositions of the tested samples are shown in Tab. 2 and Fig. 1. In the ferritic samples, only the ferrite phase was observed. Due to stronger plastic deformation, increasing amount of strain-induced martensite (α') was found in the austenitic samples with higher deformation. The two phases based on chromium ($Cr_{23}C_6$ + FeCr) were analysed in duplex samples. The trends from Fig. 1 result from the combination of strong plastic deformation ($\gamma \rightarrow \alpha'$) with annealing ($\alpha \rightarrow \gamma + Cr_{23}C_6$ + FeCr) [9]. For these reasons, the amount of austenite phase is decreasing, the ferrite phase is constant and the chromium phases is increasing with higher deformation.

The presented stresses σ (measured by X-ray diffraction) are superposition of the external stresses σ_N generated by the four-point bending, and the RS of surfaces areas of the tested samples after annealing σ_{RS}, i.e. $\sigma = \sigma_N + \sigma_{RS}$. For the external stresses calculation, Young's modules of tested materials were measured using the *ultrasonic pulse-echo method* [10]. Values of σ'_N represent the theoretical values of the measured/experimental value of σ. In ideal case, the values of σ should be equal to σ'_N.

Residual Stresses 2018 – ECRS-10   Materials Research Forum LLC
Materials Research Proceedings **6** (2018) 145-150   doi: http://dx.doi.org/10.21741/9781945291890-23

The determined stresses σ relevant for the surface layers depending on external stresses can be seen in Figs. 2. Stresses calculated using common $\sin^2\psi$ method differ from σ'_N values due to omitting of presence of texture, i.e. without taking into account the potential non-linear regression of $d(\sin^2\psi)$, and using the XEC instead of the XSF.

Tab. 2 Phase compositions in wt.% of austenitic samples

sample	strain-induced martensite	austenite
A0	2.0	98.0
A10	1.7	98.3
A20	3.2	96.8
A30	2.5	97.5
A40	4.7	95.3
A50	4.2	95.8

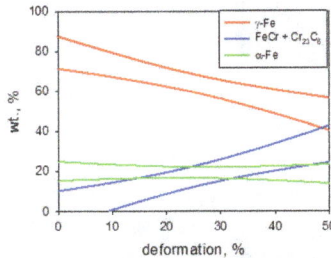

Fig. 1 Phase compositions of duplex samples with 95% confidence interval

The stresses determined by the harmonic function method have in the most cases higher values and errors in comparison with other methods. The main reason is presence of a sharp texture, see Fig. 3 and [11], which the harmonic functions are not able to describe correctly. Secondly, this method strongly depends on the accuracy of harmonic coefficients derived from calculation of ODF, which is closely related to microstructure, mainly grain size.

The new developed method uses relative intensities of pole figures of the given *{hkl}* diffraction planes. In this method, the PF or ODF are not fitted by any functions, so it is suitable for materials with very strong and sharp textures. Nevertheless, same as for the harmonic function method, the accuracy strongly depends on the ODF calculation and quality of measured data. This is the main reason, why the values of stresses determined by the new method are not equal to the σ'_N values. The good validity of the proposed method is for one-phase steel, where the function $f(I)$ could be approximated by a quadratic function, see Figs. 2a and 2b. Due to the multi-phase interaction during deformation of austenitic and duplex samples, the accuracy of the method was confirmed for stresses up to 50% of yield strength, see Figs. 2c–h. The influence of multi-phase interaction is not a part of the proposed method, that is the reason why the validity for higher deformation was not proofed. For all cases, the RS values were determined by the proposal method with smaller experimental errors than other methods.

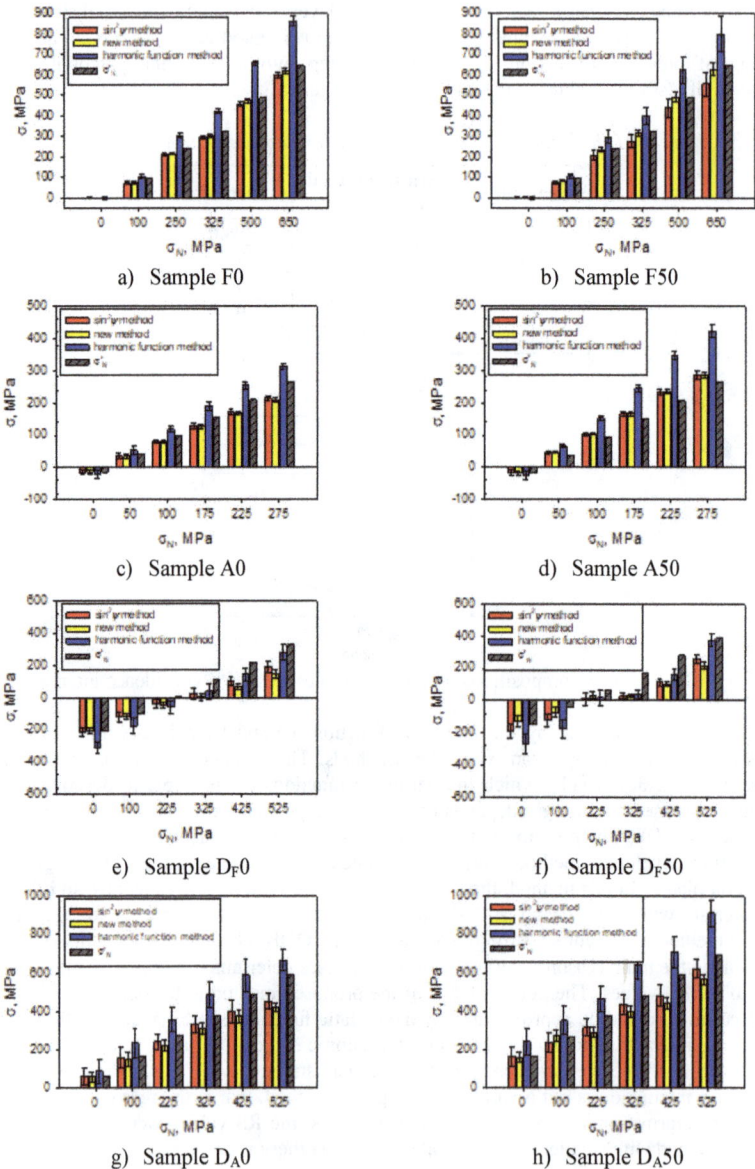

a) Sample F0

b) Sample F50

c) Sample A0

d) Sample A50

e) Sample D_F0

f) Sample D_F50

g) Sample D_A0

h) Sample D_A50

Fig. 2 Determined stresses σ depending on σ'$_N$ stress with 0 and 50% deformation

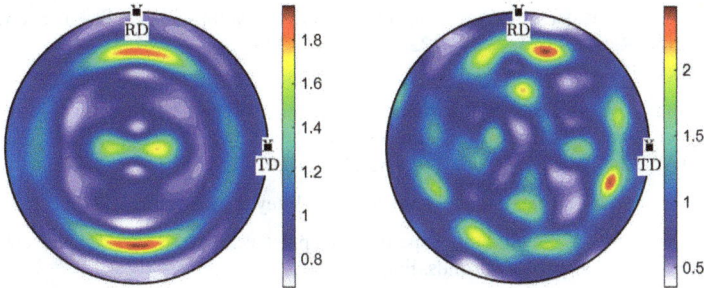

Fig. 3 Example of PF of {211} planes with sharp texture of F50 and D_F50 samples

After the final annealing, the residual stresses σ_{RS} should be approx. equal to zero. This statement is not fulfilled for the duplex samples. If the RS of each phase of duplex steel are added up, with taking into account the wt.%, the result is not zero. The possible reason is the presence of chromium phases with non-zero residual stresses.

Conclusions

The applicability of the new method of residual stress determination in textured materials was proofed for single-phase materials and for major phases of multi-phase materials up to approx. 50% yield strength, see Figs. 2. However, comparing with the standard $\sin^2\psi$ and harmonic function methods, much more accurate results were achieved. The main reason is the presence of very sharp texture, especially in duplex steel. Other reasons are using different methods of calculation RS and mainly accuracy of ODF calculation.

Acknowledgement
This work was supported by the Czech Science Foundation, grant No. 14-36566G.

References
[1] H.J. Bunge, Texture Analysis in Materials Science, Butterworth, London, 1982.

[2] U. Welzel, et al. Stress analysis of polycrystalline thin films and surface regions by X-ray diffraction, J. Appl. Crystallogr. 38 (2005) 1-29. https://doi.org/10.1107/S0021889804029516

[3] C.M. Brakman, A general treatment of X-ray (residual) macro-stress determination in textured cubic materials: General expressions, cubic invariancy and application to X-ray strain pole figures, Cryst. Res. Technol. 20.5 (1985) 593-618. https://doi.org/10.1002/crat.2170200503

[4] H. Dölle, The Influence of Multiaxial Stress States, Stress Gradients and Elastic Anisotropy on the Evaluation of (Residual) Stresses by X-rays, J. Appl. Cryst. 12 (1979) 489-501. https://doi.org/10.1107/S0021889879013169

[5] J. Capek, Z. Pala, O. Kovarik, Residual stresses determination in textured substrates for plasma sprayed coatings, IOP Conf. Ser.: Mater. Sci. Eng. 82.1 (2015) 012112.

[6] F. Bachmann, R. Hielscher, H. Schaeben, Texture Analysis with MTEX | Free and Open Source Software Toolbox, Solid State Phenomen. 60 (2010) 63-68. https://doi.org/10.4028/www.scientific.net/SSP.160.63

[7] J.D. Eshelby, The determination of the elastic field of an ellipsoidal inclusion, and related problems, Proc. R. Soc. Lond. A. 241.1226 (1957) 376-396. https://doi.org/10.1098/rspa.1957.0133

[8] R. Dakhlaoui, C. Braham, A. Baczmański, Mechanical properties of phases in austeno-ferritic duplex stainless steel – Surface stresses studied by X-ray diffraction, Mater. Sci. Eng.: A. 444 (2007) 6-17. https://doi.org/10.1016/j.msea.2006.06.074

[9] I. Alvarez-Armas, Duplex Stainless Steels, John Wiley & Sons, USA, 2009.

[10] E.P. Papadakis, T.P. Lerch, Pulse superposition, pulse-echo overlap and related techniques, Handbook of elastic properties of solids, liquids and gases. 1 (2001).

[11] J. Capek, M. Cernik, N. Ganev, K. Trojan, J. Nemecek, K. Kolarik, Comparison of rolling texture of austenite and ferrite phases of duplex steel with single-phase austenitic and ferritic steel, IOP Conf. Ser.: Mater. Sci. Eng. 375.1 (2018) 012025.

Residual Stresses 2018 – ECRS-10
Materials Research Proceedings 6 (2018) 151-156

Materials Research Forum LLC
doi: http://dx.doi.org/10.21741/9781945291890-24

Effects of Cutting Parameters on the Residual Stresses of SAE 1045 Steel after Turning

Vinícius Carvalho Pinto[1,2,a,*], Eugênio Teixeira de Carvalho Filho[2,b]
and João Telésforo Nóbrega de Medeiros[2,c]

[1]IFRN - Federal Institute of Rio Grande do Norte, 304 São Braz St., Santa Cruz, RN, Brazil, 59200-000

[2]UFRN - Federal University of Rio Grande do Norte, University Campus, Natal, RN, Brazil, 59078-970

[a]vinicius_c_pinto@hotmail.com, [b]eugenioteixeira_@hotmail.com, [c]jtelesforo@yahoo.com

Keywords: Machining Residual Stresses, Cutting Parameter, Turning

Abstract. Surface residual stresses on machined parts may be undesirable in some parts, leading to problems over their lifetime. In order to study the effects of cutting parameters on the residual stresses of SAE 1045 steel after turning, tests were performed varying cutting depth, feed rate and cutting speed. For each of these parameters, four different conditions were tested, in order to understand their influence separately from the others. The tests were performed with tungsten carbide coated tool with 80° rhomboid tip morphology. Before being used in the tests, the samples were thermally treated through the normalization process, in order to obtain a regular grain size in each sample and reduction of the residual stresses present in the billet from the manufacturing process. The $\sin^2\Psi$ method, through the X-ray diffraction technique, was used to quantify the residual stresses. The samples were divided into 4 regions for the evaluation of the residual stresses, where the analyses were performed in the longitudinal and axis direction. The analysis of the residual stress presents greater variation for the depth of cut and feed rate. With the increase of the depth of cut, the tensile residual stresses reduce, presenting compressive values in the axial direction. With the increase of the feed rate, there is increment of the tensile residual stress.

Introduction

The study of residual stresses has been of great importance in the understanding of the contact mechanics, damage mechanics and fracture mechanics, with respect to the behavior of the materials. The residual stresses are an elastic response of the material to the spatial variations of a heterogeneous microstructure [1].

The manufacture industry has been demanding an increasing productivity. In machining processes, the tools are under constant evolution to attend this growing necessity of more severe cutting parameters. Some advances in cutting tools have been made such as the development of hard coatings [2].

In machining, in the processes of milling, turning, drilling and grinding, the final state of the residual stresses will depend on factors such as machined material, cutting tool, machining parameters and coolant. In these processes, the generation and modification of stresses are given by the localized heating and the contact pressure performed by the tool, which can generate tensile or compressive stresses [3, 4].

Residual stresses increase with most machining parameters increment, including cutting speed and tool corner radius [5]. Additionally, it is worth mentioning that compressive stresses are usually desired.

Residual Stresses 2018 – ECRS-10   Materials Research Forum LLC
Materials Research Proceedings **6** (2018) 151-156   doi: http://dx.doi.org/10.21741/9781945291890-24

Some authors have investigated the influence of the cutting parameters in the residual stresses of the machined materials. Researchers found that the residual stresses tend to be more tensile for higher feed rates for a turned Iconel 718 [6, 7]. Some studies purpose a model to predict the residual stresses on a turned steel and show that the increase of cutting speed also increases the tensile residual stresses on the workpieces, according to the model [8]. Another analytical model analyzed the influence in residual stresses of machining parameters and tool parameters separately. According to this model, the depth of highest compressive residual stress increases with the increase of feed rate, and decreases with the increase of cutting speed [9].

This study aims to analyze the influence of the cutting parameters (cutting speed, feed rate and depth of cut) in the superficial residual stresses of turned pieces, in such a way that it can be used as reference when somehow it makes necessary to adjust the cutting parameters in manufacture industry in order to get higher productivity.

Experimental Procedure
Material and heat treating. The material used in the tests was the SAE 1045 steel with pearlitic/ferritic microstructure with chemical composition presented in Table 1.

Table 1 - Chemical composition of SAE 1045 steel tested

Fe [%]	C [%]	Mn [%]	Si [%]	Al [%]	S [%]
98,31	0,45	0,75	0,27	0,20	0,02

The specimens were heat treated through the normalizing process in order to standardize the grain sizes and the residual stresses.

For the normalizing heat treatment of the specimens' grain sizes, it was used as reference the ASM International Standard [10], which recommends the temperature of 880°C. The furnace was kept at this temperature for 90 minutes to stabilize the temperature and standardize the specimens. After that, the specimens were withdrawn from the furnace and arranged for air cooling.

The specimens were produced to be tested with a minimum diameter of 29 mm. The Fig. 1 shows the specimen. The diameter of 15 mm was used to fix the specimen in the chuck, furthermore it is settled by the tailstock in the opposite face.

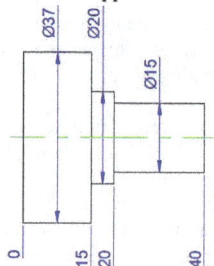

Fig. 1 - Schematic drawing of the specimen prepared for testing
(dimensions in millimeters [mm])

Inserts. For each test specimen, a new insert was used, in order to reduce the influence of the tool wear mechanisms on the results of residual stresses.

The tool and tool holder geometry is described in Table 2.

A multilayer CVD coated tool (TiCN, Al_2O_3 and TiN) was used.

Table 2 – Tool and toolholder geometry used in the tests

Insert Shape	80°
Insert Clearence Angle	0°
Insert Corner Radius	0,8 mm
Approach Angle	93°
Axial Rake Angle	-6°
Radial Rake Angle	-6°

Tests. The tests were performed without cutting fluid. The parameters of Cutting Speed (Vc), Depth of Cut (ap) and Feed Rate (f) varied.

In order to quantify the influence of each variable on the residual stress, the following parameters were used: cutting speed (60/135/190/310 [m/min]), feed rate (0.06 / 0.18 / 0.30 [mm/rev]) and depth of cut (1.0 / 2.0 / 3.0 / 4.0 [mm]).

Residual stress measurements. After the tests, the specimens were analyzed in the XRD without the necessity of a previous handling of the pieces. For this analysis, a device was developed to fix and position the sample in the equipment, presented in Fig. 2. The measurements were carried out by X-ray diffraction using the iso-inclination method, the parallel-beam geometry and the anode (target) CrKα (2.289 Å). Residual stress was measured in the cutting direction and in the feed direction, considering the state of plane stress.

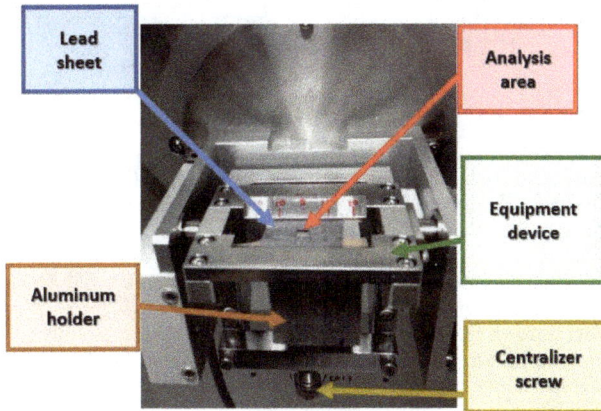

Fig. 2 – Positioning of the specimen for measurement in the XRD

Results and discussions

Results for variation of the cutting speed. The parameters of feed rate (0.18 mm/rev) and depth of cut (2.0 mm) were maintained. The results are shown in Table 3 and Fig. 3.

It is possible to observe an increase tendency of the tensile tensions with the increase of the cutting speed up to 190 m/min. For the cutting speed of 310 m/min, it was possible to observe a reduction in the tensile stress, which can be understood as a predominance of thermal loads up to the speed of 190 m/min and mechanical loads at higher speeds. An indicative that, for higher speeds, most of the thermal energy generated in the cut is coming out in the chip. Similar results were observed in the AISI 4340 steel study [11].

Table 3 – Surface residual stress results varying the cutting speed

V_c [m/min]	Residual Stress [MPa]	
	Cutting direction	Feed direction
60	260 ±15	-83 ±13
135	417 ±20	41 ±12
190	414 ±19	54 ±10
310	352 ±15	-24 ±07

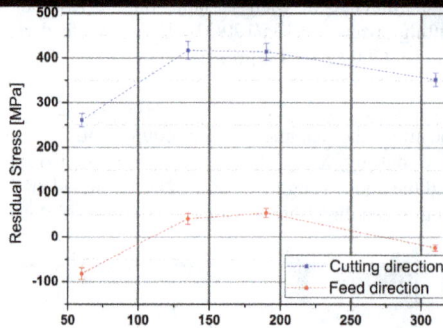

Fig.3 – Surface residual stress results, varying the cutting speed

Results for feed rate variation. The parameters of cutting speed (135 m/min) and depth of cut (2.0 mm) were maintained. The results are presented in Table 4 and Fig. 4.

The higher the feed rate, the greater the surface residual tensile stress. This gain is associated with the elevation of the cutting temperature, generated by the chip thickness increase, which boosts the heat generation related to the plastic deformations [11-13].

Results for variation of depth of cut. The parameters of cutting speed (135 m/min) and feed rate (0.18 mm/rev) were maintained. The results are shown in Table 5 and Fig. 5.

With the increase of the cutting depth, maintaining the parameters of cutting speed and feed rate, it was observed the reduction of the residual tensile surface stress (or elevation of the residual compressive surface tension), indicating elevation in the mechanical loading.

Table 4 – Surface residual stress results, varying the feed rate

f [mm/rev]	Residual Stress [MPa]	
	Cutting direction	Feed direction
0,06	286 ±13	-79 ±04
0,12	337 ±13	-83 ±09
0,18	417 ±20	41 ±12
0,30	469 ±20	182 ±10

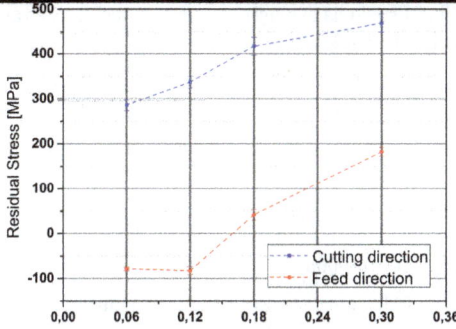

Fig. 4 – Surface residual stress results, varying the feed rate

Table 5 – Surface residual stress results, varying the depth of cut

a_p [mm]	Residual Stress [MPa]	
	Cutting direction	Feed direction
1,0	509 ±25	283 ±06
2,0	417 ±20	41 ±12
3,0	320 ±15	-164 ±12
4,0	285 ±11	-197 ±10

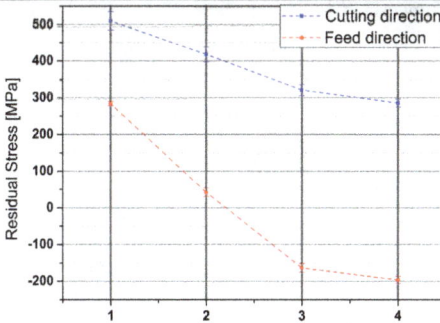

Fig. 5 – Surface residual stress results, varying the depth of cut

Summary

The parameter that influenced the variation of the residual stress the most, showing less tensile or more compressive values with its increase, was the depth of cut. The feed rate presented an inverse result, in which the increment of this parameter caused an increase in the residual tensile stress. The cutting speed was shown to be the parameter with less influence, showing an increase of the surface residual stress up to a speed of approximately 200 m/min. At higher velocities, there was a reduction of the residual tensile stress.

Residual Stresses 2018 – ECRS-10   Materials Research Forum LLC
Materials Research Proceedings 6 (2018) 151-156   doi: http://dx.doi.org/10.21741/9781945291890-24

References

[1] E. T. de Carvalho Filho, Estudo da evolução das tensões residuais através da difratometria de raios x em aço rolamento submetido a esforços cíclicos, PPGEM-UFRN, Natal, 2015.

[2] K. Bobzin, High-performance coatings for cutting tools, CIRP J. of Manuf. Sci. and Technol. 18 (2016) 1-9. https://doi.org/10.1016/j.cirpj.2016.11.004

[3] M. C. B. V. Soares, Influência das tensões residuais no comportamento em fadiga e fratura de ligas metálicas, IPEN-USP, São Paulo, 1998.

[4] J. guang Li, S. qi Wang, Distortion caused by residual stresses in machining aeronautical aluminum alloy parts: recent advances, Int. J. Adv. Manuf. Technol. 89 (2017) 997– 1012. https://doi.org/10.1007/s00170-016-9066-6

[5] C. Maranhão, J. P. Davim, Residual stresses in machining using FEM analysis - A review, Rev. Adv. Master. Sci. 30 (2012) 267-272.

[6] Y. Hua, Z. Liu, Experimental Investigation of Principal Residual Stress and Fatigue Performance for Turned Nickel-based Superalloy Iconel 718, Materials 11 (2018) 879-894. https://doi.org/10.3390/ma11060879

[7] X. Wang et. al, Experimental study of surface integrity and fatigue life in the face milling of Inconel 718, Mech. Eng. 13 (2018) 243-250. https://doi.org/10.1007/s11465-018-0479-9

[8] Z. Pan et. al, Turning induced residual stress prediction of AISI 4130 considering dynamic recrystallization, Machining Science and Technology 22 (2018) 507-521. https://doi.org/10.1080/10910344.2017.1365900

[9] H. Kun, Y. Wenyu, Analytical analysis of the mechanism of effects of machining parameter and tool parameter on residual stress based on multivariable decoupling method, International Journal of Mechanical Sciences 128-129 (2017) 659-679. https://doi.org/10.1016/j.ijmecsci.2017.05.031

[10] ASM International, ASM Metal Handbook: Heat Treating, third ed., Handbook Commitee, USA, 2001.

[11] V. G. Navas, O. Gonzalo, I. Bengoetxea, Effect of cutting parameters in the surface residual stresses generated by turning in AISI 4340 steel, Int. J. of Mac. Tools and Manuf. 61 (2012) 48– 57. https://doi.org/10.1016/j.ijmachtools.2012.05.008

[12] B. Griffiths, Manufacturing Surface Technology: Surface integrity & functional performance, New York, Butterworth-Heinemann, 2001.

[13] X. Ji. et al., The effects of minimum quantity lubrication (MQL) on machining force, temperature, and residual stress, Int. J. of Prec. Eng. and Manuf., 15 (2014) 2443–2451. https://doi.org/10.1007/s12541-014-0612-6

Heat Treatments and Phase Transformations

Residual Stresses 2018 – ECRS-10
Materials Research Proceedings 6 (2018) 159-164

Materials Research Forum LLC
doi: http://dx.doi.org/10.21741/9781945291890-25

Characterization of Residual Stresses and Retained Austenite on 416 Stainless Steel via X-Ray Diffraction Techniques

Thomas Simmons[1,a], Gabriel Grodzicki[1,b], James Pineault[2,c],
Jeffrey Taptich[1,d], Mohammed Belassel[2,e,*], Jeffrey Nantais[2,f],
Michael Brauss[1,g]

[1]Proto Manufacturing Inc., Taylor, Michigan, USA

[2]Proto Manufacturing Ltd., Ontario, Canada

[a] tsimmons@protoxrd.com, [b] ggrodzicki@protoxrd.com, [c] jpineault@protoxrd.com,
[d] jtaptich@protoxrd.com, [e] mbelassel@protoxrd.com, [f] jnantais@protoxrd.com,
[g] mbrauss@protoxrd.com

Keywords: Residual Stress, Retained Austenite, X-Ray Diffraction, Mapping, Stainless Steel

Abstract. Residual stress (RS) and retained austenite (RA) measurements performed using x-ray diffraction (XRD) techniques have been useful to assist in evaluating the quality of parts both in the field and in production environments for decades. Single spot checks at critical locations are often performed providing basic information for a given testing requirement at a specific location. Although this is useful information, the highest levels of RS or RA present in the component may be missed when performing single spot checks only if surface RS or RA gradients are present. To achieve a more in depth understanding of the range and distribution of RS and RA in a component, a relatively large area was characterized using the mapping approach on a heated and quenched 416 stainless steel sample. The results found using this characterization technique indicate that RS and RA levels are not uniform over the area examined; thus it has been demonstrated that the one point approach is not adequate in this instance since the lowest and highest levels of RS and RA would be missed without mapping. For this reason, it is recommended that the mapping approach be adopted whenever gradients are present to fully and efficiently characterize RS and RA and subsequently to help correctly assess the quality of the part.

Introduction

RS measurements using XRD techniques have been widely applied in many industries, including automotive, aerospace, power generation, medical, and others, on components that are susceptible to stress related failure. The monitoring of RS makes it possible to remove certain components from service before they suffer from cracking or other failure modes, while extending the service life of others. Typically, RS measurements are used as a spot check at failure critical locations often selected using either modeling methods or experience. When the accuracy of models are suspect, or when no experience is available for a specific application, it is possible that areas with RS that negatively impact component life cannot be easily anticipated, predicted, selected, or found. In such instances, areas with potentially harmful stresses can be characterized using RS mapping techniques. Performing a RS map over a large area can produce more useful data from failure critical regions, which can be used to make more cost effective production or component disposition decisions.

Residual Stresses 2018 – ECRS-10   Materials Research Forum LLC
Materials Research Proceedings **6** (2018) 159-164   doi: http://dx.doi.org/10.21741/9781945291890-25

The measurement of RA using XRD has also seen widespread use in quality control and quality assurance. Depending on the specific application, the presence of austenite in the steel can have either positive or negative consequences in terms of component performance and service life. For instance, the presence of RA can improve the fatigue life of bearings and gears composed of high carbon steel, as well as impact fatigue strength in bending or other more complex loading scenarios. The presence of RA may also reduce the load carrying capacity of martensitic/austenitic structures, lower the hardness and resistance to scuffing and indentation, and increase susceptibility to grinder burn and heat checking while grinding [1].

The XRD technique for mapping critical areas can be used when characterizing either RS, RA, or both. The full width at half maximum (FWHM) of the diffraction peaks can also be extracted from the RS and RA map data so that other properties such as the coherent domain size and dislocation density can be analyzed as they may also vary within the mapped region. The XRD technique may be used to collect data at a very specific location over a small region (e.g. a few square millimeters) or a very large area (e.g. several square meters) with individual measurement spot resolution ranging from 0.2 mm diameter spots up to 2 mm x 5 mm rectangular spots. As such, detailed RS, RA and/or FWHM maps can be generated without averaging results or assuming microstructural homogeneity over large areas, while capturing the inherent gradients present in failure critical regions on critical components.

Stress Measurement Using the X-Ray Diffraction Technique

The x-ray diffraction technique can only be applied to crystalline, polycrystalline and semi-crystalline materials [2]. This technique uses the distance between crystallographic planes, i.e. d-spacing, as a strain gage. When the material is in tension the d-spacing increases and when the material is in compression the d-spacing decreases. The presence of residual stresses in the material produces a shift in the x-ray diffraction peak's angular position that is directly measured by a detector [3]. For a known x-ray wavelength λ and n equal to unity, the diffraction angle 2θ is measured experimentally and the d-spacing (d) is then calculated using Bragg's law [4,5]:

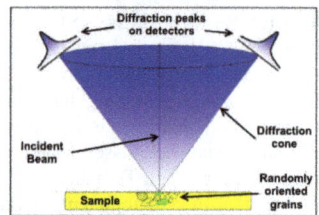

Fig. 1: X-ray diffraction configuration using 2 detectors for stress measurement. The solid angle of the diffraction cone (Debye ring) is determined by the Bragg angle of the diffraction plane.

$$n\lambda = 2d\sin\theta \tag{1}$$

Once the d-spacing is measured for unstressed (d_0) and stressed (d) conditions, the strain is calculated using the following relationship:

$$\varepsilon = (d - d_0)/d_0 \tag{2}$$

For the $\sin^2\psi$ method where a number of d-spacings are measured, stresses are calculated from an equation derived from Hooke's law for isotropic, homogeneous, fine grain materials [1]:

$$\varepsilon_{\phi\psi} = \frac{1}{2}S_2(\sigma_\phi - \sigma_{33})\sin^2\Psi + \frac{1}{2}S_2\sigma_{33} - S_1(\sigma_{11} + \sigma_{22} + \sigma_{33}) + \frac{1}{2}S_2\tau_\phi\sin2\Psi \tag{3}$$

where:

Residual Stresses 2018 – ECRS-10   Materials Research Forum LLC
Materials Research Proceedings 6 (2018) 159-164   doi: http://dx.doi.org/10.21741/9781945291890-25

½S_2 and S_1 are the x-ray elastic parameters of the material, σ_ϕ is the stress in the direction of the measurement, ψ is the angle subtended by the bisector of the incident and diffracted beam and the surface normal, and ε_ϕ is the strain at a given ψ tilt.

Two types of stresses (normal stress σ_ϕ and shear stress τ_ϕ) can be calculated from RS measurements performed in a given direction ϕ as can be seen in Eq. (3). When the d-spacing vs. $\sin^2\psi$ data are plotted, they have an elliptical form. If a shear stress is not present, the ellipse closes and becomes a straight line.

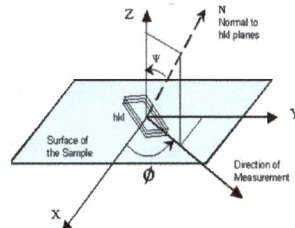

Fig. 2: Definition of the axis and the direction of measurement.

If no normal stress is present the slope of the ellipse or line is zero. The opening and slope of the ellipse or line is also dependent upon the goniometer geometry selected, i.e. if the goniometer is working in omega or psi mode or in another non-standard mode such as modified psi or grazing incidence.

Measurement of Percent Retained Austenite

A unique XRD peak pattern/fingerprint is produced by each constituent phase present in the steel under interrogation (e.g. martensite, austenite, carbides etc.). In samples with a near random grain orientation distribution, the diffracted intensity of the peaks composing each pattern will be proportional to the volume fraction of its corresponding phase [6]

$$I_\alpha^{hkl} = KR_\alpha^{hkl}V_\alpha/2\mu \tag{4}$$

where:

K is a constant which is dependent upon the selection of instrumentation geometry and radiation, but independent of the material. The R_α^{hkl} values are dependent upon the interplanar spacing, the Bragg angle θ, the diffraction {hkl} plane, the crystal structure, and the composition of the phase being measured. Each R_α^{hkl} can be calculated from basic principles. V_α is the volume fraction of the α phase , and μ is the linear mass absorption coefficient for the steel.

For steel containing only martensite (α) and austenite (γ) and no carbides, we can write for any pair of martensite or austenite hkl peaks:

$$\frac{I_\alpha^{hkl}}{I_\gamma^{hkl}} = [\left(\frac{R_\alpha^{hkl}}{R_\gamma^{hkl}}\right)\left(\frac{V_\alpha}{V_\gamma}\right)] \tag{5}$$

The ratio in Eq. 5 holds if ferrite or martensite and austenite are the only two materials present in the steel. Therefore:

$$V_\alpha + V_\gamma = 1 \tag{6}$$

Therefore, the volume fraction of austenite (V_γ) for the ratio of measured intensities of the α and γ peaks to its associated R-value is:

$$V_\gamma = \frac{\frac{I_\gamma}{R_\gamma}}{\frac{I_\alpha}{R_\alpha}+\frac{I_\gamma}{R_\gamma}} \tag{7}$$

Where more than one martensite and/or austenite peak is collected, the individual ratios of measured intensity to R-value can be summed:

$$V_\gamma = \left(\frac{1}{q}\sum_{j=1}^{q}\frac{I\gamma j}{R\gamma j}\right) / \left[\left(\frac{1}{P}\sum_{i=1}^{1}\frac{I\alpha i}{R\alpha i}\right) + \left(\frac{1}{q}\sum_{j=1}^{q}\frac{I\gamma j}{R\gamma j}\right)\right] \tag{8}$$

Experimental Procedure
In the following example a 416 stainless steel (416SS) sample was heated and oil quenched rapidly, allowing austenite to form. A total of sixty RS and RA measurements were performed utilizing a 5 point by 12 point grid encompassing a region of 12 mm in the transverse direction and 22 mm in the longitudinal direction (see Fig. 3) using a 1 x 3 mm aperture. These maps were collected at the surface, at 0.05 mm below surface, and at 0.254 mm below surface. The RS was measured in the longitudinal direction. The measurements were performed using a Proto LXRD instrument. The stress measurement parameters are as follows: Cr anode (λ = 2.291Å), hkl (211), Bragg angle 2θ = 156.4°, ½S2= $5.67.10^{-6}$ MPa^{-1}. For the RA measurements, diffraction peaks at 2θ angles of 156°, 128°, 106°, and 79° were used corresponding to the 211α, 220γ, 200α, 200γ hkl planes respectively.

Results and Discussion
The RA maps collected on the 416SS sample were plotted for each of the three depths (see Fig. 4a). It can be seen that the heating and subsequent quenching of the sample produced a localized area with a significant volume fraction of austenite. After electropolishing the sample to a depth of 0.05 mm, the maximum RA found dropped from approximately 30% at the surface to a maximum of approximately 15%. Once the sample was further electropolished to a depth of 0.254 mm, the maximum volume fraction of austenite in the peak region was found to be less than 5%, comparable to the background level prevalent throughout the sample. These maps indicate that a wide range of RA can be found over a small region in a component and that subsurface measurements are needed to accurately determine the distribution of RA as a function of depth.

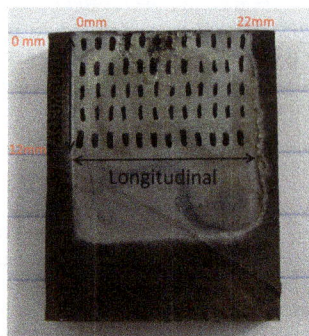

Fig. 3: Location of RS/RA maps on a 416 Stainless Steel Sample at 0.254 mm below the surface.

Three RS maps were also plotted at the same locations and depths on the same 416SS sample (see Fig. 4b). For the majority of the sampling area mapped on the 416SS sample, the RS was generally compressive. However, near the edges of the mapped region, the RS increased to near neutral, and very locally, into tension. In the center of the mapped region at 0.05 mm deep, the RS became even more compressive as compared to the surface. These maps indicate that a wide range of RS can be found over a very short distance, even on a largely apparently "uniform" sample. In such cases, it becomes useful to utilize RS mapping to capture any significant changes in RS across a sample.

Residual Stresses 2018 – ECRS-10   Materials Research Forum LLC
Materials Research Proceedings **6** (2018) 159-164   doi: http://dx.doi.org/10.21741/9781945291890-25

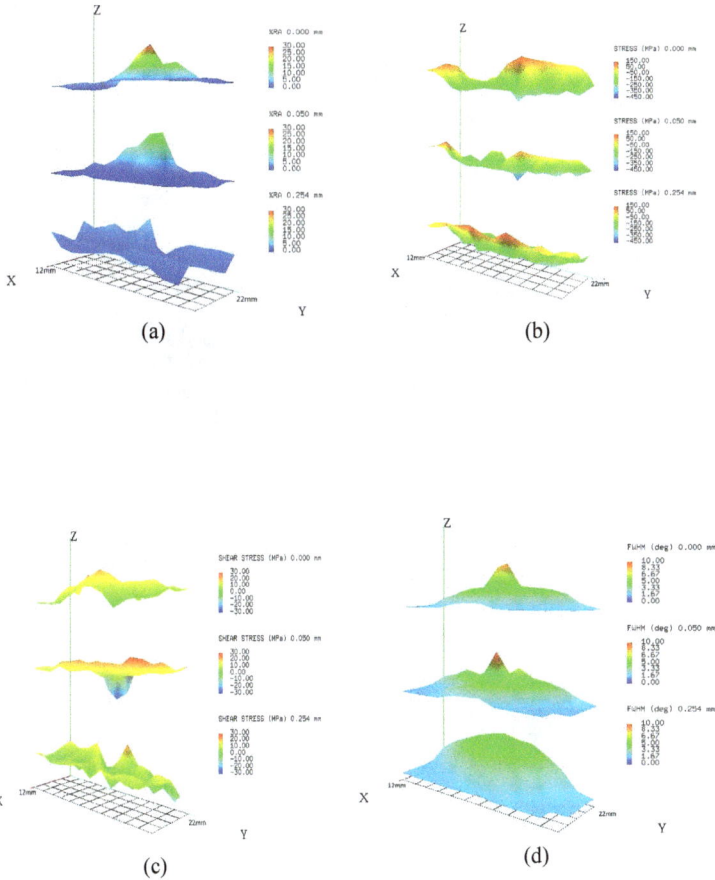

Fig. 4: (a) RA, (b) RS, (c) shear stress, and (d) FWHM maps on a 416SS sample at surface, 0.05 mm, and 0.254 mm deep.

The RS maps shown in Fig. 4b also identify the importance of performing RS measurements not only at the surface, but also at depth. A localized region that was slightly more compressive than its surroundings at the surface turned into a region where a large amount of compressive RS was present at a depth of 0.05 mm. Furthermore, the RS in that same region at a depth of 0.254 mm returned to match the RS of its surrounding area. A higher resolution depth profile at that location would provide not only this information, but would also provide a more precise indication of where this area stopped having compressive RS and matched its surroundings.

Residual Stresses 2018 – ECRS-10   Materials Research Forum LLC
Materials Research Proceedings **6** (2018) 159-164   doi: http://dx.doi.org/10.21741/9781945291890-25

These maps also indicate that the neutral/tensile regions of the mapped area were persistent with depth.

The XRD shear stress from the RS maps was also plotted over the same area and depths and varied across the sample area, as well as with depth (see Fig. 4c).

The FWHM of the diffraction peaks can be seen in Fig. 4d. The FWHM of the diffraction peaks (see Fig 4d) varied at the surface with the largest value being at a similar location as the highest value of RA. The highest FWHM wanes with increasing depth into the sample. This is consistent with the fact that the dislocation density increases with hardness [3]. Since FWHM is proportional to dislocation density, these maps thus identify regions of localized hardening at the surface and at depth, and correlate with the associated phase changes observed in the RA maps for this sample.

Conclusions

The range of RS and RA found in the 416SS sample indicate that mapping is a useful technique for characterizing RS and RA over a large area. Collecting RS and RA maps at the surface and at depth provides a more comprehensive understanding of the surface and subsurface condition of a component. In the case of RA maps, subsurface measurements provide more accurate RA characterization without having the results impacted by surface conditions. Mapping the FWHM of a sample allows an operator to identify localized regions where hardening has occurred, either by heat treatment or by cold working.   If RS or RA mapping is not performed, care and experience should be used to select appropriate measurement locations because, as can be seen in the 416SS sample maps, both RA and RS may change significantly over a distance of a few millimeters. Observing and understanding RA, RS, and FWHM maps can provide more meaningful data, which can ultimately be used to improve quality and reliability of critical components.

References

[1] C. F. Jatczak et. al, "Retained Austenite and its Measurements by X-Ray Diffraction", SAE SP-453, SAE, 1980.

[2] I. C. Noyan and J. B. Cohen, "Residual Stress - Measurement by Diffraction and Interpretation", Springer-Verlag, New York, 1987.

[3] B. D. Cullity, "Elements of X-ray Diffraction", Second Edition, Addison-Wesley, Reading Massachusetts, 1978.

[4] M.E. Hilley et.al, "Residual Stress Measurement by X-ray Diffraction" SAE HS-784, SAE, 2003.

[5] H.P. Klug, L.E. Alexander, "X-ray Diffraction Procedures for Polycrystalline and Amorphous Materials", Second Edition, Wiley-Interscience, New York, 1974.

[6] ASTM E975-03, "Standard Practice for X-Ray Determination of Retained Austenite in Steel with Near Random Crystallographic Orientation", 2008.

Residual Stresses 2018 – ECRS-10   Materials Research Forum LLC
Materials Research Proceedings 6 (2018) 165-170   doi: http://dx.doi.org/10.21741/9781945291890-26

Comparison of Nondestructive Stress Measurement Techniques for Determination of Residual Stresses in the Heat Treated Steels

Hüseyin HIZLI[1,a,*] and C. Hakan GÜR[,2,b]

[1]Special Processes and Test Technologies Department, ROKETSAN Missiles Industries Inc., Ankara, Turkey

[2]Department of Metallurgical and Materials Engineering, Middle East Technical University, Ankara, Turkey

[a]huseyin.hizli@roketsan.com.tr, [b]chgur@metu.edu.tr

Keywords: Carburizing, Residual Stress, Magnetic Barkhausen Noise, X-Ray Diffraction, Electronic Speckle Laser Interferometry

Abstract. Service life and performance of the case-hardened machine parts are greatly dependent on the residual stress state in the surface layers which directly affects the fatigue behavior. Recently, all industrial sectors have been requested for a fast and non-destructive determination of residual stress. This study aims to monitor of the variations in surface residual stress distributions in the carburized 19CrNi5H steels by means of non-destructive and semi-destructive measurement techniques, Magnetic Barkhausen Noise (MBN), X-Ray Diffraction (XRD), and Electronic Speckle Pattern Interferometry (ESPI) assisted hole drilling. Microstructural investigation by optical and scanning electron microscopy, hardness measurements, and spectroscopy analysis were also conducted. To comprehend the differences in the surface residual stress state, 19CrNi5H steel samples were carburized at 900°C for 8, 10 and 13 hours, and then, tempered in the range of 180°C and 600°C. Residual stress measurements carried out by XRD and ESPI assisted hole drilling showed that the compressive residual stress state exists for the case-hardened samples throughout the case depth regions, and the magnitude of the compressive residual stress decreases as the tempering temperature increases. MBN measurements showed that the BN activity increases with decreasing carburization time and increasing tempering temperature. It was concluded that MBN technique could be used to measure the surface residual stress distributions with a proper calibration operation.

Introduction

Carburizing is one of the most widely employed surface hardening processes to obtain high wear resistance, fatigue strength and toughness [1, 2]. Carburized steels act as a composite material that consisting of the harder surface and ductile core regions with beneficial compressive residual stress in the surface layer. The compressive residual stresses are formed by various phenomena in the carburized samples, which are the differences in the volume between the high- and low-carbon zones, the temperature gradients, and the local plastic deformation during the austenite-to-martensite phase transformation.

Several residual stress measurement techniques have been developed whether they are destructive or nondestructive. Among the nondestructive stress measurement techniques, X-ray diffraction is commonly used methods for materials possessing a crystalline structure [3]. The hole-drilling method is widely accepted [4] and ruled by a standard [5]. In this method, strain formed due to the stress relaxation measurement is performed by the strain gage rosettes [6]. Optical methods dependent on laser interferometry are promising in replacing the strain gage

Residual Stresses 2018 – ECRS-10   Materials Research Forum LLC
Materials Research Proceedings 6 (2018) 165-170   doi: http://dx.doi.org/10.21741/9781945291890-26

hole drilling method [7]. Electronic Speckle Pattern Interferometry (ESPI) has been successfully used for several years in the measurement of the residual stresses in the components [8]. Industry has been searching for methods that are capable of measuring the residual stresses accurately, quickly and easily without damaging the material under testing. A more unconventional method, the magnetic Barkhausen noise (MBN) method, is of a particular interest since it has a potential as a nondestructive industrial tool to measure residual stress and other microstructural parameters as quick and easy as possible.

Experimental Procedures

19CrNi5H steel is generally used for the production of gears, pins, and drive axles. The chemical composition and mechanical properties are given in Table 1. The steel rods were cut into the samples with the dimensions of $165 \times 36 \times 10$ mm^3. All samples were labelled in accordance with the heat treatment to be applied: the first digit indicates the heat treatment i.e., carburizing (C) whereas the second digit designates the duration of the heat treatment in hours and the third digit refers to the tempering temperature.

Table 1 Mechanical properties and chemical composition of the sample (normalized condition)

Material				19CrNi5H / SAE 3120 / En 351 (BS 970) / 20NiCr4							
Young's Modulus (E) Poisson's ratio (v)				201 – 209 GPa 0.27 – 0.29							
Yield Strength (YS) Ultimate Tensile Strength (UTS) % Elongation				350 – 550 MPa 650 – 850 MPa 8 – 25							
Chemical Composition (% weight)	C	Si	Mn	P	S	Cr	Ni	Mo	Al	V	Fe
	0.18	0.26	0.95	0.014	0.026	1.01	0.94	0.05	0.031	0.009	balanced

All samples were uniformly austenitized at 880°C for 3h; then, cooled in still-air. The residual stresses induced by previous machining operations were assumed to be relieved during this treatment. The magnitude of the surface compressive residual stress varies dependent on the heat treatment parameters. Hence, various samples were prepared by applying different carburizing and tempering operations. During carburizing operations, the samples were held in the furnace using the mixture of $C_3H_8(g)$ and the shielding gas at 900°C for 8hrs, 10hrs and 13hrs. Upon completion of carburizing process, tempering process was applied at 180°C, 240°C, and 600°C for 3 hours.

XRD stress measurements were performed using Xstress 3000 G2/G2R system and Cr-Kα radiation. Tube voltage and current were set as 30 kV and 6.7 mA, respectively. Prior to the measurements, the device was calibrated by stress-free iron reference sample. The peak of the ferrite {211} plane was used for the analysis. Five tilts for both positive and negative psi angle varying from −40° to 40° were used. The measurement directions for five equally distanced points on the sample surface was 0°, 45°, 90° with respect to the rolling direction. The retained austenite content of the samples was also determined by the four-peak XRD technique.

The PRISM ESPI system was used for residual stress measurements by the ESPI assisted hole-drilling method. The drilling trials were performed on the dummy case-hardened specimens to optimize drilling parameters as such 40,000 rpm and 0.15 mm/sec. All ESPI measurements were taken along the centerline of the samples. Measurement points were 20 mm away from the edges of sample, and they were at least 50 mm apart from each other in order to avoid the effect of the deformation area used in stress calculations. Residual stresses were calculated from the deformation data, i.e. interferograms, by using the PRISMS® software which is based on the method developed by *Nelson et al.* and *Steinzig et al.* [8].

The MBN measurements were performed using commercially available equipment and its software. The measurement parameters were determined by evaluating the series of MBN signals

Residual Stresses 2018 – ECRS-10   Materials Research Forum LLC
Materials Research Proceedings 6 (2018) 165-170   doi: http://dx.doi.org/10.21741/9781945291890-26

obtained from all samples. In order to determine the optimum measurement parameters, the reliability, sensitivity, validity indexes of MBN signal were considered. Then, the optimized measurements parameters were determined as such that magnetizing voltage and magnetizing frequency were 10 V and 250 Hz, respectively. In addition, a filter that passed frequencies from 200 to 1,000 kHz was used. The use of such a filter range allowed the measurements to be obtained at the similar depth as the XRD and ESPI method.

For microstructural investigations, the small samples were cut, ground, polished, and etched in 2% Nital solution. The microstructures were examined by an optical microscope. Effective case depth values were determined in accordance with ISO 2639. The Vickers hardness measurements were performed at 100X magnification in accordance with ISO 6507–1.

Results and Discussion

Microstructure and Hardness. Micro hardness measurements carried out on the cross-section of the carburized and tempered samples indicated that the case region were the hardest and the hardness decreased remarkably when moved from case to core region of the samples (see Figure 1). With an increase in the duration of carburization operation, the surface hardness values rose from 783 HV1 to 835 HV1. Among the same carburization condition, an increase in the tempering temperature led the surface hardness values to decrease due to the softening of

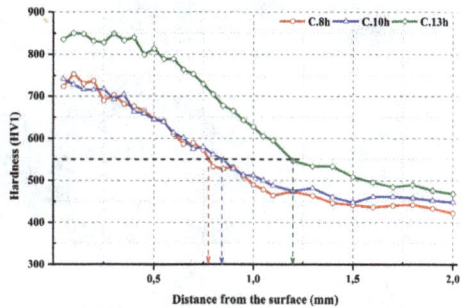

Figure 1 Effect of the carburizing time at 900°C on the hardness depth profile of the 19CrNi5H steel

martensite phase and spheroidization of the parent phase. The effective case depth of carburized samples determined as per ISO 2369 were 0.76 mm, 0.85 mm and 1.15 mm for 8hrs, 10hrs and 13hrs carburized specimens, respectively.

The micrographs seen in Figure 2 revealed that the carburized and tempered steel specimens consist of martensitic microstructure in both the core and the case regions. The needle-like shape of martensite was obtained in the quenched state, yet the application of tempering at 180°C led the needle-like martensite to be rounded at the tip (Figure 2a and b). In addition, ε-carbide precipitation occurs in the microstructure and martensite partially loses its tetragonality at this tempering temperature. Upon increasing the tempering temperature to 240°C, martensite continues to lose its tetragonality and the ε-carbide precipitations were replaced by the cementite phase (Figure 2c and d). Further increase in the tempering temperature up to 600°C, cementite coarsens and spheroidizes seen in Figure 2e and f [2]. In addition, the X-ray diffraction technique indicated that the surface regions of the samples have retained austenite. The maximum retained austenite content was observed in the 13hrs quenched specimens as 15.4±1.9%. Upon the application of tempering heat treatment, the volume percentage of the retained austenite decreased, and it was measured zero upon tempering at 240°C and 600°C since the martensite start temperature of the steel under investigation is about 190°C.

Results of XRD Measurements. Effect of tempering on the surface residual stress distribution can be seen in Figure 3 for carburized steel samples. The compressive residual stress distributions were present on the surface of the as-quenched and tempered specimens. The magnitude of the surface residual stresses was decreased with the application of the tempering operations. When the duration of the heat treatment was increased, the magnitude of the residual stress, in the same tempered condition, was also increased. This is because the magnitude of the compressive residual stress at the surface depends on the ratio between the case and the core thickness. When the other factors are the same, the surface compressive residual stress will be low unless the core is thicker than the case.

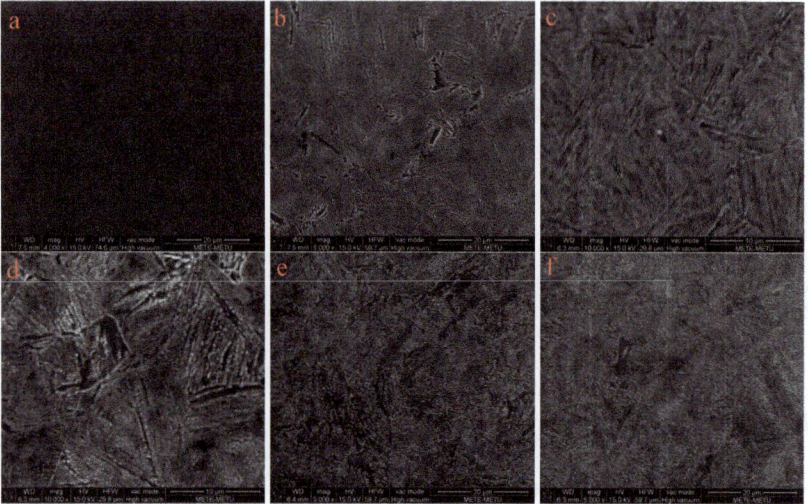

Figure 2 Representative micrographs of the carburized and tempered samples: (a) 180°C temper–case, (b) 180°C temper–core, (c) 240°C temper–case, (d) 240°C temper–core, (e) 600°C temper–case, and (f) 600°C temper–core

Results of MBN Measurements. It is well known that the MBN is affected by several microstructural features and by the stress whether applied or residual. When the magnetostriction coefficient of the material has a positive value, the MBN signal reveals an increasing trend in the direction of the applied elastic tensile stress [9]. The stress dependency of the MBN emission can be seen from Figure 4. Since the carburized and tempered steel samples having a positive magnetostriction coefficient, the

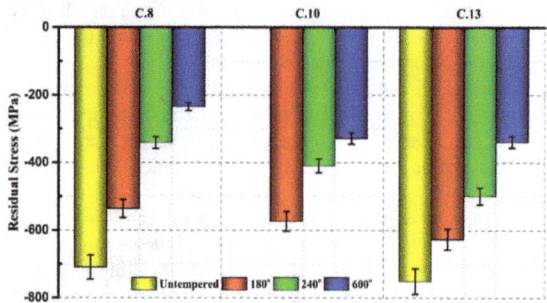

Figure 3 Effect of tempering temperature on the surface residual stress of the 19CrNi5H steel carburized at 900°C for 8 - 13 hrs (results of XRD measurements)

average MBN activity decreased when the carburizing time increased, causing more negative compressive residual stress values on the surface of the sample. In addition to this, when the samples were subjected to tempering from 180°C to 600°C, the average MBN activity showed increasing trend.

Results of ESPI Measurement. Fringe patterns recorded for various depth increments by 25 μm for the carburized sample are given in Figure 5. Since the single ESPI system was used, the fringe patterns acquired during the hole-drilling were generated on the horizontal plane, x-axis, around the hole. The brighter fringes indicate the half-wavelength increments in the component of the measured surface displacements in the sensitivity direction. The pixels adjacent to the hole within the inner dashed circle (in 100μm depth) were excluded from the stress calculation since plastic deformation zone around the hole causes de-correlation between pre- and post-hole images [10, 11]. Tikhonov regularization was used in the ESPI system to correct the zig-zag pattern of the residual stresses [5]. More than one fringe pattern were obtained and the number of the fringe images increased with increasing the depth of the hole.

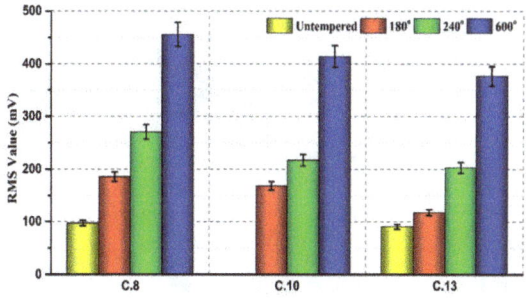

Figure 4 Effect of tempering temperature on the MBN-RMS values of the 19CrNi5H steel carburized at 900°C for 8-13hrs (the results of MBN measurements)

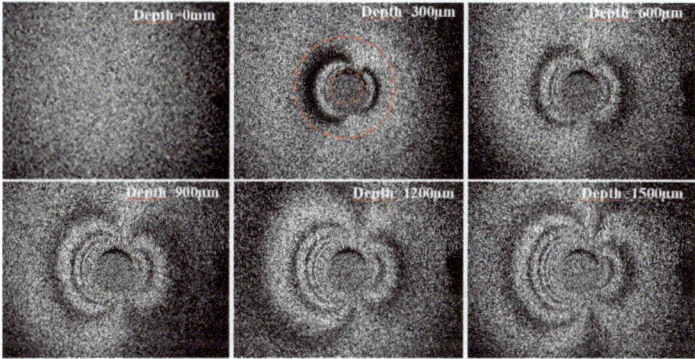

Figure 5 Fringe patterns in the carburized 19CrNi5H sample (by the ESPI method)

Surface residual stress distributions of the samples are given in Figure 6. As measured by the XRD method, the maximum compressive residual stress was observed in the 13hrs carburized sample in all tempering conditions. As observed in the other measurement techniques, the magnitude of the compressive residual stress diminished with increasing tempering temperature. Upon tempering at 600°C, the compressive residual stress transformed to the tensile residual stress in the shallow area whereas the compressive residual stress state continued to about 1.2 mm beneath the surface regions of the other samples.

Conclusion

The effectiveness of three different techniques for the measurement of residual stresses in the carburized and tempered steels was investigated. Various sample sets were prepared from steel by carburizing at 900°C for 8 to13hrs, followed by tempering at three different temperatures. The residual stress values were determined by XRD, MBN and ESPI assisted hole drilling methods.

XRD stress measurements revealed that the compressive residual stress state exists on the surfaces of all samples, and its magnitudes decrease with increasing tempering temperature. The maximum compressive residual stress exists at the surface region of the as-quenched specimen while this position shifts beneath the surface after tempering. MBN measurements gave the similar tendency. The RMS value of the MBN emission increases with increasing tempering temperature, i.e., with decreasing magnitude of the compressive residual stress. The correlation of both techniques shows very strong positive linear relationship between the root-mean-square values of MBN and the measured surface residual stress values by XRD technique. Furthermore, ESPI assisted hole-drilling method is confirmed by the XRD measurements with slight differences in the magnitude of the stresses till a depth of 0.8

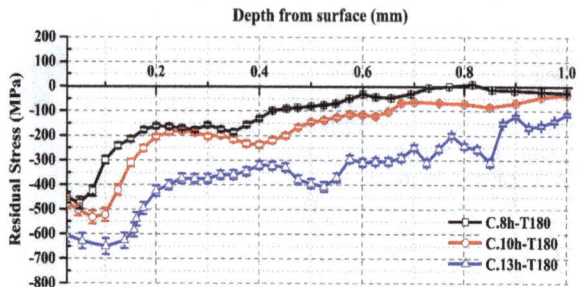

Figure 6 Effect of carburizing time on the depth profile of the residual stress in the T180 samples (by the ESPI method)

mm. For this particular study, the ESPI assisted hole-drilling method seems to be more advantageous due to its measurement speed and practicality.

References

[1] Krauss G. (1991). Microstructures and Properties of Carburized Steels, in ASM Handbook, Vol. 4, Heat Treating, ASM International 363–375.

[2] Parrish G., (1999). Carburizing: Microstructures and Properties, Materials Park, Ohio: ASM International.

[3] Rickert, T., Thomas, J., and Suominen, L. (2014). Residual Stress Measurement of Shot-Peened Steel Rings by Barkhausen Noise, ESPI Hole-Drilling and X-Ray Diffraction. AMR, 996, 380-385. https://doi.org/10.4028/www.scientific.net/AMR.996.380

[4] Casavola C., Campanelli L. S., and Pappalettere C. (2008). Experimental Analysis of Residual Stresses in the Selective Laser Melting Process, in Proceedings of the 11th International Congress and Exhibition on Experimental and Applied Mechanics,1479-1486.

[5] ASTM E837 (2008) Standard Test Method for Determining Residual Stresses by the Hole-Drilling Strain Gage Method, Annual Book of ASTM standards, Philadelphia: American Society for Testing and Materials, 747-753.

[6] Barile C., Casavola C., Pappalettera G., and Pappalettere C. (2013). Feasibility of Local Stress Relaxation by Laser Annealing and X-Ray Measurement. Strain, 49(5), 393–398. https://doi.org/10.1111/str.12045

[7] Albertazzi Jr. A., Peixoto Filho F., Suterio R., Amaral F. (2004) Evaluation of a Residual Stresses Measurement Device Combining a Radial In-plane ESPI and the Blind Hole Drilling Method, The International Society for Optical Engineering, Europe International Symposium Photonics.

[8] Steinzig, M. and Ponslet, E. (2003). Residual Stress Measurement Using the Hole Drilling Method and Laser Speckle Interferometry: Part I-IV. Experimental Techniques, 27 (3), 43-46. https://doi.org/10.1111/j.1747-1567.2003.tb00114.x

[9] Gauthier, J., Krause, T., & Atherton, D. (1998). Measurement of residual stress in steel using the magnetic Barkhausen noise technique. NDT & E International, 31(1), 23-31. https://doi.org/10.1016/S0963-8695(97)00023-6

[10] Furgiuele, F., Pagnotta, L., Poggialini, A. (1991). Measuring residual stresses by hole-drilling and coherent optics techniques: a numerical calibration. J. Eng. Mater. Technol. 113(1), 41. https://doi.org/10.1115/1.2903381

[11] Barile, C., Casavola, C., Pappalettera, G., Pappalettere, C. (2014). Analysis of the effects of process parameters in residual stress measurements on titanium plates by HDM/ESPI. Measurement 48, 220–227. https://doi.org/10.1016/j.measurement.2013.11.014

Residual Stresses 2018 – ECRS-10   Materials Research Forum LLC
Materials Research Proceedings 6 (2018) 171-176   doi: http://dx.doi.org/10.21741/9781945291890-27

Evolution of Microstructure and Residual Stress in Hot Rolled Ti-6Al-4V Plates Subjected to Different Heat Treatment Conditions

William Rae[1,2,a,*], Salah Rahimi[2,b]

[1]Design, Manufacture & Engineering Management (DMEM) Department, University of Strathclyde, 75 Montrose Street, Glasgow, UK

[2]Advanced Forming Research Centre (AFRC), University of Strathclyde, 85 Inchinnan Drive, Glasgow, UK

[a]william.rae@strath.ac.uk, [b]salah.rahimi@strath.ac.uk

Keywords: Contour Method, X-Ray Diffraction, Titanium Alloy, Phase Transformation

Abstract. Hot rolled Ti-6Al-4V plate samples were taken from three different stages of an industrial heat treatment process; one as-rolled and two heat treated. This was followed by microstructure characterization using optical microscopy. Surface and through-thickness residual stress was determined using a combination of X-ray diffraction (XRD) and the contour method. Measured residual stress distributions showed similarities in distribution with that obtained for rolled Al-7050 alloy; including compressive troughs near the outer thickness on both sides, leading towards a tensile zone around the center with a local minima at the plate center thickness. Microstructure and residual stress data was then used to draw comparisons between the investigated conditions.

Introduction

Titanium alloys are often selected as critical components in the aerospace sector due to their exceptionally high strength-to-weight ratio and ability to produce different microstructures, with tailored mechanical properties, by thermo-mechanical processing at elevated temperatures. Thermo-mechanical processing may generate residual stress which can lead to distortion out of dimensional tolerance during subsequent manufacturing processes; such as during machining [1], and can also affect material performance in service; such as reducing fatigue life [1]. The difficulty arises because Ti alloys usually have a complex, multi-phase microstructure, which varies within a component depending on process history [1]. This makes the final residual stress distribution highly dependent on microstructure and deformation history. Residual stress arises from heterogeneity in plastic deformation and during thermo-mechanical processing, phase transformation, and the differential gradient between colder outer surface and hot inner interior induced during cooling. The latter would be even more intensified by non-uniform heat exchange with the cooling environment. The deformation in Ti alloys is complex due to the inherent thermal and mechanical anisotropy of the constituent phases. Furthermore, in the temperature range of interest (600°C – 950°C), the material undergoes a reversible α-β phase transformation, which can contribute to stress relaxation [2].

Thus, it is imperative that the evolution of residual stress is understood when designing and modifying microstructure for high value, critical aerospace components. The aim of this work was to evaluate microstructure and the through-thickness residual stress distribution induced by thermo-mechanical and subsequent industrial heat treatment processing of 38 mm Ti-6Al-4V plate, using multiple residual stress measurement techniques.

Residual Stresses 2018 – ECRS-10   Materials Research Forum LLC
Materials Research Proceedings 6 (2018) 171-176   doi: http://dx.doi.org/10.21741/9781945291890-27

Materials & Methods

Materials & Processing. The material selected for this study was Ti-6Al-4V in the form of a 38 mm thickness rolled plate with dimensions of approximately $38 \times 250 \times 250$ mm^3; these sections were removed from industrially processed plates with a width of 1000 mm^3 by means of flame cutting. The nominal chemical composition of the alloy is provided in Table 1. Three plate conditions were investigated: one as-rolled (AR) condition which was deformed at approximately 950°C, and two conditions which were subjected to varying subsequent post-rolling annealing treatments (HT1 and HT2) high in the $\alpha+\beta$ phase field (>900°C) followed by air cooling. Both surfaces of HT2 were also subjected to shot blasting and its influence on residual stress was of interest. It was noted that the use of flame cutting would affect the overall residual stress distribution in the samples, however as this was carried out for all plate conditions it was deemed that the gathered data would allow for adequate comparison, and the original plate residual stress could be ascertained by further modelling which was out-with the scope of this work.

Table 1. Nominal chemical composition for Ti-6Al-4V alloy used in study (wt%).

V	Fe	Al	O	C	N	Ti
4	0.25	6	0.13	0.08	0.03	Bal.

Microstructure Characterization. Light microscopy was utilized to gain an understanding of the grain morphology and microstructural evolution in the materials. Three 10 mm^3 cubes were cut from each plate condition at a distance of over ~25 mm from the plate edge to minimize the chance of the microstructure being effected by the flame cut. Metallographic preparation was conducted until a mirror finish was achieved, followed by chemical etching using Kroll's Reagent (2% HF, 6% HNO$_3$, and 92% H$_2$O) to reveal the microstructure. A Leica DM12000M Optical Microscope was then used to acquire micrographs at a range of magnifications.

Residual Stress Measurement. Residual stress analysis was carried out on each plate using a combination of X-ray Diffraction (XRD) for surface measurements, and the contour method for through-thickness measurements; providing complementary data [3]. As it has previously been shown that the residual stress distribution parallel to the rolling direction is largely uniform [4], it was desired to investigate the residual stress distribution perpendicular to the rolling direction, along the center of the plate. The out-of-plane residual stress, σ_{zz}, was measured using both techniques.

Firstly, non-destructive XRD was carried out using a PROTO LXRD X-ray diffractometer system and the sin$^2\psi$ method. The diffraction peak at a Bragg angle of 140°, corresponding to the {211} crystallographic plane in the α phase, was used to calculate residual strain which was then converted to residual stress, assuming a Young's Modulus of 118 GPa and Poisson's Ratio of 0.342. Although Ti-6Al-4V is a dual phase alloy, it consists of approximately 92% α phase at room temperature and therefore the stress associated with the β phase is assumed to have a minimal effect on the overall residual stress magnitude [2]. Nine measurements were performed on the top surface of each plate at increments of 25 mm from the plate center position, with three measurements along the base at increments of 50 mm from the center. Nine ψ-offset angles in the range of $\pm 30°$ were employed.

A 2-D map of out-of-plane residual stress perpendicular to the rolling direction was produced using the contour method. The plates were cut through the center by electrical discharge machining (EDM) using a 250 µm brass wire. The outline and surface topography of the cut surfaces were then measured by means of a Mitutoyo Crysta Apex C coordinate measuring machine (CMM). Co-ordinate data was obtained at increments of 0.4 mm through the thickness and 1 mm along the length of the surface. The measured data was then cleaned to minimize outliers, followed by alignment of the two corresponding surface datasets to one coordinate system, which was then linearly interpolated onto a common grid and averaged to form one set of data for each plate condition. A bivariate cubic spline was fit to the data with knot spacing of 10 mm and 5.6 mm in the x (width) and y (thickness) direction, respectively. A 3-D finite element (FE) model was produced from this fitted data and the original residual stresses were computed by forcing the surface into the reverse shape of the averaged topography data.

Results & Discussion

Microstructure Characterization. The multi-orientation light micrographs obtained for each plate condition are presented in Fig. 1. The AR condition exhibited a high fraction of elongated α lamellae along the rolling direction with some globular α in a fine transformed β matrix. The HT1 condition shows a clear break-up of α lamellae in the transverse and normal planes with an increased fraction of more equiaxed, globular α with Widmenstätten α tending to replace the fine $α_s$ observed in the AR condition. The HT2 condition shows a greater degree of $α_p$ fragmentation than HT1, with less Widmenstätten α and an accompanying increase in fraction and size of globularised grains; however, a large amount of elongated α lamellae still remain.

The break-up of α lamellae by globularization as observed in the HT1 and HT2 conditions was characteristic of recrystallization annealing high in the α+β phase field (>900°C) [5]. An increased fraction and size of globular α in the HT2 condition may be due to a further annealing cycle; or increased temperature or duration than that experienced in the HT1 condition, resulting in the progression of globularization kinetics [5]. Globularization in Ti-6Al-4V has been associated with stress relaxation; due to rearrangement of dislocations which can annihilate as they move through the lattice, reducing dislocation density and therefore internal strain [6]. Thus, it was thought that the microstructural evolution observed may be indicative of stress relaxation.

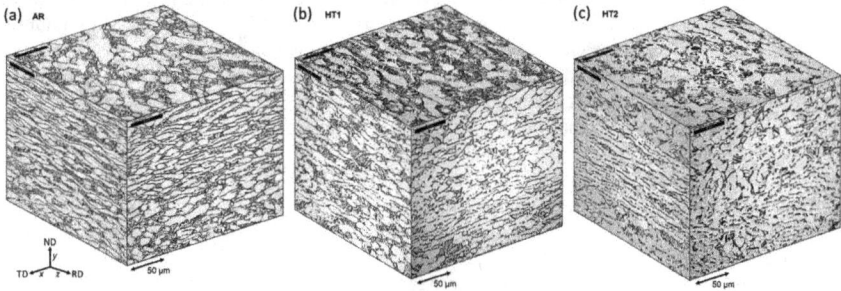

Figure 1. Multi-orientation light microscope images of (a) AR; (b) HT1; and (c) HT2 conditions.

Residual Stress Measurement. The surface residual stress measurements obtained by XRD are presented in Fig. 2 for all conditions. The AR plate exhibited a low magnitude of tensile surface residual stress for all measurement points, with a peak stress of +69 MPa on the top surface at a distance of 25 mm from the plate center. Measured stresses reduced towards +18 MPa outwards from this point. The stress distribution on the bottom was lower and ranged from +7 MPa to +25 MPa. The HT1 condition contained similar magnitudes of residual stress, however this was largely uniform along the top surface with an average of +45 MPa. The bottom surface exhibited a compressive stress of -34 MPa at -50 mm from the plate center with the other two points showing low tensile values of between 0 and +20 MPa. On the other hand, the HT2 condition exhibited much higher magnitudes of compressive residual stress for all measured points. This was largely uniform along both surfaces with an average value of approximately -300 MPa and -500 MPa for the top and bottom, respectively. The similarities in magnitudes between the AR and HT1 conditions suggest that comparable cooling conditions may have been used for both processes. The highly compressive residual stresses measured in the HT2 condition is likely to be due to shot-blasting which was applied to both sides of the material after heat treatment.

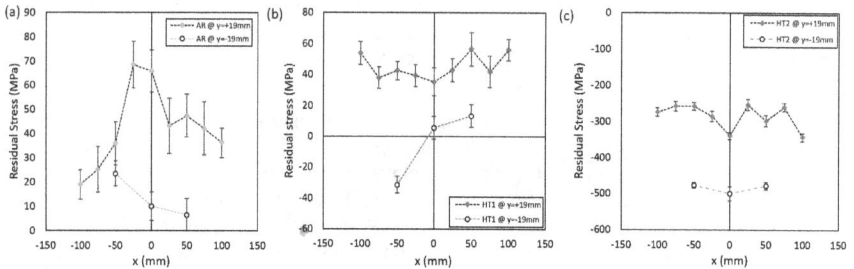

Figure 2. Surface residual stress distribution as determined by XRD for (a) AR; (b) HT1; and (c) HT2.

The contour method residual stress results for the AR condition (Fig. 3a) show a clear compressive trough on either side of the plate within the first 10 mm thickness. This region had a peak compressive residual stress of -100 MPa and is shown to elongate along the x-direction. A balancing tensile region was then observed over the following 5-10 mm on both sides with a local minima region noted at approximately the center of the plate thickness. A maximum tensile stress of +58 MPa was noted. This trend is further exemplified in the combined contour and XRD through-thickness line profiles presented in Fig. 4a; where the contour method results show good agreement with the surface data provided by XRD. On comparison with the through-thickness residual stress distribution obtained by Prime and Hill for cold-rolled aluminum (Fig. 4b), numerous similarities can be noted [4]. Both data show compressive troughs in within 20% of the outer surface followed by balancing tensile regions and a local minima at approximately 50% the plate thickness. This particular residual stress distribution is thus likely to be primarily due to heterogeneous plastic deformation during rolling as it has been observed in both hot and cold-rolled plates.

On evaluation of the HT1 contour data (Fig. 3b and 4c), a similar trend is noted including the compressive troughs and local minima around y = 0 mm, however the magnitudes of residual stress in the compressive regions are much higher than that of the AR condition. This is particularly visible in the top region between x = -50 and +50 mm. The maximum compressive stress was observed in this region with a magnitude of -220 MPa, with a comparatively lower local maximum compressive stress on the bottom side of -120 MPa. As the residual stress distribution continued to somewhat resemble that of the AR condition it is thought that this indicates the heat treatment did not lead to full relaxation of residual stress. Additionally, the increased magnitudes in the top compressive region suggests that an aspect of post rolling treatment induced increased residual stress near the top surface. On examination of the contour results for the HT2 condition (Fig. 3c and 4d), the increased intensity of the compressive trough near the top surface is immediately apparent, stretching over 200 mm in the x-direction. The maximum compressive stress in this region was noted to be -220 MPa, as with HT1, with a maximum on the bottom side of -100 MPa. The characteristic rolled plate residual stress distribution can be seen for HT2 but is less prominent than the AR or HT1 condition.

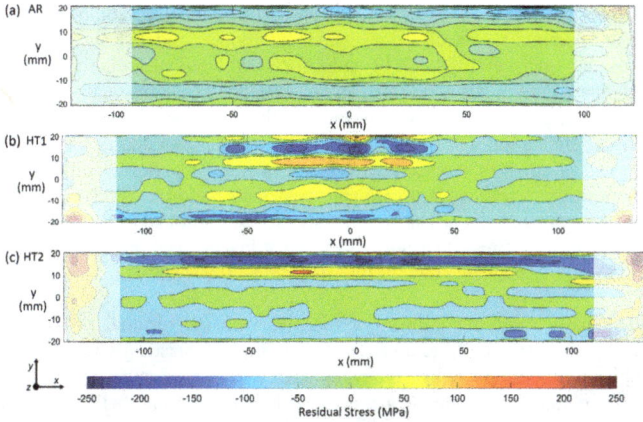

Figure 3. 2-D out-of-plane residual stress maps of samples as determined by the contour method for (a) AR, (b) HT1 and (c) HT2 plate conditions. The outer 25 mm of each plate has been shaded in grey as it was affected by the flame cut and therefore not considered in analysis.

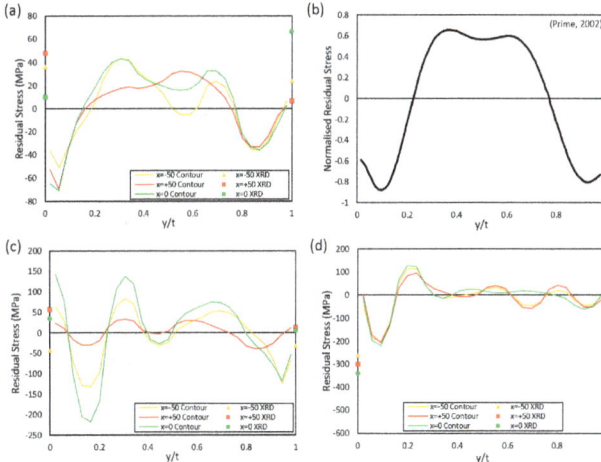

Figure 4. Combined XRD and through-thickness contour method residual stress line profile at x = -50 mm, x = 0 mm and x = +50 mm for (a) AR; (c) HT1; and (d) HT2 conditions. Normalized through-thickness residual stress line profile of cold-rolled 7050-T74 plate provided in (b) for comparison (adapted from [4]). Y-axis has been normalized for all figures.

The further decreased prominence of the rolled plate distribution in HT2, especially for y/t above 0.3, suggests that heat treatment may have relaxed initial rolling induced residual stress by a greater extent than in HT1 condition. This was likely due to either increased annealing temperature or soak duration and is supported by the observation of increased break-up of α lamellae in the HT2 microstructure.

The compressive troughs near the bottom surface of all conditions were found to be of similar magnitude (-50 to -100 MPa), whereas the troughs near the top surface showed a much increased magnitude for the HT1 and HT2 condition. Increased magnitudes of compressive near-surface residual stress are commonly associated with higher thermal gradients in the region [3]. This indicates that the top surfaces of the HT1 and HT2 conditions were likely exposed to higher cooling rates compared to the AR condition. The consistency of the residual stress distribution of the bottom surfaces suggests slower cooling rates in this area for all conditions; likely through conductive cooling, indicating that the bottom surface was not exposed to the air. This suggests that although annealing acts to relieve original residual stresses and provide a more homogeneous microstructure, the final residual stress distribution is highly dependent on post-heat treatment cooling rates [5]. Further work should be completed to quantify the effect of cooling rate on the residual stress distributions using the inverse heat transfer coefficient method.

Conclusions
- All hot-rolled plate conditions showed similarities with the residual stress distribution for cold-rolled Al plate reported in literature, indicating that the plastically induced residual stress from rolling is not fully relieved by processing at elevated temperature.
- Heat treated conditions showed a clear break-up of α lamellae but this is not thought to have greatly affected the residual stress magnitudes or distribution.
- Increased cooling rates may have caused increased compressive residual stress near the top surface of the heat treated conditions.

Acknowledgements
This work was supported by the Scottish Association for Metals (SAM) and Engineering and Physical Sciences Research Council (EPSRC) grant (EP/1015698/1). The authors would like to thank TIMET for the provision of materials, and also Aubert & Duval for their support on the project. The research was performed at the Advanced Forming Research Centre (AFRC), which receives partial financial support from the UK's High Value Manufacturing Catapult.

References
[1] M. G. Glavicic, D. U. Furrer, and G. Shen, "A Rolls-Royce Corporation industrial perspective of titanium process modelling and optimization: current capabilities and future needs," J. Strain Anal. Eng., vol. 45, no. 5, 2010, pp. 329-335. https://doi.org/10.1243/03093247JSA577

[2] E. Alabort, P. Kontis, D. Barba, K. Dragnevski, and R. C. Reed, "On the mechanisms of superplasticity in Ti–6Al–4V," Acta Mater., vol. 105, 2016, pp. 449-463. https://doi.org/10.1016/j.actamat.2015.12.003

[3] Rae, W., Lomas, Z., Jackson, M., & Rahimi, S., "Measurements of residual stress and microstructural evolution in electron beam welded Ti-6Al-4V using multiple techniques," Mat. Char., vol. 132, 2017, pp. 10-19. https://doi.org/10.1016/j.matchar.2017.07.042

[4] Prime, M.B. and Hill, M.R., "Residual stress, stress relief, and inhomogeneity in aluminum plate". Scripta Materialia, 2002, vol. 46, no. 1, pp.77-82. https://doi.org/10.1016/S1359-6462(01)01201-5

[5] Stefansson, N., and S. L. Semiatin. "Mechanisms of globularization of Ti-6Al-4V during static heat treatment," Metall. Mat. Trans. A, 2003, vol. 34, no.3, pp. 691-698. https://doi.org/10.1007/s11661-003-0103-3

[6] Babu, B., & Lindgren, L. E., "Dislocation density based model for plastic deformation and globularization of Ti-6Al-4V". Int. J. Plasticity, 2013, 50, pp. 94-108. https://doi.org/10.1016/j.ijplas.2013.04.003

Residual Stresses 2018 – ECRS-10
Materials Research Proceedings 6 (2018) 177-182

Materials Research Forum LLC
doi: http://dx.doi.org/10.21741/9781945291890-28

Influence of Surface Pretreatment on Residual Stress Field of Heat-Treated Steel Induced by Laser Local Quenching

Yoshihisa Sakaida[1,a,*], Yuki Sasaki[2,b] and Haruo Owashi[3,c]

[1] Dept. of Mech. Eng., Shizuoka University, Naka-ku, Hamamatsu 432-8561, Japan

[2] Grad School of Eng., Shizuoka University, Japan

[3] Atsumi Kogyo Co. Ltd., Higashi-ku Hamamatsu 435-0042, Japan

[a] sakaida.yoshihisa@shizuoka.ac.jp, [b] yuki.sasaki.16@gmail.com, [c] owashi@atsumi-kogyo.co.jp

Keywords: Laser Quenching, Carbon Steel, Residual Stress Field, Surface Pretreatment, X-Ray Stress Measurement, Quenching Temperature

Abstract. In order to clarify an influence of the surface pretreatment condition on the residual stress field after laser local quenching, three kinds of surface pretreatment were applied to the steel specimens. In this study, the mirror-polished, matte-black-painted and mechanical-polished surfaces were prepared, and were heat-treated by two times laser irradiation. After quenching, the irradiation marks and hardened layers were observed. The irradiation mark on the mirror-polished surface was insufficient. On the other hand, the hardened layers under the matte-black painted and mechanical-polished surfaces were formed enough. The residual stress fields of the latter two specimens were measured. As a result, surface pretreatment condition was found to make a big difference in both the 2θ-$\sin^2\psi$ diagram and the residual stress field after laser quenching. Therefore, in order to investigate these causes, surface temperature histories during laser irradiation were monitored using a radiation thermometer. From these results, it was clarified that the difference of the first heat-treatment temperature directly affects the presence or absence of residual stress gradient and the depth of hardened layer under the laser-irradiated surface. Finally, adequate surfaces for the steel parts before laser local quenching could be proposed.

Introduction

Laser local quenching [1] is very useful technique to improve the surface strengths of the local area, such as hardness, fatigue and fracture strengths, of steel parts having a complicated shape. The strengthening effect due to laser quenching depends directly on the hardness and compressive residual stress distributions near the quenched surface. However an influence that laser and surface pretreatment conditions give to intensities of hardness and compressive residual stress field is not sufficiently clear. In this study, the surface pretreatment condition is thought to give a maximum influence on the improved surface strength after quenching. Three kinds of pretreatment having different heat-absorptivity were applied to the surface of the steel specimens. These pretreated specimens were heat-treated by two times laser irradiation. After second laser irradiation, the residual stress field of each specimen was measured experimentally using x-ray stress measurement. An influence of surface pretreatment condition on the residual stress field and the hardness distribution under the quenched surface were examined.

Experimental Procedure

Materials and Specimens. The initial material was a carbon steel with 0.45 mass% C, 0.21 mass% Si, and 0.81 mass % Mn. At first, the material was normalized twice from 900°C in air, and was quenched from 870°C in oil at 80°C. Before laser quenching, the material was changed from martensite to fine pearlite by tempering from 550°C in air. The plate specimens with a

Residual Stresses 2018 – ECRS-10   Materials Research Forum LLC
Materials Research Proceedings 6 (2018) 177-182   doi: http://dx.doi.org/10.21741/9781945291890-28

width of 25mm and a thickness of 10mm was used. In this study, three kinds of different surfaces were prepared. One is mirror-polished surface. Another is matte-black-painted surface with soluble acrylic resin. The other is mechanical-polished surface with #600 sandpaper.

Laser Local Quenching. These specimens were heat-treated in air by two times laser irradiation as shown in Fig. 1. Each laser irradiation was conducted at a laser power, P_L = 1kW, on the local area of the pretreated surface using a semiconductor laser processing machine with a direct diode laser (L1, Enshu Ltd., Japan). In experiment, a rectangular laser beam about W_L = 3.9mm and W_T = 8.8mm was used. An incident angle, δ, and a scanning speed, V_F, were 15 or 45 degrees and 1.0 m/min, respectively. By scanning the laser beam, the local area was heated and quenched immediately by self-cooling. After first laser irradiation, the laser beam was returned to the starting point. The local area was heated and quenched again by second laser irradiation.

X-ray Stress Measurement. After second laser irradiation, the 2θ-$\sin^2\psi$ diagrams were measured experimentally from the irradiation mark using an x-ray stress measurement equipment with Ω- goniometer (MSF-2M-PSPC, Rigaku Co. Ltd., Japan). Residual stress field of each specimen was estimated from the 2θ-$\sin^2\psi$ diagrams at several φ-angles, as shown in Fig.2. Table 1 summarizes x-ray stress measurement condition. In this study, the diffraction from 211 plane of martensite by Cr-Kα radiation was used. The x-ray Young's modulus, E_X, and Poisson's ratio, v_X, of 211 plane of martensite were 186GPa and 0.30, respectively [2]. The existences of ψ-splitting phenomenon and stress gradients near the quenched surface were examined from the 2θ-$\sin^2\psi$ diagrams.

Table 1 X-ray stress measurement condition for laser local quenching.

Characteristic x-ray	Cr-Kα
Tube voltage (kV)	30
Tube current (mA)	8
Diffraction	Fe-211
Diffraction angle (deg)	155.57
Range of $\sin^2\psi$	0-0.75
Step in $\sin^2\psi$	0.05
φ angle (rad)	0, $\pi/4$, $\pi/2$, $3\pi/4$, π, $5\pi/4$, $3\pi/2$, $7\pi/4$
Incident collimator (mm^2)	2 x 2

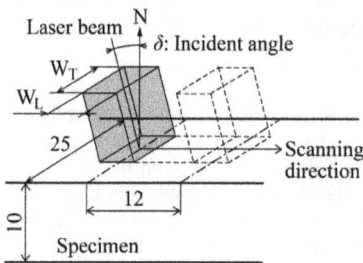

Fig.1 Local area heating and quenching by scanning a semiconductor laser beam.

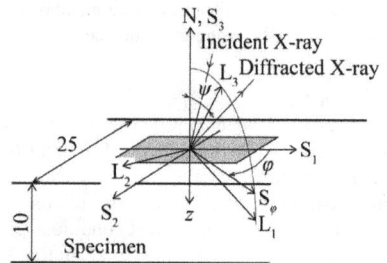

Fig.2 Laboratory and sample coordinate systems, L_i and S_i, and ψ and φ-angles.

Results and Discussion

Irradiation Mark of Laser Beam and Hardened Layer Formed near the Surface.

Figures 3(a), 3(b) and 3(c) show images of the irradiation mark of the laser beam formed on the different surfaces. On the mirror-polished surface, the irradiation mark at δ=45deg resembles a teardrop shape. The hardened layer below the teardrop area was insufficient because the incident laser beam scattered on the mirror surface and the heat-treatment temperature near the surface did not rise to austenite phase. On the other hand, other irradiation mark shapes are rectangular, and are larger than that of the mirror polished surface at δ=45deg. Therefore, matte-black painting and mechanical polishing were more adequate pretreatments than mirror

Residual Stresses 2018 – ECRS-10   Materials Research Forum LLC
Materials Research Proceedings 6 (2018) 177-182   doi: http://dx.doi.org/10.21741/9781945291890-28

polishing, because the laser quenched mark was stable even though the incidence angle of laser beam changed. Next, using a wire-electrical discharge machine, the matte-black-painted and mechanical-polished specimens after second laser irradiation were cut to observe the hardened layer.

Figures 4(a) and 4(b) show the cross-sectional micrographs of two different pretreated specimens. At first, the hardened layer of the matte-black-painted specimen was found to be deeper than that of the mechanical–polished specimen. Furthermore, the depth of hardened layer at δ=15deg was found to be deeper than that at δ=45deg. Therefore, two kinds of pretreated specimen quenched at δ=15deg were picked up. X-ray stress measurements were carried out on the two quenched marks at the next step.

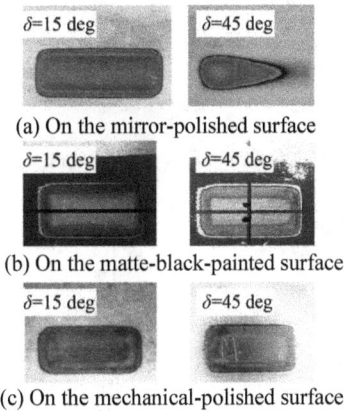

(a) On the mirror-polished surface

(b) On the matte-black-painted surface

(c) On the mechanical-polished surface

Fig.3 Photographs of irradiation mark of laser beam formed on the surfaces.

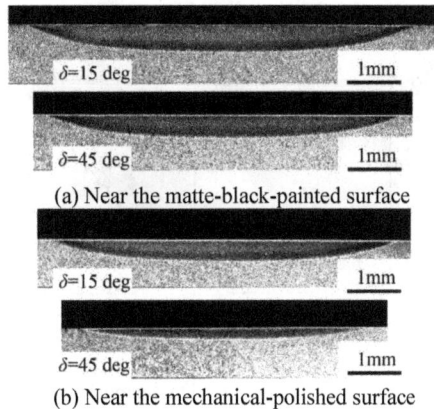

(a) Near the matte-black-painted surface

(b) Near the mechanical-polished surface

Fig.4 Cross-sectional micrographs of hardened layer below the irradiation marks.

Typical 2θ-$\sin^2\psi$ diagrams after Laser Quenching. Figures 5(a) and 5(b) show the typical 2θ-$\sin^2\psi$ diagrams at δ=15deg from the matte-black-painted and mechanical-polished specimens, respectively. In each figure, the first, second, third and fourth figures from the left show the results for $\pm\psi$ at φ=0, $\pi/4$, $\pi/2$ and $3\pi/4$, respectively. These four pair figures were used to judge the existences of ψ-splitting phenomenon and stress gradient.

In the case of the matte-black-painted specimen as shown in Fig. 5(a), each pair figure almost overlaps, and the overlapped 2θ-$\sin^2\psi$ diagram exhibits a convex upward quadratic function. On the other hand, in the case of the mechanical-polished specimen as shown in Fig. 5(b), each pair figure almost overlaps too, but the overlapped 2θ-$\sin^2\psi$ diagram exhibits a linear function. Therefore, the ψ-splitting could not be observed in all 2θ-$\sin^2\psi$ diagrams. After two times laser quenching, the matte-black- painted specimen was found to have a stress gradient below the quenched mark, while mechanical- polished specimen did not have a stress gradient.

Estimation of Residual Stress Fields after Laser Quenching. The residual stress field near the quenched mark of the matte-black-painted specimen was first estimated from four pair figures 5(a). In this case, the biaxial principal compressive residual stresses were thought to be generated near the quenched surface and to have constant stress gradients because these 2θ-$\sin^2\psi$ diagrams could be approximated as quadratic functions. In this study, three residual stress components, normal stresses, $\sigma_{11}(z)$, $\sigma_{22}(z)$ and shear stress, $\sigma_{12}(z)$, near the local quenched surface were defined as,

Residual Stresses 2018 – ECRS-10   Materials Research Forum LLC
Materials Research Proceedings **6** (2018) 177-182   doi: http://dx.doi.org/10.21741/9781945291890-28

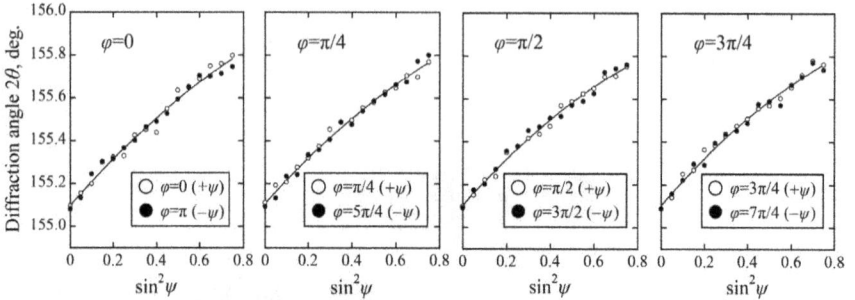

(a) On the matte-black-painted specimen.

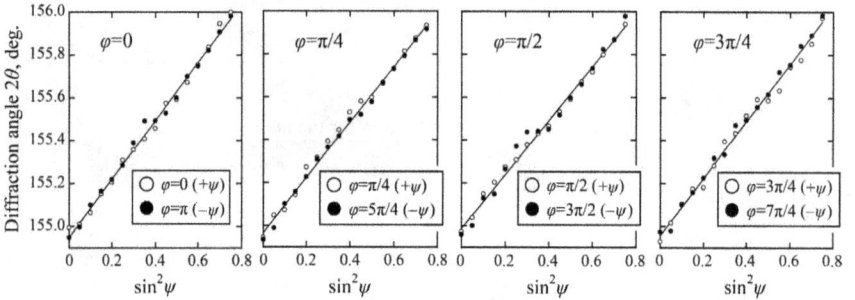

(b) On the mechanical-polished specimen.

Fig.5 The 2θ-sin²ψ diagrams from the quenched marks by laser beam at δ=15 deg.

$$\sigma_{ij}(z) = \begin{pmatrix} \sigma_{11}(0) & \sigma_{12}(0) \\ \sigma_{12}(0) & \sigma_{22}(0) \end{pmatrix} + \begin{pmatrix} A_{11} & A_{12} \\ A_{12} & A_{22} \end{pmatrix} z \, , \tag{1}$$

where $\sigma_{ij}(0)$ is residual stress component on the surface and A_{ij} is a constant stress gradient. In the case of stress field having stress gradients, the measured stress parallel to the S_φ-direction is the weighted average stress $<\sigma_{ij}>$, as follows [3, 4],

$$<\sigma_{ij}> = \left[\int_0^t \sigma_{ij}(z)\cdot\exp\left(-\frac{z}{T_0}\right)dz\right] \Big/ \left[\int_0^t \exp\left(-\frac{z}{T_0}\right)dz\right] = \sigma_{ij}(0) + A_{ij}T_0 W \, , \tag{2}$$

where W is weighted coefficient. W is equal to 1.0 at $t=\infty$[4]. T_0 is the effective x-ray penetration depth with Ω-goniometer is expressed as [3, 4],

$$T_0 = (\sin^2\theta_0 - \sin^2\psi)/(2\mu\sin\theta_0\cos\psi) = T_0 + T_1\sin^2\psi \, , \tag{3}$$

where μ is a linear absorption coefficient, $2\theta_0$ is stress-free diffraction angle estimated from the measured 2θ-$\sin^2\psi$ diagrams, T_1 and T_0 are regression coefficient and intercept that are determined by a least square method. In this case, $\mu=844$ cm^{-1}, $2\theta_0=155.57$ deg., $T_0=5.95$ μm and $T_1=-4.30$ μm, respectively.

Next, four overlapped 2θ-sin2ψ diagrams for φ=0, π/4, π/2 and 3π/4 were fitted as quadratic functions by a least square method. Three residual stress components on the surface and constant stress gradients were determined experimentally [3, 4]. As a result, the residual stress field was obtained as follows,

$$\sigma_{ij}(z) = \begin{pmatrix} -151.9 & -6.8 \\ -6.8 & -124.1 \end{pmatrix} + \begin{pmatrix} -19.1 & 2.5 \\ 2.5 & -24.1 \end{pmatrix} z \text{ , [units of } \sigma_{ij}(0)\text{: MPa, } A_{ij}\text{: MPa/}\mu\text{m]. (4)}$$

The compressive residual stresses about $\sigma_{11}(0) = -152$ MPa and $\sigma_{22}(0) = -124$ MPa were generated parallel to the S_1 and S_2 directions on the matte-black-painted surface. The residual stress gradients abut $A_{11} = -19$ MPa/μm and $A_{22} = -24$ MPa/μm were also generated near the quenched surface. The compression residual normal stress components were found to increase with an increase in the depth from the local quenched surface.

On the other hand, the residual stress field near the quenched surface of the mechanical-polished specimen was estimated from four pair figures 5(b). In that case, the constant biaxial principal compressive residual stresses were thought to be generated from the local quenched surface to the effective x-ray penetration depth because those 2θ-$\sin^2\psi$ diagrams could be approximated as linear functions. Therefore, four overlapped 2θ-$\sin^2\psi$ diagrams for $\varphi=0$, $\pi/4$, $\pi/2$ and $3\pi/4$ were fitted as linear functions by a least square method. Three residual stress components were determined experimentally. As a result, the residual stress field was obtained as follows,

$$\sigma_{ij} = \begin{pmatrix} -367.3 & 5.7 \\ 5.7 & -339.6 \end{pmatrix} \text{, [unit of } \sigma_{ij}\text{: MPa].} \tag{5}$$

The large compressive residual stresses about $\sigma_{11} = -367$ MPa and $\sigma_{22} = -340$ MPa were generated parallel to the S_1 and S_2 directions on the mechanical-polished surface.

Heat-treatment Temperature during Irradiating Laser Beam. Surface pretreatment condition before laser quenching was found to make a big difference in both the 2θ-$\sin^2\psi$ diagram and the residual stress field after two laser quenching. Therefore, surface temperatures, T_1 and T_2, at the center were monitored immediately after two times laser irradiation using a radiation thermometer with an emissivity of 0.8. Figure 6 shows the comparison of apparent temperature changes of two kinds of pretreated specimens. It was found that T_1 of the matte-black -painted specimen was much higher than that of the mechanical-polished specimen. It is thought that difference of T_1 results from high heat-absorptivity of the matte-black-painted surface. On the other hand, T_2 of two specimens was about 1070°C. The mat-black coating on the

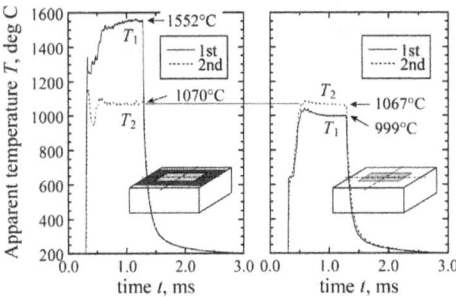

Fig.6 Comparison of surface temperatures, T_1 and T_2, after 1st and 2nd laser irradiation at δ=15 deg.

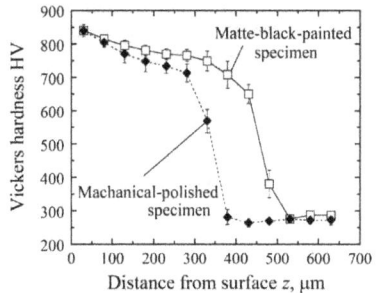

Fig.7 Vickers hardness distributions of hardened layers in two specimens.

surface was sublimated during first laser irradiation, so that T_2 dropped to 1070°C due to the reduced heat-absorptivity of the first quenched mark. In contrast, the first quenched mark also put on the surface of the mechanical- polished specimen, and then T_2 slightly rose to 1067°C due to the enlarged heat-absorptivity.

Figure 7 shows Vickers hardness distributions of hardened layers in both matte-black-painted and mechanical-polished specimens. The surface hardness of two specimens was equal to 840HV because of same T_2 before second self-cooling. However, both first-heating temperature T_1 and total amount of heat, Q, absorbed on the matte-black painted surface were higher and larger than T_1 and Q on the mechanical-polished surface. Higher heat-treatment temperature, T_1, and larger amount of absorbed heat were found to give rise to compressive residual stress field having a constant stress gradient and the deeper hardened layer near the matte-black-painted surface. Therefore, it is concluded that matte-black painting is more adequate pretreatment from the viewpoint that the depth of hardened layer is preferably deeper. In contrast, it is also concluded that mechanical polishing is more adequate pretreatment from the viewpoint that the compressive residual stress on the surface is preferably higher for the surface strength of mechanical parts.

Conclusions

Three kinds of pretreated specimens having different surfaces were heat-treated by two times laser irradiation. The residual stress field near the irradiation mark and hardness distribution were measured experimentally. The results are summarized as follows:

(1) Matte-black painting and mechanical polishing were more adequate surface-pretreatments than mirror polishing, because the irradiation marks on the matte-black-painted and mechanical-polished surfaces were stable even though the incidence angle of laser beam changed.

(2) Surface pretreatment condition was found to make a big difference in the 2θ-$\sin^2\psi$ diagram and the residual stress field after laser quenching. The 2θ-$\sin^2\psi$ diagram of the matte-black-painted specimen exhibits a quadratic function. The biaxial compressive residual stress field having a constant stress gradient was found to be generated. On the other hand, the 2θ-$\sin^2\psi$ diagram of mechanical-polished specimen exhibits a linear function. The constant biaxial compressive residual stress field was found to be generated within the effective x-ray penetration depth.

(3) By measuring the heat-treatment temperature, higher temperature and larger amount of absorbed heat were found to give rise to compressive residual stress field having a stress gradient and the deeper hardened layer near the matte-black painted surface. Therefore, matte-black painting is found to be more adequate pretreatment from the viewpoint that the depth of hardened layer is preferably deeper. In contrast, mechanical polishing is found to be more adequate pretreatment from the viewpoint that the compressive residual stress on the surface is preferably higher.

References

[1] Edited by the Japan Society of Mech. Eng., Thermomechanical behavior of materials during transformation, theory and numerical simulation, Corona Pubishing. Co., Ltd: Tokyo, (1991), 142-162.

[2] Y. Sakaida and K. Tanaka, Change of x-ray elastic constants of carbon steel caused by quenching and effect of its change on x-ray stress measurement, J. the Society of Materials Science, Japan, 56-7 (2007) 602-608. https://doi.org/10.2472/jsms.56.602

[3] I. C. Noyan and J. B. Cohen, Residual stress; measurement by diffraction and interpretation, Springer-Verlag: New York, (1987), p. 134. https://doi.org/10.1007/978-1-4613-9570-6

[4] K. Suzuki, K. Tanaka and Y. Sakaida, Penetration depth in x-ray analysis of steep stress gradient near surface, J. the Society of Materials Science, Japan, 45-7 (1996) 759-765. https://doi.org/10.2472/jsms.45.759

Welding, Fatigue and Fracture

Residual Stresses 2018 – ECRS-10
Materials Research Proceedings 6 (2018) 185-190

Materials Research Forum LLC
doi: http://dx.doi.org/10.21741/9781945291890-29

Residual Stress Formation in Component Related Stress Relief Cracking Tests of a Welded Creep-Resistant Steel

Michael Rhode[1,2,a,*], Arne Kromm[1,b], Dirk Schroepfer[1,c],
Joerg Steger[1,d], Thomas Kannengiesser[1,2,e]

[1]Bundesanstalt für Materialforschung und -prüfung (BAM), Unter den Eichen 87, 12205 Berlin, Germany

[2]Otto-von-Guericke University, Universitätsplatz 2, 39106 Magdeburg, Germany

[a]michael.rhode@bam.de, [b]arne.kromm@bam.de, [c]dirk.schroepfer@bam.de, [d]joergsteger@gmx.de, [e]thomas.kannengiesser@bam.de

Keywords: Welding, Residual Stresses, Creep-Resistant Steel, Post Weld Heat Treatment, Stress Relief Cracking, 13CrMoV9-10

Abstract. Submerged arc welded (SAW) components of creep-resistant low-alloyed Cr-Mo-V steels are used for thick-walled heavy petrochemical reactors (wall-thickness up to 475 mm) as well as employed in construction of modern high-efficient fossil fired power plants. These large components are accompanied by significant restraints during welding fabrication, especially at positions of different thicknesses like welding of nozzles. As a result, residual stresses occur, playing a dominant role concerning so-called stress relief cracking (SRC) typically during post weld heat treatment (PWHT). Besides specific metallurgical factors (like secondary hardening due to re-precipitation), high tensile residual stresses are a considerable influence factor on SRC. For the assessment of SRC susceptibility of certain materials mostly mechanical tests are applied which are isolated from the welding process. Conclusions regarding the influence of mechanical factors are rare so far. The present research follows an approach to reproduce loads, which occur during welding of real thick-walled components scaled to laboratory conditions by using tests designed on different measures. A large-scale slit specimen giving a high restraint in 3 dimensions by high stiffness was compared to a medium-scale multi-pass welding U-profile specimen showing a high degree of restraint in longitudinal direction and a small-scale TIG-re-melted specimen. The small-scale specimens were additionally subjected to mechanical bending to induce loads that are found during fabrication on the real-scale in heavy components. Results show for all three cases comparable high tensile residual stresses up to yield strength with high gradients in the weld metal and the heat affected zone. Those high tensile stresses can be significant for cracking during further PWHT.

Introduction

Safety and economical use of welded components depend to a large extent on reliable welding processing. Submerged arc welding (SAW) is the most important welding technique for creep-resistant steels like the low-alloyed 13CrMoV9-10 for thick-walled components in heavy petrochemical reactors (wall-thickness up to 475 mm) as well as fossil-fired power plants [1, 2]. Those large components are accompanied by significant restraints during welding fabrication, especially at positions of different thicknesses like nozzles. The resulting residual stress state is of immense importance, since highly restrained components tend to stress relief (or relaxation) cracks (SRC) during the final post weld heat treatment (PWHT). SRC phenomenon is determined by the interaction of the design, metallurgical and thermal factors as shown in Fig. 1.

Residual Stresses 2018 – ECRS-10   Materials Research Forum LLC
Materials Research Proceedings 6 (2018) 185-190   doi: http://dx.doi.org/10.21741/9781945291890-29

Metallurgical reasons for SRC are differences in strength within the matrix (re-precipitation of carbides at PAGs and the associated secondary hardening) and sufficient tensile stresses which lead to material failure. Those cracks are characterized by intergranular transverse and longitudinal cracks within the weld metal (WM) and the heat affected zone (HAZ). During the PWHT of the welded component, free expansion and shrinkage (relaxation) occur with corresponding local and global stressing of the weld seams. This final manufacturing step is indispensable to set the required material properties for later high-temperature use [3, 4]. Besides specific metallurgical factors, especially high tensile residual stresses are a considerable influence factor on SRC. For assessment of SRC susceptibility, mostly mechanical tests are applied which are isolated from the welding process. Conclusions regarding the influence of mechanical factors are rare so far. For that reason, it is currently not possible to realistically depict the interaction between the mechanical, thermal and metallurgical factors on a laboratory scale. For that purpose, geometry of laboratory specimens was adapted that residual stress state of realistic component-like weld can be simulated [5, 6]. Different stress conditions were set by varying the geometry and welding parameters. The residual stress distributions of a heavy slit sample (Spec A. - 3D-hindered shrinkage due to high restraint, cf. [6]), a medium U-shape notch specimen (Spec. B - high longitudinal shrinkage inhibition) and a small-scale specimen (Spec. C - external loading) were compared. The aim was to investigate if smaller specimens are suitable to represent longitudinal residual stresses as they occur in realistic SAW-manufactured components.

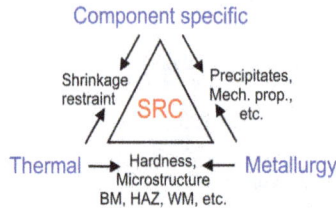

Fig. 1: Influencing factors on SRC [5]

Experimental
Submerged arc welding (SAW) experiments were carried out using the low-alloyed creep-resistant steel 13CrMoV9-10 [7]. The delivered plates had a thickness of 50 mm and three different specimen types Spec. A to C were machined. Specimen dimensions are shown in Fig. 2.

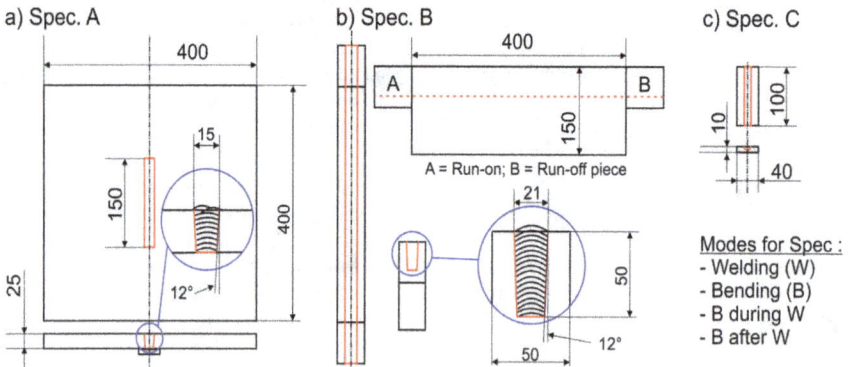

Fig. 2: Specimen types: (a) Spec A: Slit specimen, (b) Spec B: U-shape notch, (c) Spec C: Small-scale specimen

Residual Stresses 2018 – ECRS-10   Materials Research Forum LLC
Materials Research Proceedings 6 (2018) 185-190   doi: http://dx.doi.org/10.21741/9781945291890-29

The chemical composition of the base material (BM) and the deposited filler/weld metal (WM) are shown in Table 1 [5]. Additionally, it contains the mechanical properties. The corresponding welding parameters are shown in Table 2.

Table 1: Chemical composition of BM, deposited WM and mechanical properties [5]

	Content [wt.-%], Rest Fe									$R_{p0.2}$ [MPa]	R_m [MPa]	A_5 [%]
	C	Si	Mn	Cr	Cu	Mo	Ni	V	Nb			
BM	0.12	0.08	0.51	2.29	0.11	0.98	0.12	0.30	0.03	607	725	22
WM	0.07	0.08	1.23	2.27	0.04	0.96	0.16	0.23	0.01	415*	600*	18*

* Specifications from welding consumable supplier after PWHT at 705 °C for 4 h

Table 2: Layer sequence and welding parameters of Spec. A to C

Sample	Layer number	Load condition	Weld speed [mm/s]	Current [A]	Voltage [V]	Heat input [kJ/mm]
Spec. A	10	Self-restraint	6.0	600	30	~ 3.0
Spec. B	18	Self-restraint	6.0	650/500	34/31	~ 6.8
Spec. C	1	External load	2.0	210	15	~ 1.5

Spec. A and B were joint by SAW process (wire-Ø 4 mm: Spec. A: Single DC-mode and Spec. B: Tandem process DC+AC) with multi-layer technique and represent an approach for self-restraint weld joint samples in accordance to [5, 6]. A disadvantage for a self-restraint specimen is the necessary thickness. Hence, Spec. C was focused on inducing comparable mechanical stresses. For that reason, the weld heat input was simulated by high-energy TIG (tungsten inert gas) weld arc and an adapted test facility [8]. This facility offers a superimposed mechanical bending of the specimen perpendicular to the weld direction [8]. Four different conditions of Spec. C have been examined: (I) pure bending, (II) pure welding, (III) bending during welding and (IV) bending after welding. Conditions I and II were the reference conditions III and IV were used to simulate similar load conditions as occurred in self-restraint Spec. A and B but with significantly smaller specimen size (see Fig. 2). All specimens were in as-welded condition without further PWHT. The reason is that the initial residual stresses are important for SRC susceptibility. Hence, the as-welded condition must be compared prior to further PWHT.

The local longitudinal residual stresses were determined in each case by X-ray diffractometry (*Stresstech G3)* using the sin²ψ-technique [9]. The stresses were measured on the specimen's top surface across the weld (Spec. C only), the HAZ and the base material using a step size of 1 mm, further parameters can be found in [10].

Results and Discussion

Spec. A - Slit specimen. The longitudinal residual stress is shown in Fig. 3 (left). Tensile residual stresses up to 400 MPa were found in the HAZ approximately 3 mm from the fusion line. In transition to the weld metal the stresses tended to decrease but remained tensile. Note, that due to microstructural reasons no reliable stress determination was possible in the weld metal itself. Adjacent to the HAZ large stress gradients were found. With increasing distance to the HAZ the stresses quickly drop into compression around -100 MPa. With larger distance compressive values about -200 MPa were present in the base material. The same compressive stress level was found in transverse direction as indicated in Fig. 3 (right). In the HAZ still a compressive stress level was present. The transition into the weld metal was just characterized by -50 MPa, respectively 0. The stress distribution was not symmetric to the weld centre as the layers were deposited eccentrically. As a result, high tensile residual stress peaks were present in longitudinal direction in the HAZ. The transverse residual stresses are much lower [5, 11].

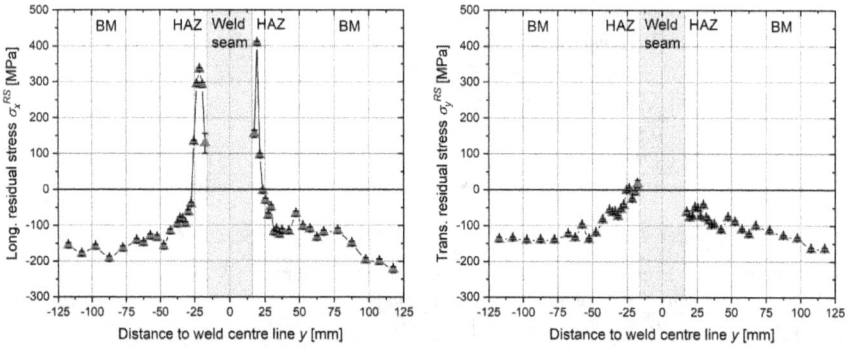

Fig. 3: Longitudinal (left) and transverse residual stresses (right) in Spec. A

Spec. B - U-shape notched specimen. The U-shape notched specimen had only a small width of 42 mm. The longitudinal residual stresses shown in Fig. 4 (left) indicate that the high tensile stresses are located on the edges of the fillets. The HAZ extends here about the whole width of the specimen. The stress level is comparable to the slit specimen with maximum values around 400 MPa. Towards the weld metal the stresses drop into compression. This gradient is larger than observed for the slit specimen. The low restraint acting in transverse direction causes compressive stresses in transverse direction, see Fig. 4 (right). Their level is even lower compared to the slit specimen. The special shape of the U-notched specimen leads to comparable longitudinal residual stresses as found in the heavy slit specimen. However, the transverse residual stresses are neglectable. High tensile residual stress maxima located in the HAZ are characteristic for transformation affected welds. Similar residual stress distributions are known from high-strength steel welds [6, 12, 13].

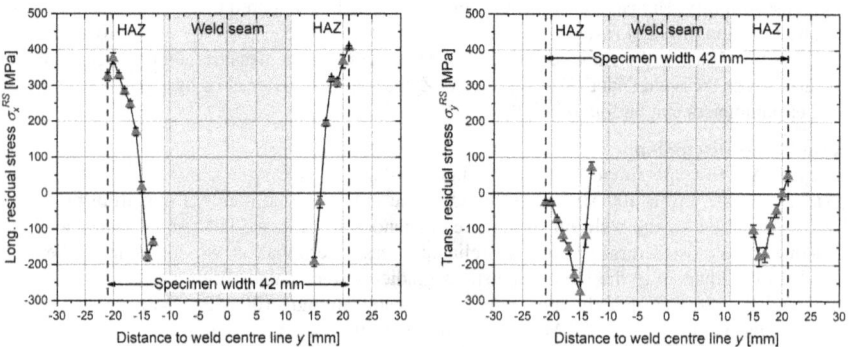

Fig. 4: Longitudinal (left) and transverse residual stresses (right) in Spec. B

Residual Stresses 2018 – ECRS-10   Materials Research Forum LLC
Materials Research Proceedings 6 (2018) 185-190   doi: http://dx.doi.org/10.21741/9781945291890-29

Spec. C - Small-scale Specimens. For testing of the small-scale specimens, a mechanical bending was additionally applied longitudinal to the welding direction. Fig. 5 shows the longitudinal residual stresses for the four different states derived from the experiments.

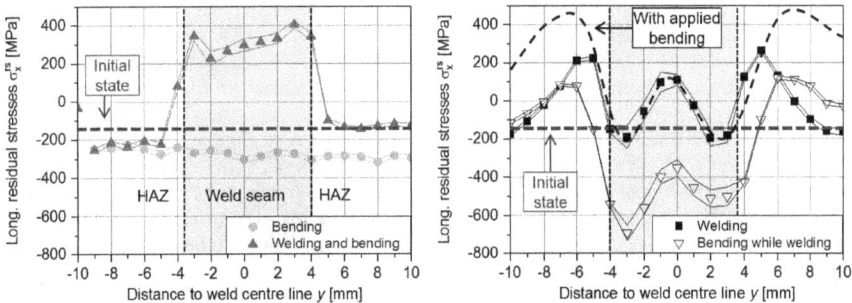

Fig. 5: Spec C - Longitudinal residual stress distributions for the four experimental conditions

While simply bending of the specimen, a plastic deformation of the top side of the specimen is observable. After release of the specimen from the testing machine a spring back occurs, which leads to higher compressive residual stresses (approx. -300 MPa) at the top side compared to the initial state of about -150 MPa.

Solitary welding of the specimen affects longitudinal stress distributions as typical for steel with undergoing phase transformation [12, 13]. The surface of the weld seam shows a stress maximum of 100 MPa at weld center and compressive stresses down to -200 MPa near the fusion line, which is due to the volume expansion of the phase transformation while cooling of the weld. The restraint of the cooler weld vicinity leads therefore to an increase of the residual stresses in the heat effected zone due to shoring effects. Tensile residual stress maxima of over 200 MPa can be observed in the HAZ.

The highest tensile residual stresses (400 MPa) can be observed in the weld seam, if the specimen is welded and subsequently bending is applied while cooling of the specimen. The transformation stresses are completely suppressed due to the superimposing bending stresses while cooling. Therefore, no shoring effects towards the weld vicinity were observed and the residual stresses in the HAZ are at the amount of the initial state.

In the last case the superposition of stresses at the weld top side due to additional bending at high temperatures while welding causes a plastic deformation with almost no resistance of the whole weld zone. After welding, subsequent cooling of the bended specimen leads to major evolution of welding residual stresses. At room temperature, the specimen is released from the testing machine as are the superimposed bending stresses, which affects a spring back of the specimen and a translation of the weld topside tensile residual stresses to compressive stress values from -700 to -350 MPa. Compared to the weld seam stresses without additional bending, the resulting residual stresses are approx. 400 MPa lower in the weld seam. The weld vicinity shows only stress differences from -100 to 100 MPa. However, if it is presumed that the translation of the stresses due to the spring back are constant over the whole specimen top side, the result would be high tensile residual stresses of 400 to 500 MPa in the HAZ due to the shoring effects, restraint shrinkage and bending stresses. This shows an important aspect for component relevant testing using small-scale specimen, especially in terms of weld cracking investigation.

Summary

The presence of high tensile residual stresses is a prerequisite for SRC during PWHT of creep-resistant CrMoV steels. The residual stresses resulting from the welding process are strongly determined by the boundary conditions resulting from the production of large welding components. Testing such scenarios on the laboratory scale is elaborate. In the present study, three different welding tests (Spec. A to C) were evaluated using XRD regarding the residual stresses that occur. The tests differ in size and complexity. In the heavy slit specimen (Spec. A) high tensile residual stresses were found in the HAZ in longitudinal direction. A special U-shaped notch specimen (Spec. B) also provided comparable residual stress levels in the HAZ. On the other hand, the transverse residual stresses in both tests were negligible. Small-scale bead-on-plate welds (Spec. C) gave similar residual stress distributions, but with lower peak values in the HAZ. Additional mechanical bending raised the stresses over the entire seam area to a level comparable to the other two tests. Thus, even this small sample can be used to a limited extent to simulate corresponding residual stresses with justifiable effort. Further modifications are planned.

References

[1] L. Antalffy, Metallurgical, design & fabrication aspects of modern hydroprocessing reactors, Weld. Res. Counc. Bull. 524 (2009) 77-115.

[2] K. Park et al., Post-weld heat treatment cracking susceptibility of T23 weld metals for fossil fuel applications, Mater. Des. 34 (2012) 699-706. https://doi.org/10.1016/j.matdes.2011.05.029

[3] K. Tamaki, J. Suzuki, H. Kawakami, Metallurgical factors affecting reheat cracking in HAZ of Cr-Mo steels. Proceedings of the Finnish-German-Japanese Joint International Seminar on Mechanical Approaches to New Joining Process (2004), 155-168.

[4] P. Nevasmaa, J. Salonen, Reheat Cracking Susceptibility and Toughness of 2% CrMoWVNb P23 Steel Welds, Weld. World. 52(3) (2008) 68-78. https://doi.org/10.1007/BF03266633

[5] T. Lausch, Zum Einfluss der Wärmeführung auf die Rissbildung beim Spannungsarmglühen dickwandiger Bauteile aus 13CrMoV9-10, BAM Dissertationsreihe 134, Berlin, 2015.

[6] D. Schroepfer, A. Kromm, T. Kannengiesser, Load analyses of welded high-strength steel structures using image correlation and diffraction techniques, Weld. World. 62(3) (2018) 459–469. https://doi.org/10.1007/s40194-018-0566-x

[7] DIN EN 10028-2, Flat products made of steels for pressure purposes - Part 2: Non-alloy and alloy steels with specified elevated temperature properties, German version (2017).

[8] T. Kannengiesser, T. Boellinghaus, Hot cracking tests-an overview of present technologies and applications, Weld. World. 58(3) (2014) 397-421. https://doi.org/10.1007/s40194-014-0126-y

[9] E. Macherauch, P. Müller, Das sin²ψ - Verfahren der röntgenografischen Spannungsmessung, Zeitschrift für angewandte Physik 13 (1961) 305-312.

[10] M. Rhode, A. Kromm, T. Kannengiesser, Residual stresses in multi-layer component welds, in: T. DebRoy et al. (Eds.), Trends in Welding Research: Proceedings of the 9th International Conference, ASM International, Materials Park (Ohio), 2013, pp. 48-54. https://doi.org/10.1016/j.jmatprotec.2013.01.008

[11] T. Lausch, T. Kannengiesser, M. Schmitz-Niederau, Multi-axial load analysis of thick-walled component welds made of 13CrMoV9-10, J. Mater. Process. Tech. 213(7) (2013) 1234-1240. https://doi.org/10.1016/j.jmatprotec.2013.01.008

[12] T. Nitschke-Pagel, H. Wohlfahrt, Residual stresses in welded joints - Sources and consequences, ECRS 6: Proceedings of the 6th European Conference on Residual Stresses, 2002, pp. 215-224.

[13] M. Farajian, R.C. Wimpory, T. Nitschke-Pagel, Relaxation and Stability of Welding Residual Stresses in High Strength Steel under Mechanical Loading, Steel. Res. Int. 81(12) (2010) 1137-1143. https://doi.org/10.1002/srin.201000194

Residual Stresses 2018 – ECRS-10
Materials Research Proceedings 6 (2018) 191-196

Materials Research Forum LLC
doi: http://dx.doi.org/10.21741/9781945291890-30

In-Situ Determination of Critical Welding Stresses During Assembly of Thick-Walled Components made of High-Strength Steel

Dirk Schroepfer[1,a,*], Arne Kromm[1,b], Andreas Hannemann[1,c] and Thomas Kannengiesser[1,d]

[1]Bundesanstalt für Materialforschung und -prüfung (BAM), Unter den Eichen 87, 12205 Berlin, Germany.

[a]dirk.schroepfer@bam.de, [b]arne.kromm@bam.de, [c]andreas.hannemann.de, [d]thomas.kannengiesser.de

Keywords: Residual Stress, Welding, Large-Scale Test, X-Ray Diffraction, HSLA Steel

Abstract. The performance and safety of welded high-strength low-alloyed steel (HSLA) components are substantially affected by the stresses occurring during and after welding fabrication, especially if welding shrinkage and distortion are severely restrained. The surrounding structure of the whole component affects loads in the far-field superimposing with welding stresses in the near-field of the weld. In this study a unique testing facility was used to restrain shrinkage and bending while analyse multiaxial far-field loads (max. 2 MN) during assembly of thick-walled component. A novel approach for the assessment of the in-situ-measured far-field data in combination with the actual weld geometry was elaborated. For the first time, analyses of the global bending moments of restrained welds based on the neutral axis of the actual weld load bearing section were achieved. Hence, far-field measurements offered the possibility to determine critical near-field stresses of the weld crosssections for the entire joining process. This work presents the approach for far-to-near field in-situ determination of stresses in detail for the 2-MN-testing system based on an extensive experimental work on HSLA steel welds, which demonstrates sources and consequences of these high local welding stresses. Thus, it was clarified, why the first weld beads are crucial regarding welding stresses and cold cracking, which is well known, but has never been measured so far. Accompanying analyses using X-ray diffraction (XRD) after welding show effects on local residual stress distributions. These analyses indicated viable prospects for stress reduction during assembly of thick-walled HSLA steel components.

Introduction

Residual stresses in welds are the result of the inhomogeneous volume changes due to welding process, cooling and material behaviour [1,2]. Especially in HSLA steels high tensile residual stresses are crucial due to the high yield ratio and low plastic strain reserves as well as the challenging and limited parameter range (i.e. cooling time) for the welding processes. An improved knowledge of evolution, distribution and magnitude of the arising residual stresses may obviate failure during manufacturing or even service. Welding process optimization considering stressing even enables enhancements of safety and performance (strength, toughness, fatigue, cracking resistance etc.) of HSLA steel welds and a sustainable, economic utilization of the material [2,3]. Studies and standardized weld tests involving stresses predominantly focus on local residual stresses in free shrinking laboratory size specimens [4]. Realistic heat conduction and the rigidity of the oftentimes thick-walled component welds are usually not taken into consideration. Generally, during component welding, shrinkage is hindered additionally to the local restraint in the weld and its vicinity as a result of global shoring effects and stiffness of the surrounding structure [5]. The arising far-field reaction forces and stresses by reasons of global restraint superimpose with local restraint stresses due to welding process and material behaviour, see Fig. 1a. Current studies reveal that the far-field loads may increase dependent on the stiffness of the structure, the welding process conditions and the applied filler and base materials, and critically prepossess also the near-field residual stress level in the weld seam [3,6].

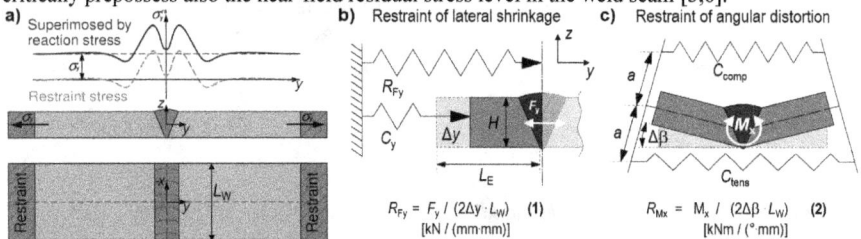

$$R_{F_y} = F_y / (2\Delta y \cdot L_W) \quad (1)$$
$$[kN / (mm \cdot mm)]$$

$$R_{Mx} = M_x / (2\Delta\beta \cdot L_W) \quad (2)$$
$$[kNm / (° \cdot mm)]$$

Fig. 1. Schematic of superposition of near and far-field loads (a) due to restraint of lateral shrinkage (b)/ angular distortion (c) with formula for restraint intensity factors R_{Fy} (Eq. 1)/ R_{Mx} (Eq. 2) [3].

For quantification for the stiffness of the weld seam towards the circumjacent construction, the restraint intensity factors R_{Fy} (lateral restraint, Fig. 1b) and R_{Mx} (bending restraint, Fig. 1c) based on the seam length L_W were established by [5]. Recently, numeric and experimental works dealt with the impact of restraint intensity on the residual stress state after welding and cooling of HSLA steel components [6–8]. However, numerical analysis of residual stress evolution in component-related welds are challenging and may lead to rash misinterpretation [9]. Besides, severe effects are also expected for the weld seam during the assembly, especially if highly restrained angular distortion occurs. Primarily, the first passes of HSLA steel welds are critical in the face of crack initiation due to high welding stresses. This phenomenon is well-known. Systematic measurements and detailed analyses to quantify these stresses and the influence of the boundary process parameters, material and realistic restraint conditions are absent, since in-situ analyses of local stresses while welding are not feasible directly by conventional testing methods. At BAM, a unique testing facility was developed [3,10], with which a defined restraint of both the lateral shrinkage and the angular distortion is achievable at the amount of real HSLA steel components. Simultaneously to the restraint, the multiaxial far-field loads can be analysed in-situ during welding experiments dependent on the applied welding heat control.

Experimental
Material and Welding Parameters. For the experiments, $H = 20$ mm thick plates of HSLA S690QL (EN 10025-6 [11]) were multilayer MAG-welded with matching solid wire of same type using typical parameters for HSLA steel application (i.e. mobile cranes), see Table 1. A pair of plates were clamped hydraulically into the testing facility and tack welded. Additional to the welding data, temperatures of the weld and weld vicinity were measured using thermocouples and a two-colour pyrometer. For comparison of XRD data and verification of required weld properties, free shrinking samples were welded as well. Cooling times ($t_{8/5}$) were appropriate to steel producer recommendations for test no. 1 to 4 ($t_{8/5} = 11$ to 13 s). All tested free shrinking specimens with adequate weld heat control had suitable weld properties, except test no. 5 due to $t_{8/5}$-times being much higher than recommended ($t_{8/5} > 20$ s).

Restraint Intensities. Fig. 2 shows a schematic of the testing facility and the XRD measurements at the clamped specimens. With the formulas for restraint intensity factors (Eq. 1 and Eq. 2) the rigidity of the test weld could be quantified and compared to typical values for restraint conditions in case of HSLA steel application, Table 2 [5,8]. The restraint intensity R_{Fy}

was discretised as shown in Eq. 3 for the test system $R_{Fy,2MN}$, from the experimental determined spring rigidity of the three-legged test frame (C_y) [10], and the plate/seam geometry $R_{Fy,12}$ calculated from the specimen stiffness [8]. The bending restraint R_{Mx} is the result of Eq. 4 with lever arms of $a = 230$ mm from the neutral axis of the specimen to the upper and lower piston rods of the testing facility. For the free shrinking specimens, the restraint intensity factors are both zero, since there is no stiffness of the weld against any surrounding structure. Note that due to bisect of the weld length L_W the lateral restraint intensity of the test setup would increase by 30 %, whereas the bending restraint intensity would almost double.

Table 1. Chemical composition (FES, Fe balanced), mechanical properties of the 20 mm thick test plates and welding parameters of the weld tests.

Element in %	C	Si	Mn	Cr	Mo	Ni	V	Nb	Ti
S690QL	0.14	0.71	1.15	0.03	0.17	0.07	0.009	0.005	0.010
Property	$R_{p0.2}$ in MPa		R_m in MPa		A_5 in %		Av in J at -40 °C		HV10
S690QL	768		821		19		198		270
Seam geometry	W.current	W.voltage		W.speed		Wire feed speed	W.process		Preheating T_p
V-type butt joint, 45°	255 ± 25 A	25 ± 5 V		245 to 340 mm min⁻¹		6.5 to 8.0 m min⁻¹	MAG		120 °C
Weld test no.				1	2 (repeat)		3	4	5 (high $t_{8/5}$)
Interpass temp. T_i [°C] / Heat input E [kJ mm⁻¹]				150 / 1.4	150 / 1.4		100 / 1.7	200 / 1.1	200 / 1.7

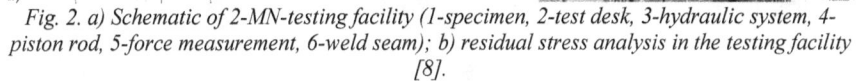

Fig. 2. a) Schematic of 2-MN-testing facility (1-specimen, 2-test desk, 3-hydraulic system, 4-piston rod, 5-force measurement, 6-weld seam); b) residual stress analysis in the testing facility [8].

Table 2. Restraint intensity factors for different test setups ($C_{y,tens,cyl}$ = 407.3 kN mm⁻¹; $C_{y,comp,cyl}$ = 459.0 kN mm⁻¹).

Test setup	Weld length	Plate length	System restraint	Specimen restraint	Lateral restraint	Bending restraint
-	L_W [mm]	L_E [mm]	$R_{Fy,2MN}$[kN(mmmm)⁻¹]	$R_{Fy,12}$[kN(mmmm)⁻¹]	R_{Fy}[kN(mmmm)⁻¹]	R_{Fy}[kNm/(°mm)⁻¹]
Free shrinking	200	150	0	0	0	0
2-MN-test fac.	200	330	6.1	5.8	3.0	8

$$R_{Fy} = (R_{Fy,2MN} \cdot R_{Fy,12}) / (R_{Fy,2MN} + R_{Fy,12}). \qquad (3)$$

$$R_{Mx} = \{a^2 \cdot \tan\Delta\beta \cdot (C_{y,comp,cyl} + C_{y,tens,cyl})\} / \{2\Delta\beta \cdot L_W\}. \qquad (4)$$

Results

Global Forces. Fig. 3a depicts reaction forces $F_y(t)$ while preheating, welding and cooling of the two weld tests no. 1 and 2, welded under the same boundary conditions, welding parameters and recommended heat control (medium T_i and E). As a result, both graphs are qualitatively and quantitatively almost equal with a little scatter range in-between. After some compression while preheating, during root welding, the already solidified inserted weld metal produces transversal shrinking forces, which increase further to a first maximum of $F_y = 62$ kN and $M_x = 0.5$ kNm during cooling to $T_i = 150$ °C. Forces decrease while welding of the second layer due to the local heat input combined with stress relief. Subsequent cooling to T_i leads to a growth of reaction force, the next weld run to a transient reduction. This evolution repeats for each weld sequence. Reaction force amplitudes increase with each weld layer due to welding heat input. Subsequent

Residual Stresses 2018 – ECRS-10   Materials Research Forum LLC
Materials Research Proceedings 6 (2018) 191-196   doi: http://dx.doi.org/10.21741/9781945291890-30

cooling to room temperature (RT) after the cap bead leads to a maximum force of $F_{y,end}$ = 535 kN.

Bending Moments. Restraint angular distortion of welds cause a disequilibrium of reaction forces over the plate thickness direction, resulting in a reaction moment M_x around the neutral axis of the weld in x- direction [8].

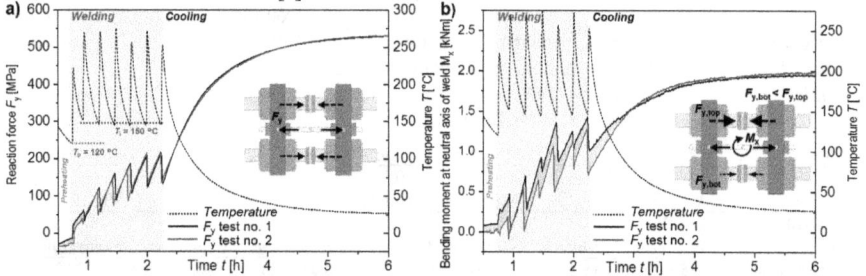

Fig. 3. Reaction forces $F_y(t)$, temperature $T(t)$ (a) and Bending moments $M_x(t)$ (b) of two weld tests with same parameters, 2-MN-testing facility (T_p = 120 °C, T_i = 150 °C, E = 1.4 kJ mm^{-1}).

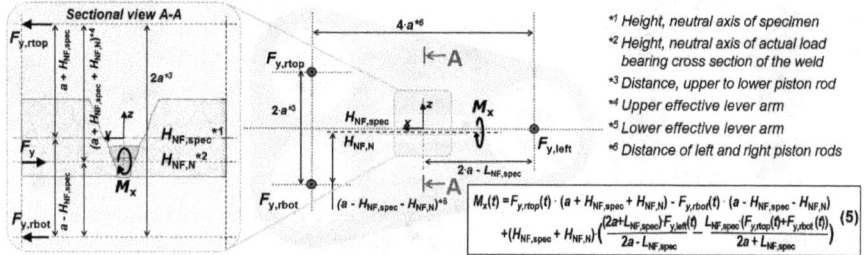

Fig. 4. Determination of the bending moments $M_x(t)$ based on the neutral axis of the actual load bearing cross section of the weld in the 2-MN-testing facility [8].

For detection of this bending moment M_x while welding, the height of the neutral axis of the specimen $H_{NF,spec}$ and of the actual load bearing cross section $H_{NF,N}$ of the weld have to be known, see Fig. 4. With the determined upper and lower lever arms of the proof frame and a compensation regarding the neutral axis of the specimen towards the left piston rod and force measurements of the hydraulic system, finally, the actual bending moment $M_x(t)$ can be calculated, using Eq. 5, shown exemplarily for the two medium parameter weld tests (no. 1 and 2) in Fig. 3b. It is obvious that the $M_x(t)$-graphs are qualitatively equal to the $F_y(t)$-graphs, since the force above the neutral axis of the weld is ever higher than below. The scatter band between the two $M_x(t)$-graphs of weld test no. 1 and 2 is slightly broader compared to the $F_y(t)$-graphs, which reflects the higher susceptibility to errors, i.e. due to metallographic analyses of the weld layers at metallographic cross section. The maximum bending moment of both weld tests occur after cooling to RT at approx. $M_{x,end}$ = 2 kNm.

Far-field Loads. Reaction forces due to welding result in global stresses at the weld seam, which can be calculated and in-situ analysed, see Fig. 5, cf. Fig. 3. Constant forces over the weld seam thickness lead to normal reaction stress. The magnitude depends on the load bearing section of the weld, see Fig. 5a. Transient stress increases significantly while cooling to interpass temperature already after root welding due to the comparable small load bearing cross section area. It could be observed that the heat input and phase transformation of the weld metal at the next weld runs leads to a transient stress reduction and a new stress increase during cooling to T_i.

Residual Stresses 2018 – ECRS-10   Materials Research Forum LLC
Materials Research Proceedings **6** (2018) 191-196   doi: http://dx.doi.org/10.21741/9781945291890-30

Compared to the $F_y(t)$-graphs, there is, hence, a smaller difference between the stress levels while welding and after cooling to RT, especially if low interpass temperatures were applied [3,8].

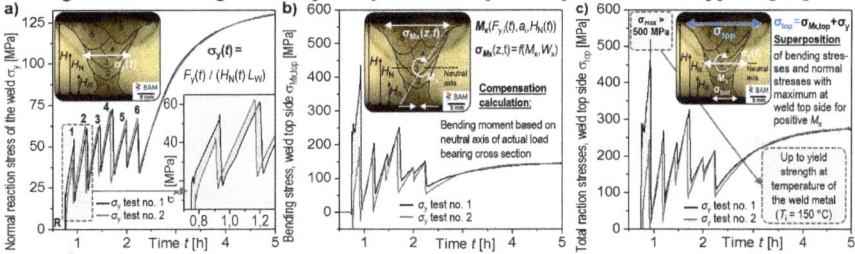

Fig. 5. Normal stress $\sigma_y(t)$ (a), bending stresses $\sigma_{Mx}(t)$ (b), total stress at weld top side $\sigma_{top}(t)$ (c) of two weld tests with same parameters, 2-MN-testing facility ($T_p = 120\ °C$, $T_i = 150\ °C$, $E = 1.4\ kJ\ mm^{-1}$).

$$\sigma_{Mx}(t) = M_x(t) / W_x \qquad \text{with } W_{x} = L_W H^2/6. \tag{6}$$

Near-field Loads. A positive bending moment arising during welding and cooling of the restrained weld, causes bending stresses, which concentrate on the top side of the weld. With in-situ bending moment analyses based on the neutral axis of the load bearing cross section the calculation of the actual local bending stresses on the weld top side $\sigma_{Mx}(t)$ was possible for the first time for the whole welding process of HSLA steels, see Eq. 6 and Fig. 5b. A superposition with the normal reaction stresses leads to total reaction stresses at the top side of the weld, Fig 5c. It was observed that especially cooling of the root weld already at medium interpass temperature leads to high stresses exceeding 500 MPa, the yield strength at temperature for the microstructure of the weld metal, and the total stresses of the complete weld after cooling to RT. Also, the superimposing bending stresses of approx. $\sigma_{Mx} = 150$ MPa cause almost 300 MPa total end reaction stresses $\sigma_{top,end}$.

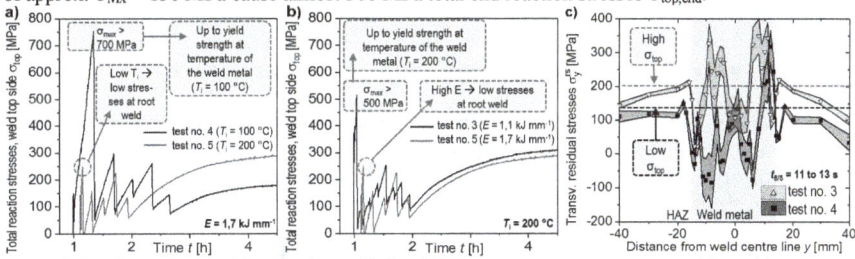

Fig. 6. Total stress at weld top side $\sigma_{top}(t)$ for different interpass temperature (a) and heat input (b); comparison of transverse residual stress distributions of two weld tests with different total stress levels.

Heat control affects the arising welding stresses significantly, see Fig. 6. A low interpass temperature leads to lower end reaction stresses, as observed also in [3,8,10], b ut high total stresses at the top of the root weld due to longer cooling phases compared to higher T_i, with stresses exceeding the yield strength of the weld metal. A lower heat input at high T_i results also in higher root stresses reaching the yield strength at temperature due to a smaller root section area. Somewhat higher total end reaction stresses occur. XRD analysis of the local residual stresses depicts Fig. 6c for the two welds with varied heat control and appropriate $t_{8/5}$-time (11 to 13 s, test no. 3 and 4). The welding stress level is significantly increased if high T_i and low E (test no. 3) were applied. Especially in the HAZ the increase is at the amount of the difference between the total end reaction stresses at the top of the weld (approx. 100MPa), cf. Fig. 6a and b.

Residual Stresses 2018 – ECRS-10   Materials Research Forum LLC
Materials Research Proceedings 6 (2018) 191-196   doi: http://dx.doi.org/10.21741/9781945291890-30

Summary

With a unique testing facility at BAM, component related weld test of a HSLA steel were performed using application related parameters and boundary conditions. Multiaxial far-field loads (max. 2 MN) during multilayer welding could be analysed. Special compensation calculation based on the neutral axis of the actual load bearing section allowed for the first time in-situ analyses of local welding stresses on the top of the weld. Coming from medium values for heat control, heat input and working temperature should be adapted together to reach recommended cooling time values of the weld. It could be shown that low heat and high working temperature lead to significantly higher welding stresses. Low working temperature with increased heat input enables low welding stresses and low local residual stresses especially in the HAZ, but leads to high welding stresses during root weld's cooling to low interpass temperature, which is crucial for the whole welding process and, hence, weld integrity. Special procedures for root welding to prevent high root stresses could be: ductile filler materials, adapted working temperature strategy or higher deposition rate as well as adapted seam geometry to increase the cross section area and section modulus of the root [8]. Further investigations of such adapted concepts would help to enhance performance of HSLA steel welds and further utilization of the materials.

References

[1] T. Nitschke-Pagel, H. Wohlfahrt, The Generation of Residual Stresses due to Joining Processes, in: V. Hauk, H. Hougardy, E. Macherauch (Eds.), Residual Stress. - Meas. Calc. Eval., DGM Informationsgesellschaft mbH, 1991, pp. 121–133.

[2] P.J. Withers, H.K.D.H. Bhadeshia, Residual Stress. Part 2 - Nature and Origins, Mater. Sci. Technol. 17 (2001) 366–375. https://doi.org/10.1179/026708301101510087

[3] D. Schroepfer, A. Kromm, T. Kannengiesser, Engineering approach to assess residual stresses in welded components, Weld. World. 61 (2017) 91–106. https://doi.org/10.1007/s40194-016-0394-9

[4] T. Kannengiesser, T. Boellinghaus, Cold cracking tests - An overview of present technologies and applications, Weld. World. 57 (2013) 3–37. https://doi.org/10.1007/s40194-012-0001-7

[5] K. Satoh, Y. Ueda, H. Kihara, Recent Trends of Research into Restraint Stresses and Strains in Relation to Weld Cracking, Weld. World. 11 (1973) 133–156.

[6] D. Schroepfer, A. Kromm, T. Kannengiesser, Load analyses of welded high-strength steel structures using image correlation and diffraction techniques, 62 (2018) 459–469.

[7] M. Hirohata, Y. Itoh, Effect of Restraint on Residual Stress Generated by Butt-welding for Thin Steel Plates, in: 9th Ger. Bridg. Symp. Kyoto, Japan, 2012, pp. 1–6.

[8] D. Schröpfer, Adaptierte Wärmeführung zur Optimierung schweißbedingter Beanspruchungen und Eigenschaften höherfester Verbindungen, Dissertation, OVGU Magdeburg, Shaker Verlag Aachen, 2017.

[9] J. Klassen, T. Nitschke-Pagel, K. Dilger, Challenges in the Calculation of Residual Stresses in Thick-walled Components, in: Mater. Res. Proceedings, Residual Stress. 2016 ICRS-10, Materials Research Forum, 2016: pp. 299–304.

[10]T. Lausch, T. Kannengiesser, M. Schmitz-Niederau, Multi-Axial Load Analysis of Thick-Walled Component Welds made of 13CrMoV9-10, J. Mater. Process. Technol. 213 (2013) 1234–1240. https://doi.org/10.1016/j.jmatprotec.2013.01.008

[11]EN 10025-6: Hot rolled products of structural steels – Part 6: Technical delivery conditions for flat products of high yield strength structural steels in the quenched and tempered conditions, 2018.

Residual Stresses 2018 – ECRS-10   Materials Research Forum LLC
Materials Research Proceedings 6 (2018) 197-202   doi: http://dx.doi.org/10.21741/9781945291890-31

Reviewing the Influence of Welding Setup on FE-Simulated Welding Residual Stresses

Stefanos Gkatzogiannis[1,a,*], Peter Knoedel[1,b] and Thomas Ummenhofer[1,c]

[1]KIT Steel & Lightweight Structures, Research Center for Steel, Timber & Masonry, Otto-Ammann-Platz 1, D-76131 Karlsruhe, Germany

[a]stefanos.gkatzogiannis@kit.edu, [b]peter.knoedel@kit.edu, [c]thomas.ummenhofer@kit.edu

Keywords: Welding Residual Stresses, FE Simulation, Boundary Conditions, Heat Input, Material Model, Aluminum, Steel

Abstract. Previous and new simulations of welding residual stresses with the finite element method are reviewed in the present study. The influence of modelling mechanical boundary conditions, erroneous prediction of the weld heat source coefficient and the influence of microstructural changes in aluminum welds are investigated. The results are analyzed so that concrete suggestions regarding the investigated factors, acting as guidance to the practitioner, can be presented.

Introduction

Finite element (FE) simulation of welding has evolved rapidly in the last decades, although the major goal has ever since remained the same, i.e. the calculation of welding residual stresses (WRS) and plastic strains. The possibility to take into consideration even more effects of real welding in the simulation increased therewith as well. Nevertheless, welding is a complex multiphysics process and many different phenomena (electromagnetic, thermal, mechanical etc.) take place, as it was thoroughly described by Francis et al. [1]. A new challenge has therefore arisen for modern engineers: take into consideration only those factors who exhibit a non-negligible influence on welding residual stresses, depending on the desired preciseness level of each investigated case, in order to balance computational cost and enable better overview of the modelling approach by the practitioner. Straightforward modelling approaches can provide precise results, as long a proper choice of the factors to be taken into consideration is made [2].

Lindgren proposed a concept of filtering simulated factors based on the desired precision [3], from down to basic up to very accurate levels of simulation, by making, nevertheless, a qualitative and not quantitative evaluation. The influence of each simulated factor on the welding residual stresses can differ in each investigated case. For example, the influence of clamped edges, which is proven to have a significant effect on the transverse WRS [4] of butt-welds, is expected to be less significant in the case of fillet welds. Therefore, an investigation of the influence of individual factors on WRS is necessary for a further improvement of welding simulation techniques.

The present study reviews previous and new FE simulations in order to evaluate the influence of selected factors of welding simulation. A review of selected welding parameters and modelling approaches, regarding their influence on WRS, is carried out, with the objective to offer guidance to the practitioner.

Theoretical Background

As already previously discussed, welding is a multiphysics problem, but simulation of WRS requires modelling only of some aspects of the welding process. The thermal behavior and possible microstructural changes are investigated along with the mechanical behavior of a component when WRS are under the scope. These three fields, which are presented in Fig. 1 are interacting bidirectionally with each other. Practical models, predicting the WRS with satisfying preciseness by only simulating the interaction of thermal on microstructural and mechanical behavior and of microstructural field on the mechanical behavior have been developed [2]. The considered interactions are presented in Fig. 1. Ignoring the reverse influence of microstructural and mechanical behavior enables solution of the problem with a sequential unidirectionally coupled thermal–mechanical analysis [2]. A description of the numerical background is avoided in the present case for sake of space, but it is thoroughly provided in previous work of the authors [2].

Modelling of the thermal field includes the simulation of the weld heat source, the heat transfer inside the investigated component and heat losses to the surrounding environment (boundary conditions). The Goldak's double ellipsoidal is the state-of-the-art approach for modelling the weld heat source [5]. The effective heat input is predicted by multiplying the power of the heat source with a coefficient of the weld heat source. Proposed values of this coefficient for various weld types are proposed in [6]. Nevertheless, different values usually in a range of ±10 % for same welding types are found elsewhere (see [7] etc.). Heat transfer inside the component is governed by three dimensional Fourier's law of heat conduction. Heat losses are modelled according to Newton's law of cooling. A transient thermal analysis is carried out, whereby the temperature history of the nodes of the FE model are saved at each solution step.

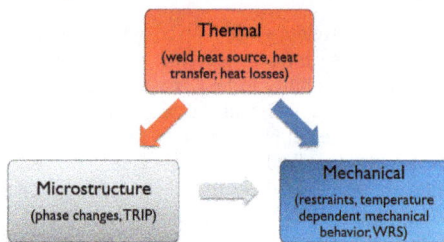

Fig. 1: Investigated fields and respective interactions in an engineering approach for arc welding simulation, TRIP stands for transformed induced plasticity [2]

Regarding the modelling of the microstructural changes that take place, different models have been proposed in the past. A straightforward engineering approach, which was proposed by the authors of the present study [2], provided results with sufficient preciseness and can be applied for the simulation of various materials [8], [9], [10]. Main feature of the approach lies on the assignment of material models to the finite elements inside the fusion zone (FZ) and the heat-affected zone (HAZ) during cooling down based on predominant parameters of the thermal cycle, in order to simulate the modified mechanical behavior of the transformed microstructure. The selection of material models is based on predictions of the transformed microstructure according to continuous cooling transformation (CCT) diagram or assessment of microstructure through measurements. The importance or negligibility of microstructural changes during modelling of ferritic or austenitic steels respectively has been confirmed in the past [2], [9]. Softening in the HAZ of aluminum alloys was as well successfully modelled elsewhere [8]. Nevertheless, a quantification of this influence on WRS is not known to the authors of the present study. The same problem arises as well when high strength steels are regarded. Liu et al. [11] and Lee and Chang [12] simulated weldments by ASTM A514 (yield strength of 717 MPa) with phase changes and a high strength carbon steel with yield strength of 790 MPa by neglecting them respectively.

Residual Stresses 2018 – ECRS-10   Materials Research Forum LLC
Materials Research Proceedings 6 (2018) 197-202   doi: http://dx.doi.org/10.21741/9781945291890-31

Finally, during modelling of the mechanical field, the nodal temperature history from the transient thermal analysis is applied as external thermal strains. Solution takes place in a static structural analysis, whereby temperature dependent material parameters and the restraints applied to the real welded component are modelled [10]. Usually, components are clamped down during welding. Fixing of the respective nodes in the FE models is a common approach of modelling but deviates from physical restraining reality. An alternative approach applied elsewhere [10] is the use of linear spring elements, restraining the respective nodes (see Fig. 4). Further steps of the mechanical solution are based on classical finite element theory for nonlinear materials.

Investigated Factors

Results from previous and new FE analyses, which were all carried out according to the above-described theoretical background, are presented in the current study in order to review specific aspects of weld simulation. FE commercial software ANSYS [13] was applied in all cases. Solid 8 node elements "solid 90" and "solid 180" were applied for the transient thermal and the static structural analyses respectively in all cases. The following aspects are investigated:

- Modelling of mechanical boundary conditions (BC): A single-pass weld component of Swedish steel HT36 (equivalent to S355) with dimensions of 2000 mm x 1000 mm x 15 mm was simulated in [10] (component A in the present study). Geometry of the weld section is presented in Fig. 2. The component was welded with submerged-arc welding (three electrodes welding consecutively), an electric power of 98 kW and a welding speed of 25 mm/s (150 cm/min; 3.92 kJ/mm heat input). Applied material parameters are given in [2]. Clampers were modelled either by fixing the respective nodes in all directions or by applying linear spring elements with stiffness of 10^6 N/mm to all nodes in the clamped area in the longitudinal and transverse direction and fixing their vertical displacement. The results are compared with measurements for an identical component found elsewhere [7].

- Thermal input and welding sequence of multi-pass welds: Influence on the WRS, which can be caused by possible erroneous modelling of thermal heat input is investigated in [9]. The 5-pass X-grooved butt-weld with dimensions 300 mm x 300 mm x 10 mm by AISI 316L, which is presented in Fig. 2, was simulated (component B in the present study) and different cases of differentiating heat input were modelled. The component was initially considered to be welded with an electric power of 4.3 kW, a welding speed of 5 mm/s (30 cm/min; 0.86 kJ/mm heat input) and the welding sequence A-B-C-D-E. Applied material parameters are given in [9]. Effective heat input was changed by either increasing the welding speed, or reducing the heat input (see Table 1). The results are reviewed from a different scope in the present study in order to estimate the possible error of calculated WRS under the assumption that an under- or overestimation of ± 10 % of the weld heat source's efficiency has taken place.

- Microstructural modelling: Numerical studies regarding welding of Aluminum alloy initially presented at [8], are repeated in the present study, by neglecting this time the phase transformations, in order to evaluate qualitatively and quantitatively the influence of recrystallization in the HAZ of aluminum alloys on the simulated WRS. A single-pass V grooved component of EN AW-6060 welded with an electric power of 3.1 kW and a welding speed of 10 mm/s (60 cm/min; 0.31 kJ/mm heat input) is modelled.

Results and Discussion

The calculated longitudinal WRS for the HT36 component A are presented in Fig. 4. "Meas." stands for measured WRS found in [7], while "Fixed" and "Springs" are referring to the respective applied modelling approach of BC. In both cases, tensile longitudinal stresses, higher than the nominal yield limit of the material at room temperature are met in the FZ and HAZ and

Residual Stresses 2018 – ECRS-10   Materials Research Forum LLC
Materials Research Proceedings **6** (2018) 197-202   doi: http://dx.doi.org/10.21741/9781945291890-31

compressive stresses in the areas away from the weld. In the case of modelling with spring elements, tensile stresses of even up to 700 MPa are met (yield strength in the weld area is higher than the nominal 355 MPa due to the phase changes taken place – increase of bainitic and martensitic phases), while for the fixed case the maximum tensile stress is not higher than 450 MPa. Although both approaches produce similar results qualitatively, the agreement of the spring model with the measurements inside the weld section is clearly better. Similar improvement due to use of spring elements was observed as well in the case of calculated transverse WRS [10].

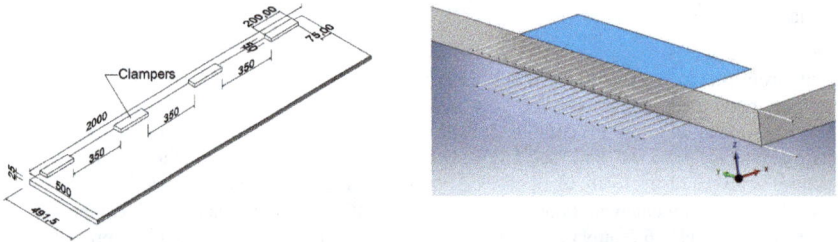

Fig. 2: Investigated component A – left: Geometry and clampers, dimensions are given in mm – right: Applied spring elements for the simulation of clampers

The calculated longitudinal WRS for the AISI 316L component B are presented in Fig. 4. Similar profiles of longitudinal WRS are calculated in all investigated cases (see Table 1). Tensile and compressive WRS are met near and away from the weld respectively. The increase of the heat input rate, either by increase of heat input or reduction of welding speed, shifts down the calculated profile of WRS. Nevertheless, the largest deviation from the initial profile (H1) is met in the case of reduced welding speed down to 50 % (case V2). A difference of up to 150 MPa or 28 % between the peak tensile stresses of those two cases is observed. Similar but not so significant differentiation of the transverse WRS profiles was observed as well [9].

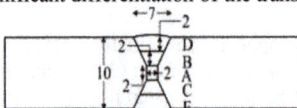

Fig. 3: Investigated component B, dimensions are given in mm

Table 1: Reviewed investigated cases found in [9]		
Investigated case	Model	Investigated variable
welding parameters	H1	initial welding parameters
	V1	welding speed – 25 % reduced
	V2	welding speed – 50 % reduced
	Q1	heat input – 25 % increased
	Q2	heat input – 50 % increased

The calculated WRS for the EN AW 6060 component C are presented in Fig. 5 for the cases of taking into consideration and neglecting the recrystallization in the HAZ. In the case of the longitudinal WRS the calculated profile differs significantly, both qualitatively and quantitatively, near the weld. Neglecting the microstructural changes, leads to a lower calculated peak stress and a shift of this peak stress from the boundaries between HAZ and parent material in the middle of the component. In the case of the transverse WRS, neglecting recrystallization causes a shift up of the calculated profile of up to 100 MPa.

Conclusions

The following conclusions were drawn, based on the above-presented results:

- Applied approach for modelling of clampers during a weld simulation can have a significant effect on the calculated WRS, at least in the case of butt-welds. The use of spring elements for modelling the longitudinal and transverse restraints in the clamped area produces results

Residual Stresses 2018 – ECRS-10   Materials Research Forum LLC
Materials Research Proceedings 6 (2018) 197-202   doi: http://dx.doi.org/10.21741/9781945291890-31

that show better agreement with experimentally measured WRS. Moreover, as higher tensile WRS are calculated, this approach lies on the safe side. This modelling approach is therefore suggested for adoption in practical applications as well.

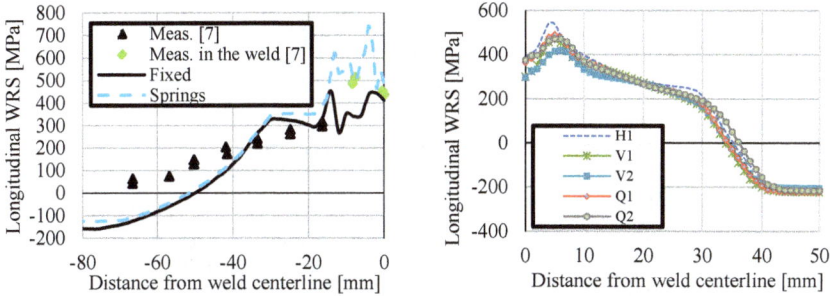

Fig. 4: Longitudinal WRS – left: Component A – right: Component B, resulting WRS on the top of the component, for the investigated cases presentes in Table 1

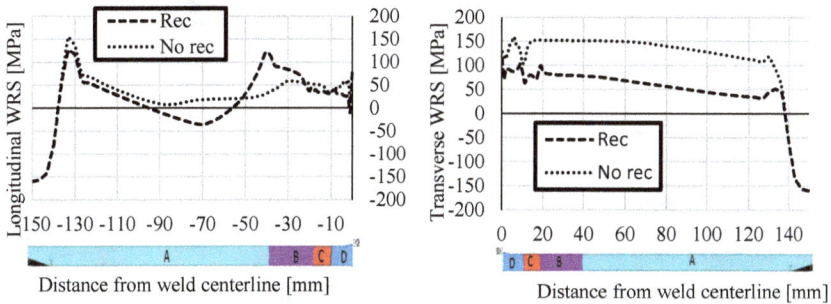

Fig. 5: WRS of the alluminum component C with and without modelling of recrystilisation in the HAZ, collored areas of the component show modelling of the microstructural changes, A: parent material, B: HAZ 50 % recrystalized, C: HAZ 100 % recrystlized, D: FZ.

- A possible erroneous modelling of the weld heat source could lead to an underestimation of the WRS and therefore non-conservative results. An increase of 50 % of the heat input rate led to a reduction of 28 % of the peak tensile stress (worst case scenario). Assuming a linear behavior, an overestimation of 10 % of the heat source coefficient would lead to an underestimation of 5.6 % of the peak tensile stresses. This deviation lies in the boundaries of the acceptable numerical error, of practical weld simulations (± 10 % [2]). Therefore, applying values for the coefficient of weld heat source from literature or previous measurements during the simulation of WRS, is considered valid.
- Aluminum welding simulations neglecting the recrystallization in the HAZ, produce erroneous results. A lower longitudinal peak stress is calculated and is present at the middle of the component. Simulations considering the microstructural changes show that the peak stress is met on the boundaries of the HAZ, exhibiting a much more crucial case, regarding fatigue strength of the investigated component. Therefore, neglecting recrystallization in the HAZ of aluminum weld during weld simulation is not conservative and should be avoided.

Acknowledgement
The above-presented work was carried out as a part of the framework of a PhD thesis at Karlsruhe Institute of Technology [16].

References

[1] J.A. Francis, H.K.D.H. Bhadeshia, P.J. Withers, Welding residual stresses in ferritic power plant steels, Material Science and Technology 23 (2007) 1009-1020. https://doi.org/10.1179/174328407X213116

[2] P. Knoedel, S. Gkatzogiannis, T. Ummenhofer, Practical aspects of welding residual stress simulation, Journal of Constructional Steel Research 132 (2017) 83–96. https://doi.org/10.1016/j.jcsr.2017.01.010

[3] L.-E. Lindgren, Computational Welding Mechanics - Thermomechanical and Microstructural Simulations, Woodhead Publishing in Materials, first ed., Cambridge England, 2007. https://doi.org/10.1201/9781439824092

[4] S. Kou, Welding Metallurgy, John Wiley & Sons, Inc., second ed., Hoboken, New Jersey, 2003.

[5] J.A. Goldak, A. Chakravarti, M. Bibby, A new finite element model for welding heat sources, Metall. Trans. B 15 (1984) 299–305. https://doi.org/10.1007/BF02667333

[6] J.N. Dupont, A.R. Marder, Thermal efficiency of arc welding processes, Weld. J. 74 (1995) 406–416.

[7] B. Andersson, Thermal Stresses in a submerged-arc welded joint considering phase transformations, Trans. ASME 100 (1978) 356–362. https://doi.org/10.1115/1.3443504

[8] P. Knoedel, S. Gkatzogiannis, T. Ummenhofer, FE simulation of residual welding stresses: Aluminum and steel structural components, Key Engineering Materials 710 (2016) 268-274. https://doi.org/10.4028/www.scientific.net/KEM.710.268

[9] S. Gkatzogiannis, P. Knoedel, T. Ummenhofer, Influence of welding parameters on the welding residual stresses, Proceedings of the VII International Conference on Coupled Problems in Science and Engineering, Rhodes Island, Greece, June 12–14 (2017) 767–778.

[10] S. Gkatzogiannis, P. Knoedel, T. Ummenhofer, FE welding residual stress simulation - Influence of boundary conditions and material models. EUROSTEEL 2017, September 13–15, 2017, Copenhagen, Denmark, (2017), Ernst & Sohn Verlag für Architektur und technische Wissenschaften GmbH & Co. KG, Berlin.

[11] W. Liu, J. Ma, F. Kong, S. Liu, R. Kovacevic, Numerical modeling and experimental verification of residual stress in autogenous laser welding of high-strength steel, Lasers Manuf. Mater. Process. 2 (2015) 24–42. https://doi.org/10.1007/s40516-015-0005-4

[12] C.H. Lee, K.H. Chang, Prediction of residual stresses in high strength carbon steel pipe weld considering solid-state phase transformation effects, Computers and Structures 89 (2011) 256–265. https://doi.org/10.1016/j.compstruc.2010.10.005

[13] ANSYS® Academic Research, Release 18.2, Help System, ANSYS, Inc., (2018).

[14] S. Gkatzogiannis, Finite Element Simulation of High Frequency Hammer Peening, Ph.D. thesis (in progress), KIT, Karlsruhe Institute of Technology, Department of Civil, Geo and Environmental Sciences, KIT Steel & Lightweight Structures, 2018.

Residual Stresses 2018 – ECRS-10
Materials Research Proceedings 6 (2018) 203-208

Materials Research Forum LLC
doi: http://dx.doi.org/10.21741/9781945291890-32

Effect of Dengeling on Bending Fatigue Behaviour of Al Alloy 7050 and Comparison with Milling and Shot Peening

Ru Lin Peng1[,a,*], Linnéa Selegård[2,b], Mattias Jonsson[2,c], Markus Ess[3,d]
Gert Petersén[2,e]

[1]Dept of Management and Engineering, Linköping University, SE-581 83 Linköping, Sweden

[2]Saab, S-58188 Linköping, Sweden

[3]Starrag AG, Seebleichestrasse 61, 9404 Rorschacherberg, Switzerland

[a]ru.peng@liu.se, [b]linnea.selegard@saabgroup.com, [c]mattias.jonsson2@saabgroup.com, [d]markus.ess@starrag.com, [e]gert.petersen@saabgroup.com

Keywords: Dengeling, Shot Peening, Milling, Bending Fatigue, Al Alloy

Abstract. Dengeling is a new surface mechanical treatment developed as an alternative to the shot peening method used for enhancing the fatigue resistance of metallic materials. In this work, Dengeling is compared with milling and shot peening with regard to the effect on bending fatigue behavior of aluminium alloy AA 7050 T7651. In addition, the influence of certain Dengeling process parameters on the fatigue resistance is studied. Flat bar samples were milled and then subjected to the respective surface treatments. The induced surface integrity changes, namely residual stresses and surface deformation, were characterized by X-Ray diffraction measurement. Four-point bending fatigue tests with a stress ratio of R=0.1 were performed. The results show that all the surface treatments in general improve the fatigue performance of the milled samples but the samples treated by the Dengeling process with similar Almen intensity as the shot peening treatment perform best.

Introduction

Surface mechanical treatments especially shot peening has been developed and widely used by various industries including automative and aerospace to enhance fatigue properties of metallic components [1,2]. The increased fatigue resistance is attributed to beneficial compressive residual stresses and strain hardening induced in a surface layer due to impact of hard shots. Dengeling is a new surface mechanical treatment developed as an alternative method to shot peening. The treatment is carried out by striking the metal surface with a hard indenter to induce surface plastic deformation and compressive residual stresses. Dengeling can be performed on the same machine that is used for machining the component and the operator can control exactly the location and magnitude of the residual stress.

In a previous study [3], it was shown that significant compressive residual stresses can be introduced to a large depth in aluminum alloy 7050 by Dengeling. In the current project, the potential of Dengeling as an effective way to improve the fatigue resistance of milled aluminium alloys was investigated. The fatigue behavior of Dengeling treated AA7050 was studied by the four-point bending fatigue testing and the results were interpreted with respect to the induced changes in residual stresses. Comparison with shot peening was also made.

Experiment details

Rectangular bars of 10 mm x 10 mm x 80 mm with 1 mm chamfer on both sides of the surface to be tested in bending fatigue were machined from AA7050 T7651. In addition to the milled samples used as reference, four other groups were treated to different surface conditions, as listed in Table 1. For shot peening (T6) a common process for the alloy was used to compare with a

Residual Stresses 2018 – ECRS-10   Materials Research Forum LLC
Materials Research Proceedings **6** (2018) 203-208   doi: http://dx.doi.org/10.21741/9781945291890-32

Dengeling treatment (T1) of similar Almen intensity as the shot peening. Two other groups (T2 and T3) were selected to investigate the effect of Dengeling process parameters. The dimple overlap is calculated from the percentage of diameter overlapping between two neighboring dimples.

Table 1 Dengeling and shot peening process parameters

Sample group	Treatment	Indenter size, stroke distance, dimple size, dimple overlap and line feed direction
T1	Dengeling	Φ3 mm, 0.2 mm, 0.215 mm, 25%, parallel
T2	Dengeling	Φ8 mm, 0.5 mm, 0.72 mm, 50%, parallel
T3	Dengeling	Φ8 mm, 0.5 mm, 0.72mm, 0%, parallel
T6	Shot peening	Shots: S230H, Φ0.59 mm; intensity: 0.2 mmA; coverage: 125%
M	Milling	

The 4-point bending fatigue testing was carried out with a stress ratio of 0.1 and in a frequency of 15 Hz. Samples survived more than 3×10^6 cycles are considered to be runout.

Residual stresses were measured using the X-Ray diffraction technique. The Cr-Kα radiation was used to measure elastic strains for Al-(311) planes. Upon the assumption of biaxial stress state, in-plane residual stresses were calculated using the sin2ψ method and an elastic constant ½S_2 of 19.54×10^{-6}/MPa. In order to obtain the depth residual stress profile, stepwise layer removal by electrolytic polishing was employed.

Fractographic analysis was performed in SEM to identify the fatigue initiation points and possible origins.

Results and discussion

Surface morphology. The surface morphology is compared in Fig. 1 between the Dengeling (T1) and shot peening (T6) with similar Almen intensity. As can be seen, a regular pattern of indents was observed for the Dengeling but a random pattern consisting of larger and smaller crats was found for the shot peening. Some small, untreated areas could also be seen for the shot peened surface while full coverage was observed for the Dengeling group. The corresponding roughness is Ra = 2.3±0.3 µm for T6 and Ra = 3.1±0.8 µm for T1 measured along the line direction and Ra= 3.1±0.5 µm for measurement along the line feed direction. For Dengeling using the larger indenter (T2 and T3), the surface impression was less obvious and marks from the milling operation were still visible.

Residual stresses. Residual stress profiles for all the sample conditions are presented in Fig. 2. Low residual stresses are observed in the milled sample, as shown in Fig. 2a. As expected, the shot peening process introduced a compression layer of about 0.3 mm with maximum compressive stress slightly over 300 MPa and significant plastic deformation as indicated by the diffraction peak broadening (Fig. 2b). Comparison between Fig. 2b and c reveals that the corresponding Dengeling process generated a larger compression layer (about 0.4 mm), lower surface plastic deformation and similar maximum subsurface compressive residual stress in the axial direction while the transverse compressive stresses were much larger.

For the two Dengeling treatments using the large indenter (Fig. 2d and e) the compression layer thickness was about doubled as compared to shot peening, the maximum compressive stresses and surface deformation are however smaller than for both shot peening and Dengeling with the small indenter. The lowest surface deformation and compressive stresses were found with the treatment with no dimple overlap (T3).

Residual Stresses 2018 – ECRS-10   Materials Research Forum LLC
Materials Research Proceedings 6 (2018) 203-208   doi: http://dx.doi.org/10.21741/9781945291890-32

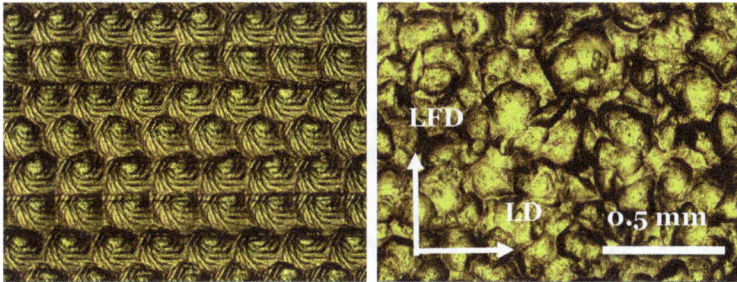

Figure 1. Micrograph showing surface indents on the shot peened (left) and Dengeling sample
(T1) (right). LD: line direction; LFD: line feed direction.

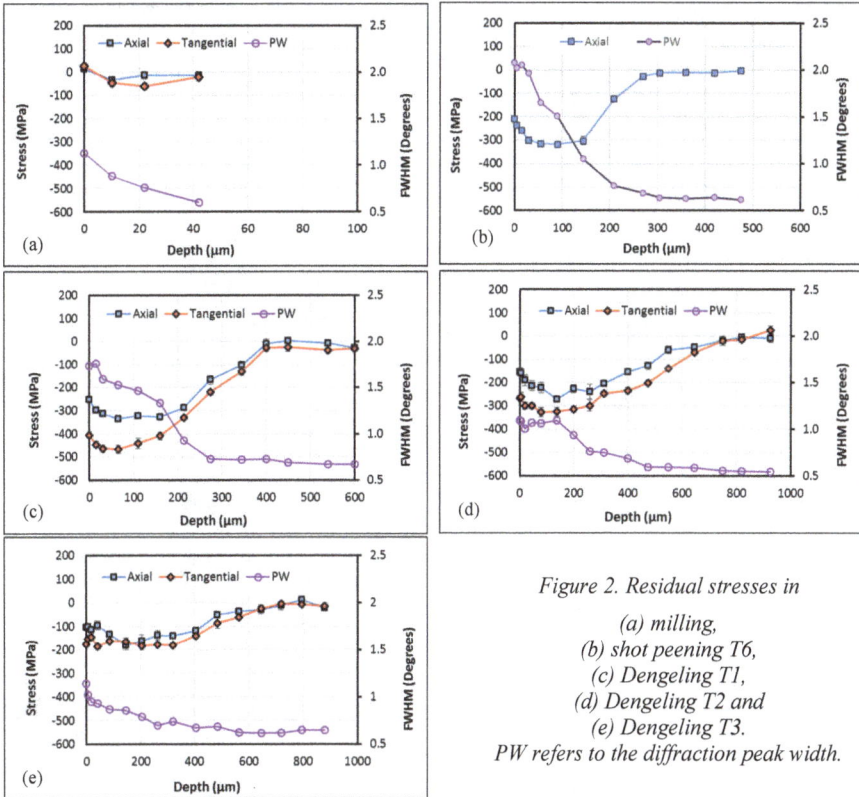

Figure 2. Residual stresses in

(a) milling,
(b) shot peening T6,
(c) Dengeling T1,
(d) Dengeling T2 and
(e) Dengeling T3.
PW refers to the diffraction peak width.

Fatigue. The fatigue testing results are presented as S-N plots in Fig. 3 and 4. On the
assumption of elastic loading, the maximum stress in the surface was calculated according to the

bending beam theory. XRD measurements were made on Dengeling treated and shot peened samples that fractured from a fatigue load of 436 MPa. The obtained residual stresses were similar to the as treated samples, indicating no global yielding in the surface of the samples during fatigue loading at and below 436 MPa.

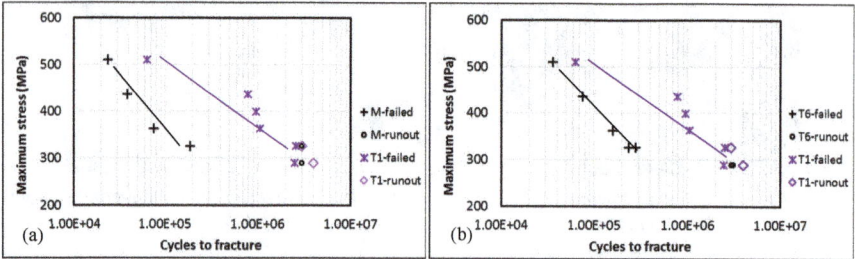

Figure 3 Comparison of Dengeling (T1) with milling (M) (a) and with shot peening (SP) (b). The lines serve as a guide to the eye.

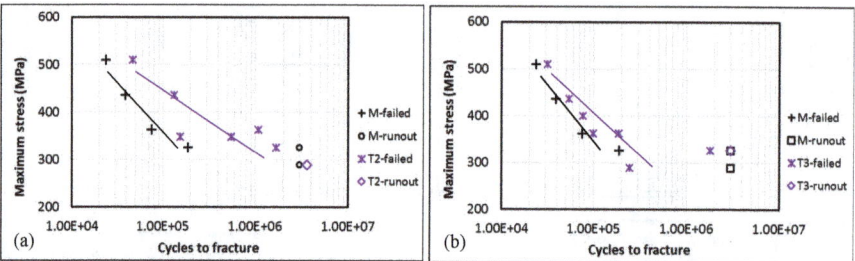

Figure 4. Comparison of Dengeling treatment using Φ8 mm indenter and 50% dimple overlapping (T2) and 0% overlapping (T3) with milling (M). The lines serve as a guide to the eye.

The Dengeling group (T1) is compared with the milling group (M) in Fig. 3a. As can be seen, application of the Dengeling treatment greatly improved the fatigue performance of milled samples. More than one order of magnitude better in fatigue life was observed for the slope region. At the lower stress, 326 MPa, both runout and failure were observed. Nonetheless, the failed Dengeling sample had a much longer life than the milled sample. Comparison for the lowest stress level is difficult as only one milled sample was tested.

An increase of fatigue life up to 100% in the slope region was obtained by the shot peening. The improvement is however much less in comparison with the Dengeling treatment as Fig. 3b illustrates. At the lowest stress level, the two treatments seem to show similar behavior. One Dengeling sample and two shot peened samples survived 3 million loading cycles while the Dengeling samples failed with a fatigue life close to 3 million cycles. Actually the two "Runout" samples of shot peening failed shortly after continued loading beyond 3 million cycles.

The fatigue life of T2 samples in Fig. 4a was twice to three times of the milled samples in the upper part of the slope region. However, in the lower part of the slope region the fatigue data scattered largely, from about 1.55×10^5 to over 3×10^6 cycles. The T3 group presented in Fig. 4b

Residual Stresses 2018 – ECRS-10   Materials Research Forum LLC
Materials Research Proceedings 6 (2018) 203-208   doi: http://dx.doi.org/10.21741/9781945291890-32

was treated with same parameters as T2 but with no dimple overlap. For this group, the improvement in fatigue life was marginal in the slope region and varied in the low stress region.

Fractography. Fractographic analysis reveals that surface crack initiation, preferably at precipitates, and its propagation resulted in fatigue failure of the milled samples. For the Dengeling samples, the fatigue damage always started below the surface and the initiation site moved closer to the surface with increasing applied stress. For loading at and below 363 MPa, the crack initiation site was located about 500 to 560 μm below surface, i.e. outside the compression zone which is about 400 μm. For larger applied stresses, the fatigue cracking started inside the compression zone. It can be concluded that the Dengeling treatment successfully suppressed the surface crack initiation. Examples of fracture surface revealing the crack initiation sites (indicated by arrows) are given in Fig. 5a for milling and Fig. 5b for Dengeling treated samples.

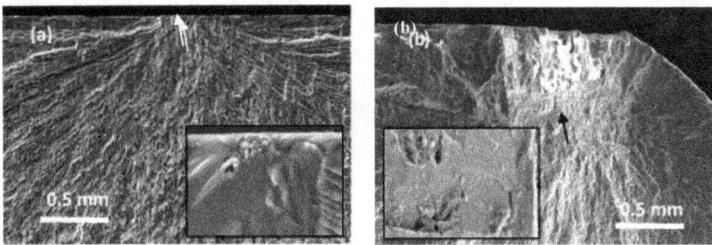

Figure 5. Crack initiation from (a) the surface of an M sample that failed at 363 MPa and (b) subsurface (about 560 μm below surface) of a T1 sample that failed under 326 MPa. The insets reveal the respective crack initiation site.

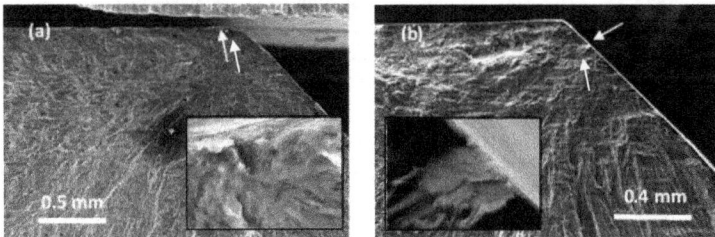

Figure 6 Surface crack initiation in (a) a T6 sample failed at 326 MPa and (b) a T2 sample failed at 363 MPa. The insets show the respective fatigue initiation site at the surface.

In spite of the surface compressive residual stresses, fatigue cracks tend to start from surface or very near to the surface, see Fig. 6a, in the shot peened group. In another word, the shot peening process employed is less effective in suppressing surface crack initiation in comparison with the Dengeling treatment.

For the T2 samples, fatigue crack origins were observed mostly on the chamfer surface, see Fig. 6b, although crack initiation near the edge of the top flat surface was also observed. The treatment of the chamfer surfaces was made separately from the top flat surface. The chamfer surfaces are more prone to fatigue, which means that the strengthening was not as effective as on the top flat surface. It could be due to the indenter diameter that is much larger than the width of the chamfer surface. Similar to the milled group, surface crack initiation often resulted in final failure in T3 samples. The 0% overlap means that about 21.5% of the surface was not covered by the indents. Such bared areas could become preferable sites for crack initiation.

Concluding remarks

The effect of a surface mechanical treatment on fatigue originates from the induced changes in surface integrity especially surface roughness, strain hardening and compressive residual stresses [1]. The results in the previous section reveal that for the Dengeling and shot peening treatments using similar Almen intensity, the magnitude of compressive residual stresses in the fatigue loading direction is similar and better surface roughness with a higher degree of strain hardening was found in the shot peened samples. However, the Dengeling treatment is much more effective for fatigue resistance reinforcement in the slope region of the S-N graph. This is likely attributed to that the Dengeling treatment successfully suppressed surface initiation of fatigue cracks where the shot peening failed. For the shot peened samples subjected to higher loading stresses, microcracks might exist or quickly develop from other types of surface defects and the positive effect of shot peening can be explained by the retardation of crack growth by the compressive residual stresses. For the Dengeling samples in the same stress region, fatigue cracks initiated below the surface, which is a much slower process and the more significant improvement may come from the delay of fatigue crack initiation. At the low stress region, surface crack initiation may become difficult for the shot peened samples and therefore the behavior of both groups are similar. It should be pointed out that the fatigue crack initiation sites tend to be located close to the edge near the chamfer especially for the shot peened samples.

Dengeling treatments using the large indenter generated a much deeper compression zone. However, as fatigue crack initiates often from the weaker chamfer surface in T2 or untreated surface areas in T3, both treatments are less effective and the fatigue data are scattered in the low stress region.

Large intermetallic particles are common in the alloy. Those located at or near surface may result in microcracks during shot peening or Dengeling. EDS analysis of crack initiation sites also reveals such precipitates at crack origins in a number of failed samples in all conditions: milling, shot peening and Dengeling. These particles and their distribution could be responsible for scattered data near the runout stress region.

Acknowledgement

This project is carried out within the Strategic Innovation Program "Metallic Materials", a joint venture of Vinnova, Formas and Energy Agency of Sweden.

Reference

[1] K. A. Soady, "Life assessment methodologies incorporating shot peening process effects: Mechanistic consideration of residual stresses and strain hardening: Part 1 - Effect of shot peening on fatigue resistance," *Mater. Sci. Technol. (United Kingdom)*, vol. 29, no. 6, pp. 637–651, 2013. https://doi.org/10.1179/1743284713Y.0000000222

[2] A. Azhari, S. Sulaiman, and A. K. P. Rao, "A review on the application of peening processes for surface treatment," in *IOP Conference Series: Materials Science and Engineering*, 2016, vol. 114, no. 1. https://doi.org/10.1088/1757-899X/114/1/012002

[3] L. Selegård, R. Lin Peng, A. Billenius, G. Petersén, M. Ess, M. Jonsson, "Residual Stresses in Dengeling-Treated Aluminum Alloy AA 7050," in *Residual Stresses 2016: ICRS-10*, 2016, pp. 425–430.

Residual Stresses 2018 – ECRS-10   Materials Research Forum LLC
Materials Research Proceedings 6 (2018) 209-214   doi: http://dx.doi.org/10.21741/9781945291890-33

Analysis of Residual Stress State in Deep-Rolled HT-Bolts

Julian Unglaub[1,a,*], Jonas Hensel[2,b], Robert Wimpory[3,c], Thomas Nitschke-Pagel[2]
Klaus Dilger[2] and Klaus Thiele[1]

[1]Inst. of Steel Structures, TU Braunschweig, Germany

[2]Institute of Joining and Welding. TU Braunschweig, Germany

[3]Helmholtz-Zentrum Berlin für Materialien und Energie, Germany

[a]j.unglaub@stahlbau.tu-braunschweig.de, [b]j.hensel@tu-braunschweig.de,
[c]robert.wimpory@helmholtz-berlin.de

Keywords: HT-Bolts, Deep Rolling, Residual Stress Measurement, Neutron Diffraction

Abstract. Results of residual stress measurements of HT-bolts gained by neutron diffraction at HZB will be presented. The in-depth residual stress of different conditions of "rolled-before heat-treatment and galvanized" M24 HT-bolts made from 33MnCrB5-2 will be shown: As-manufactured, pre-stressed and fatigue loaded. Additionally, results of an unloaded, "rolled-after heat-treatment and hot-dipped" galvanized M36 HT-bolt will be presented. It will be shown that the manufacturing sequence "rolled-before heat-treatment and galvanized" can develop compressive residual stresses due to heat-treatment close to the surface and that "rolled-after heat-treatment and hot-dipped" shows a contraire residual stress path as "rolled-before heat-treatment and galvanized" with a maximum below the surface.

Introduction

Bolted joints are generally the most frequently used joining elements in mechanical and plant engineering. High-tensile (HT) bolts with large diameters (M30 up to M72) are heavily used in wind power plants both onshore and offshore. Bolted joints are made from hardened steels (for instance 33MnCrB5-2) and show high stress concentration in the thread. One of the fatigue strength governing parameters is the production method used to manufacture those threads, shown in Fig. 1. During the cold-rolling process compressive residual stresses are generated at the root of the thread in combination with strain hardening. The residual stress state can be significantly influenced by heat-treatment or loading.

The in-depth residual stresses of bolts in different conditions of "rolled before heat-treatment" and galvanized HT-bolts M24x130 10.9 tZn made from 33MnCrB5-2 have been investigated by neutron diffraction: As-manufactured, pre-stressed in tension and fatigue loaded. Additionally, an unloaded, "rolled after heat-treatment and hot-dipped galvanized" M36x235 10.9 tZn HT bolt has been examined. The mechanical properties are listed in Tab. 1.

The as-manufactured sample represents the residual stress condition induced by the manufacturing process. The pre-stressed condition reflects the in-situ stress state of a mounted bolt under pre-tension (about 70% of Rp0.2). This was achieved by mounting the bolt in an artificial restraint device. The pre-loaded bolt was pre-stressed and fatigue loaded at TU Braunschweig in the finite life regime until N = 100.000 load cycles which is assumed to correspond to a cyclically stabilized state after residual stress relaxation. The residual stress history can be used to further interpret given fatigue test results.

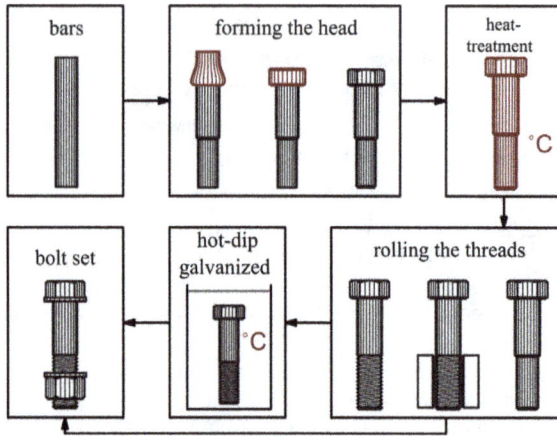

Figure 1: Manufacturing process of rolled after heat-treatment and hot dipped galvanized HT bolts, [4]

Table 1 Mechanical properties

	$R_{p0,2}$ [MPa]	R_m [MPa]	E [MPa]	v -	θ_0 [rad]
M24	969	1081	198400	0,28	77,882
M36	912	1050	194400	0,28	77,882

Setup

The measurement setup at experiment E3 at Helmholtz-Zentrum Berlin für Materialien und Energie (Berlin) for the M24 bolts is shown in Fig. 2 and Fig. 3. The setup shown in Fig. 2 was used for the measurement of the radial, tangential directions for the M24 bolts. Two sections of the bolts have been examined: the first load bearing thread and the shank. In order to cancel out the surface effect the measurements have been carried out in pairs 180° to each other.

The measurement of the axial direction is shown in Fig. 3. With this setup the shank and the first load bearing thread was investigated.

The gauge volume was $2x2x2$ mm^3 which was defined by the primary slit of the neutron beam and radial collimator as a secondary optic. Along the cross section of the measuring path, from the center of the bolt to the surface, the volume distance was equidistant with overlapping volumes close to the surface.

The corresponding measurement setup for the M36 bolts is shown in Fig. 4. and follows the same principles as the setup for the M24 bolts. Because of the sample size and density only the unloaded sample of the M36 bold has been scanned. Due to time constraints solely measurements close to the surface have been made.

The reference Bragg-angel θ_0 of the base material was determined from a cut out sample of the shank of an unloaded bolt. The same material reference sample was used for both bolt diameters. The measurement uncertainties were estimated by the method proposed by Wimpory et. al. [3].

Residual Stresses 2018 – ECRS-10   Materials Research Forum LLC
Materials Research Proceedings **6** (2018) 209-214   doi: http://dx.doi.org/10.21741/9781945291890-33

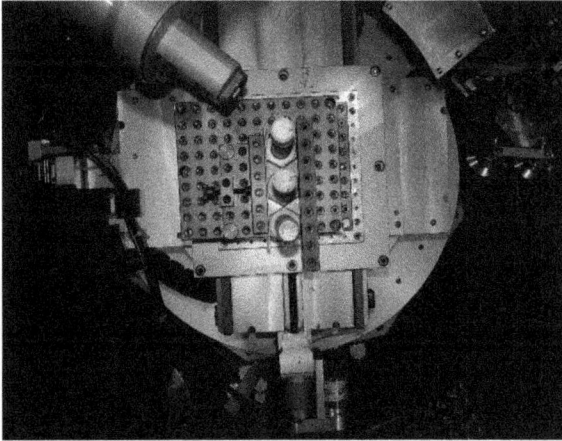

Figure 2: Setup M24, top view for radial, tangential, in measurement pairs '0° and 180°'

Figure 3: Setup M24, side view for axial, in measurement pairs 0° and 180°

Results

The results of the measurement of the "rolled after heat-treatment", hot dipped galvanized M36 HT-bolt is shown in Fig. 4. As expected, compressive residual stresses in the axial and tangential direction are present in the thread root. The maximum of the compressive residual stresses is located below the surface.

Residual Stresses 2018 – ECRS-10   Materials Research Forum LLC
Materials Research Proceedings 6 (2018) 209-214   doi: http://dx.doi.org/10.21741/9781945291890-33

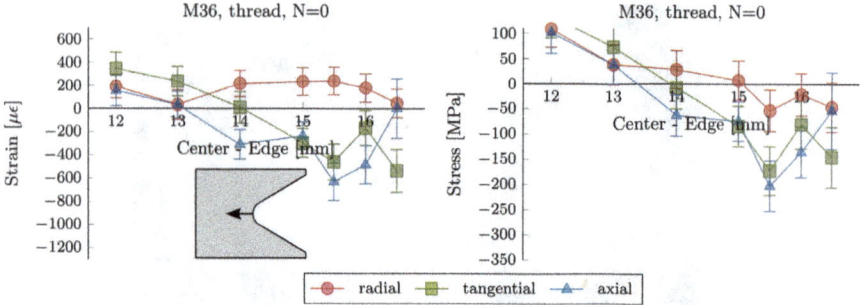

Figure 4: Residual Strain and Stresses of "rolled after heat-treatment" and hot dipped galvanized M36x235 10.9 tZn HT-bolt, threat root, unloaded

The results of the measurements of the M24 HT-bolts are shown in Fig. 5. The unloaded, "rolled after heat-treatment", hot dipped galvanized M24 HT-bolt shows minor compressive residual stresses very close to the surface. The axial stress component of the Sample M24, N=1 is calculated from the radial and tangential strain component under the assumption of zero radial stress. This stress state is influenced by the loading and can be seen in the results of the measurements of the loaded sample with N=1 and N=100.000 load cycle.

Discussion

Unfortunately, there are only a few publications with in depth residual stress measurements on hot-dip galvanized bolts available. These investigations mainly investigate small bolt diameters.
Stephens et. al. [2] investigates the influence of "rolled after heat-treatment" and "rolled before heat-treatment". A residual stress measurement by help of x-ray diffraction shows residual stresses close to the surface of the thread root up to 500 MPa (compression) in axial and -1000 MPa (tension) in tangential direction for rolled after heat-treatment bolts. Residual stresses in "rolled before heat-treatment" bolts are close to zero.

The measured residual stresses of M 36 bolts differ from those in the literature. However, the general path of the residual stresses resembles those from Stephens et. al. [2] quite well. A smaller distance between the measured volumes could capture the maximum better.

The results for the unloaded M24 shows contrary behavior to the investigations of Stephens et. al. [2] for "rolled before heat-treatment" bolts. However, after the heat-treatment process (900 °C) of the bolts the temperature drops rapidly, compressive residual stresses can develop due to the change of the material from cubic face-centred to body-centred cubic.

The change of stress from N=1 to N=100.000 under loading conditions can be led back to cyclic softening, as described by Panic et. al. [1].

Materials Research Forum LLC
doi: http://dx.doi.org/10.21741/9781945291890-33

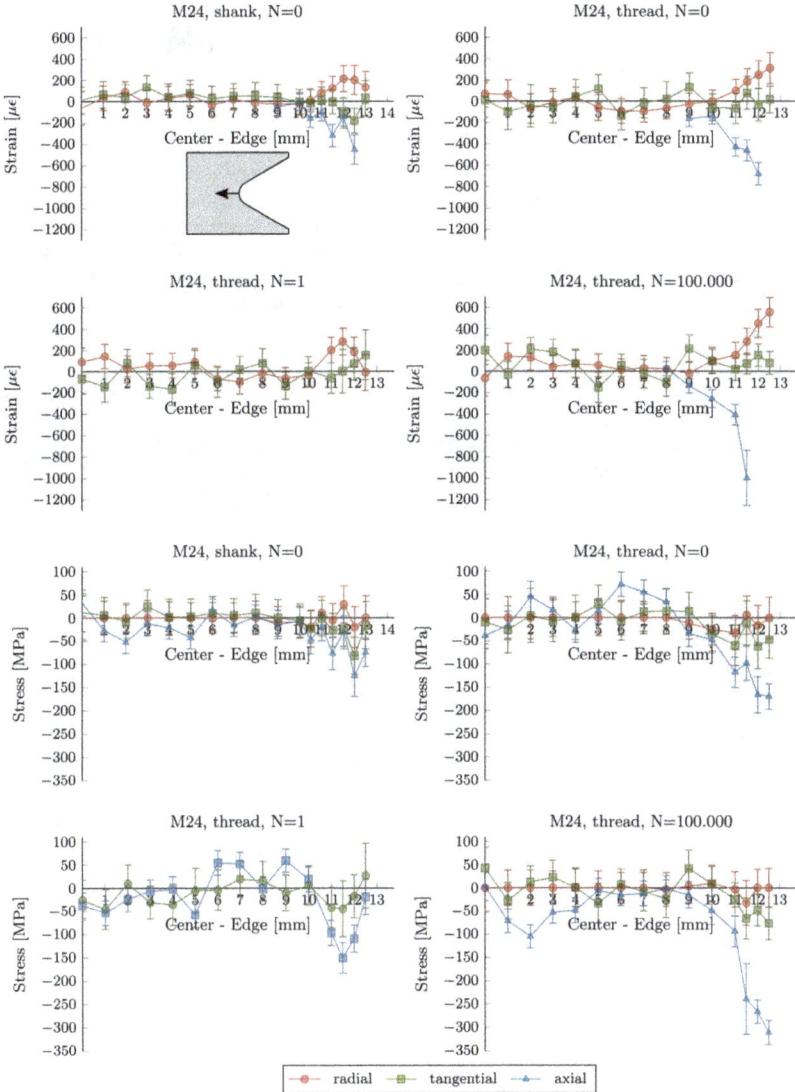

Figure 5: Residual Strain and Stresses of "rolled before heat-treatment" and hot dipped galvanized M24x120 10.9 tZn HT-bolts, threat root, unloaded, load N=1 and load N=100.000.

Residual Stresses 2018 – ECRS-10   Materials Research Forum LLC
Materials Research Proceedings **6** (2018) 209-214   doi: http://dx.doi.org/10.21741/9781945291890-33

Summary
For the first time, "rolled after heat-treatment", hot-dipped galvanized M24 HT-bolts with different loading situations have been investigated with neutron diffraction. It can be seen that the heat-treatment process can produce small compressive residual stresses close to the surface. Theses stresses are subject to cyclic softening. Additionally, "rolled after heat-treatment", hot-dipped galvanized M36 HT-bolts have been examined and show promising results for the unloaded sample. It is evident that the maximum of the compressive residual stresses is below the surface. Future research has to be performed to describe the residual stress relaxation due to loading. For this purpose, all three strain components need to be measured.

The measurements have been performed within a subproject of the DFG Graduiertenkolleg 2075 Models for the description of aging of construction materials and structures. One topic is the investigation of the influence of residual stresses on the fatigue life of large bolt diameters M24, M36 and M48.

Acknowledgements
Measurements were carried out at the E3 instrument station at the Helmholtz-Zentrum Berlin. We thank HZB for the allocation of neutron beam time.

We thank Peiner Umformtechnik GmbH und August Friedberg GmbH for providing test samples.

References

[1] Panic, D., Beier, T., Vormwald, M.: Damage Assessment of Threaded Connections based on an Advanced Material Model and Local Concepts. Procedia Engineering 74 (2014), S. 119–128. https://doi.org/10.1016/j.proeng.2014.06.235

[2] Stephens, R., Bradley, C., Horn, N., Gradman, J., Arkema, J., Borgwardt, C.: Fatigue of High Strength Bolts Rolled Before or After Heat Treatment with Five Different Preload Levels. 2005 SAE World Congress (2005). https://doi.org/10.4271/2005-01-1321

[3] Wimpory, R. C.; Ohms, C.; Hofmann, M.; Schneider, R.; Youtsos, A. G. (2009): Statistical analysis of residual stress determinations using neutron diffraction. In: International Journal of Pressure Vessels and Piping 86 (1), S. 48–62. https://doi.org/10.1016/j.ijpvp.2008.11.003

[4] Unglaub, J., Reininghaus, M., Thiele, K.: The fatigue behaviour of bolts with large diameters under overloading. In: The Eighth International Conference on Low Cycle Fatigue (LCF8) 27.- 29.06 2017 Dresden (2017).

Residual Stresses 2018 – ECRS-10   Materials Research Forum LLC
Materials Research Proceedings 6 (2018) 215-220   doi: http://dx.doi.org/10.21741/9781945291890-34

Comparing the Influence of Residual Stresses in Bearing Fatigue Life at Line and Point Contact

Timm Coors[1,*], Florian Pape[1] and Gerhard Poll[1]

[1] Institute of Machine Design and Tribology, Leibniz University Hannover, Welfengarten 1 A, 30167 Hannover, Germany

* coors@imkt.uni-hannover.de

Keywords: Residual Stress, Bearing Fatigue Life, Tailored Forming

Abstract. The targeted insertion of compressive residual stresses can positively influence the fatigue life of rolling element bearings. Adapted manufacturing processes such as hard turning and deep rolling can help optimising the subsurface residual stress state of these machine elements, which also improves the surface quality of the bearings raceway. In this contribution, a numerical calculation method was developed to predict the influence of residual stresses on bearing fatigue life. By means of a finite element analysis, the component stresses due to the rolling contact load can be determined. The resulting shear stresses find input in a bearing fatigue life calculation based on the approach of IOANNIDES, BERGLING and GABELLI. This statistically based method refers a material-dependent stress fatigue limit to a local stress related fatigue criterion, which is influenced by the residual stress condition. On this basis, the influence of residual stresses on two different bearing types is investigated. Line contact is represented by a cylindrical roller bearing and an angular contact ball bearing is chosen to investigate the point contact. For angular contact ball bearings, a rolling motion is superimposed by a drilling movement perpendicular to the contact plane, which is caused by the kinematics of the rolling element. The calculation method is used for bearings made of classic bearing steel and bearings made of two different steels by tailored forming in order to regard the residual stress conditions of different manufacturing types. It can be shown that the influence of residual stresses on bearing fatigue life is higher for bearings with line contact than for bearings with point contact.

Introduction

Rolling bearings are widely used machine elements, which used in mechanical and plant engineering, drive technology or energy industry. They are used to fix shafts, transmitting forces between shaft and housing and at the same time allowing the shaft to rotate. Rolling bearings typically consist of an inner and an outer raceway, between which several rolling elements (Fig. 1: yellow) roll with low frictional losses. Rolling bearings can be differentiated according to the shape of their rolling elements in roller bearings and ball bearings, for which the contact area between rolling element and raceway is different. In the case of roller bearings (Fig. 1a), the contact is in the form of a line under conditions of zero load. With ball bearings (Fig. 1b), the contact area can be idealized by a point contact. Due to the elastic material behavior under the influence of an external force F, elastic deformation occurs in the contact. The idealized contacts then form two-dimensional surfaces. Previous studies showed a positive influence of pre-induced residual stresses by manufacturing related boundary zone properties [2] or the targeted use of high-strength materials [3] on bearing fatigue life, which therefore has to be regarded at rolling contact fatigue conditions [4][5][6]. In the presented paper, the following rolling bearings are considered exemplarily to compare the influence of residual stresses in bearing fatigue life at line and point contact in order to calculate the load-bearing capacity and the fatigue life under cyclic rolling loads:

Residual Stresses 2018 – ECRS-10   Materials Research Forum LLC
Materials Research Proceedings **6** (2018) 215-220   doi: http://dx.doi.org/10.21741/9781945291890-34

a)

b)

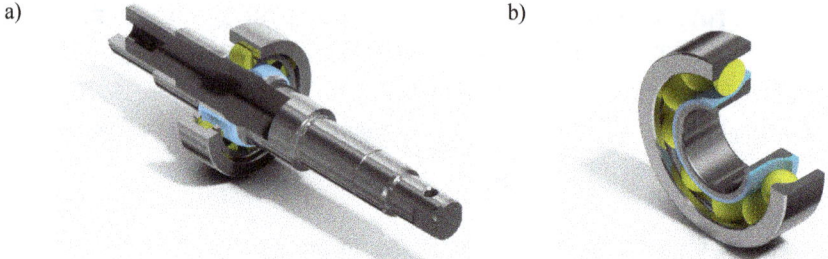

Fig. 1: Contact conditions for two exemplary Tailored Forming demonstrator components: a) cylindrical roller bearing with line contact; b) angular contact ball bearing with point contact.

A cylindrical roller bearing type RNU 204 represents a line contact. The inner ring was substituted by a shaft (Fig. 1a), which in turn was designed and manufactured within the Collaborative Research Centre 1153 "Tailored Forming". In this process chain, a hybrid semi-finished product is produced from two different steel alloys by cladding through plasma transferred arc welding and then formed subsequently by cross-wedge rolling [7]. This offers material and process-related advantages, because a high strength steel (material number 1.7035) can be bonded to an inexpensive steel (1.0402). The cladding steel is commonly considered to be poorly weldable, but offers a good hardness and fatigue resistance [8].

The angular contact ball bearing type 7306 shown in Fig. 1b) is also cladded with a high performance material on which the rolling elements run. This second tailored forming part is used to represent a highly loaded point contact.

Method of calculation

The functional capability and fatigue life of a rolling bearing are primarily determined by the boundary zone properties of the finished component, as they are directly exposed to mechanical loads. In order to calculate the stress state under load, a finite elements (FE) model was set up for each component using Ansys Mechanical APDL (see Fig. 2a and b). For the fatigue life calculation described below, only the most heavily loaded material volume is assumed to be critical for damage. Focus has therefore been set on the inner raceway and the symmetry of geometry was used by modeling a 9°-segment of each component. The contact area is discretised with a regularly mapped fine mesh. The remaining supporting structure is meshed with tetrahedron-shaped elements (see Fig. 2). As input parameters, local material properties of both, the high strength cladding and the base material, as well as the geometry of the contact zone (rolling element profiling) are taken into account. According to the HERTZian contact theory [9], the external load is transformed into a three-dimensional pressure distribution, which is then applied as initial stress to the inner rings surface. For angular contact ball bearings (Fig. 2b) the rolling motion is superimposed by a drilling movement perpendicular to the contact plane, whereby sliding friction occurs in the contacting area [10].

Pre-induced residual stresses, which may result from the manufacturing process, e.g. tempering, hard turning or deep rolling, are also applied as initial stress state regarding the residual stress depth profile in circumferential direction. The subsurface residual stress state was measured based on experimental x-ray diffractometry. In this work, a residual stress state induced by deep rolling is applied to the model (cf. Fig. 2). Due to deep rolling, residual stresses of up to -1000 MPa with a maximum depth of 100 μm could be achieved in previous measurements on cylindrical roller bearings [11].

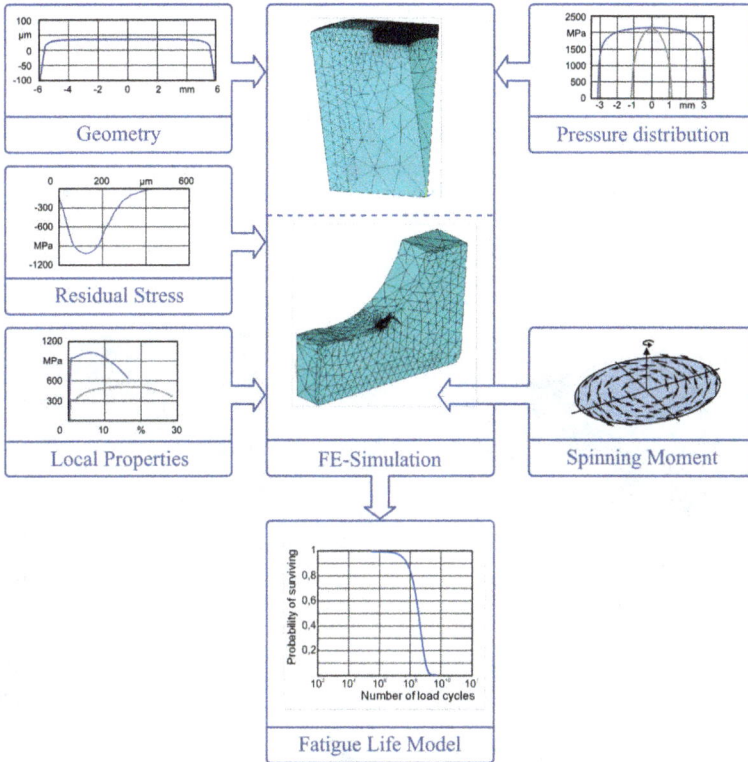

Fig. 2: Flow chart of fatigue life calculation: a) line contact; b) point contact.

Solving the FE simulation provides the 3D shear stresses and hydrostatic stresses in the inner rings resulting from the superimposition of the loading pressure distribution and the residual stress state. Peak value and location of this stress field are used to calculate the bearing fatigue life in Mathworks Matlab for the parts volume under the influence of rolling contact fatigue based on the approach of Ioannides, Bergling and Gabelli [12][13]. This statistically based method refers a material-dependent stress fatigue limit to a local stress related fatigue criterion, which is again influenced by the residual stress condition. Here, the Dang Van stress criterion is used, which sets the orthogonal shear stress in relation to the hydrostatic stress [14]. To cause a damage a limiting value resulting from a linear combination of both conditions has to be exceeded.

Computation results

Based on the calculation approach described above, the FE model was first solved to analyse the component stress state for the examples of lines and point contact. For comparability, a resulting Hertzian pressure of about 2.1 GPa for each component was applied. This represents a medium to high load on the rolling bearing, which are used in fatigue tests [15]. The input parameters for the simulation can be found in Table 1 and Fig. 2. The height of the cladding layer was 0.5 mm, which only has a slight effect on the fatigue life [16].

217

Table 1: Input parameters

Material number	Young's modulus E (GPa)	Assumed shear stress fatigue limit τ_u (MPa)
1.7035	210	300
1.0402	209	210

Fig. 3 shows the equivalent stresses according to the VON MISES yield criterion without consideration of residual stresses. It is obvious that the maximum stresses in the material should occur within the higher strength part volume in order to prevent fatigue damage in the base material. According to the HERTZian theory, the line contact (Fig. 3a) loads a larger component volume with the same maximum pressure as the point contact (Fig. 3b).

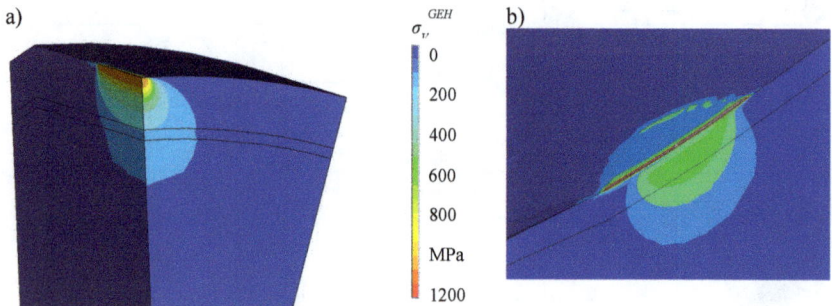

Fig. 3: Induced VON MISES stresses at a HERTZian pressure of 2.1 GPa without consideration of residual stresses: a) line contact; b) point contact.

The critical fatigue stress is highest in the component volume of the highest load-induced stress below the surface. By applying compressive residual stresses at this depth, the damage relevant stresses are reduced (cf. Fig. 4) and bearing life can be positively influenced. With pre-induced residual stresses, the maximum stress is also shifted towards the surface so that a thinner cladding thickness is sufficient.

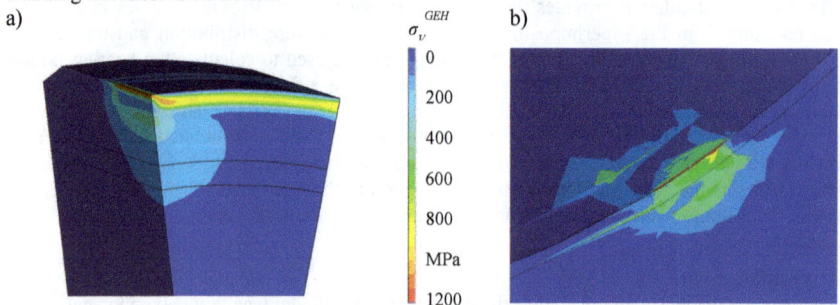

Fig. 4: Induced VON MISES stresses at a HERTZian pressure of 2 GPa with residual stress depth profile through deep rolling (see Fig. 2): a) line contact; b) point contact.

Fig. 5 depicts the calculated bearing fatigue life for the exemplary bearings with line and point contact. To verify the approach, fatigue tests were carried out on a test bench with specimen made form monolithic high strength steel. The calculation and bench tests show good agreement.

Residual Stresses 2018 – ECRS-10   Materials Research Forum LLC
Materials Research Proceedings **6** (2018) 215-220   doi: http://dx.doi.org/10.21741/9781945291890-34

It becomes clear, that a pre-induced residual stress state affects the fatigue life positively by a factor of 1.6 (line contact) respectively 1.3 (point contact). The influence is higher for line contacts, because the damage-critical volume is larger, which increases the cumulated probability of fatigue damage. Note that the higher fatigue life of the angular contact ball bearing results from lower overall external load on the bearing, which is sufficient for the selected surface pressure.

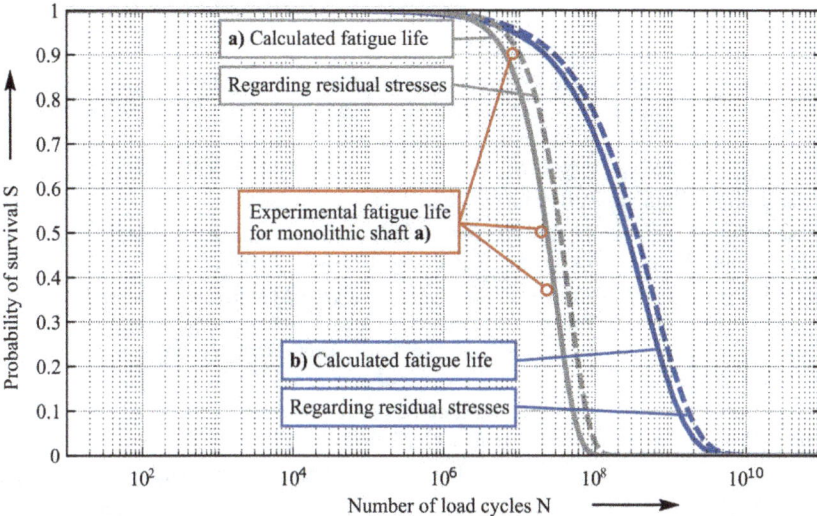

Fig. 5: Comparing the influence of residual stresses in bearing fatigue life for specimen with a HERTZian contact pressure of 2 GPa: a) line contact and b) point contact.

Summary
This paper presents the difference between line and point contact regarding bearing fatigue life, in case of pre-induced residual stresses. It can be shown that the influence of residual stresses on bearing fatigue life is considerably greater for bearings with line contact than for bearings with point contact. This is due to the larger volume in damage risk, which determines the risk for fatigue. For the manufacturing through Tailored Forming, inducing residual stresses can optimize the use of materials and help saving resources.

In order to verify the presented findings, further bearing fatigue life tests should be conducted. For the hybrid shaft (Fig. 1a), this has already been started and first results show good accordance to the calculation results. Accordingly, adapted manufacturing processes can be developed in order to utilize the mentioned advantages for large-scale applications. This will extend the presented approach and help optimising commonly used machine elements under rolling contact fatigue.

Acknowledgment
The results presented in this paper were obtained within the Collaborative Research Centre 1153 "Process chain to produce hybrid high performance components by Tailored Forming" in the subproject C3. The authors would like to thank the German Research Foundation (DFG) for the financial and organisational support of this project.

References

[1] W. Steinhilper, B. Sauer (Eds.), Konstruktionselemente des Maschinenbaus 2, 7th ed., Springer Vieweg, Berlin Heidelberg, (2012).

[2] T. Coors, F. Pape, G. Poll, Concept for enhancing machine elements by residual stresses and tailored forming, Proceedings of IAMOT 2017, ISBN: 978-3-200-04986-4, (2017) 2040 – 2050.

[3] F. Pape, T. Neubauer, O. Maiß , B. Denkena, G. Poll, Influence of Residual Stresses Introduced by Manufacturing Processes on Bearing Endurance Time. Tribology Letters, 66(205), (2017) 65 – 70. https://doi.org/10.1007/s11249-017-0855-3

[4] A.P. Voskamp, Rolling Contact Fatigue and the Significance of Residual Stresses, Deutsche Gesellschaft für Metallkunde, Proceedings: Residual Stresses in Science and Technology (1987) 713-720.

[5] T.A. Harris, M.A. Ragen, R.F. Spitzer, The Effect of Hoop and Material Residual Stresses on the Fatigue Life of High Speed, Rolling Bearings, Tribology Transactions, 35:1, (1992) 194-198.

[6] T. Neubauer, Betriebs- und Lebensdauerverhalten hartgedrehter und festgewalzter Zylinderrollenlager, Phd. Thesis, Leibniz Universität Hannover, (2016).

[7] Mildebrath, M.; Blohm, T.; Hassel, T.; Stonis, M.; Langner, J.; Maier, H. J.; Behrens, B.-A. Influence of Cross Wedge Rolling on the Coating Quality of Plasma-Transferred Arc Deposition Welded Hybrid Steel Parts, International Journal of Emerging Technology and Advanced Engineering, (2017).

[8] H.J. Fahrenwaldt, V. Schuler (Eds.), Praxiswissen Schweißtechnik. Werkstoffe, Prozesse, Fertigung, Vieweg+Teubner Verlag, Wiesbaden, (2009).

[9] H. Hertz, Über die Berührung fester elastischer Körper. Journal für die reine und angewandte Mathematik, Berlin, (1881) 156-171.

[10] G. Poll, L. Deters, Lagerungen, Gleitlager, Wälzlager. in Konstruktionselemente des Maschinenbaus 2, W. Steinhilper, B. Sauer (eds.), Springer (2008).

[11] F. Pape, O. Maiß, G. Poll, B. Denkena, Reibungsminderung bei Wälzlagern und Gleichlaufgelenken durch eine innovative Hartbearbeitung. In: Sonderband Abschlußkolloquium „Ressourceneffiziente Konstruktionselemente" SPP 1551, 58. Tribologie-Fach-tagung 2017, Goettingen, (2017) 21 – 37.

[12] E. Ioannides, T.A. Harris, A New Fatigue Life Model for Rolling Bearings. ASME Journal of Tribology, 107(3), (1985) 367–377. https://doi.org/10.1115/1.3261081

[13] E. Ioannides, G. Bergling, A. Gabelli, An analytical formulation for the life of rolling bearings, Acta Polytechnica Scandinavica, Mechanical engineering series No. 137, (1999).

[14] K. Dang Van, B. Griveau, O. Message, On a New Mulitiaxial Fatigue Criterion: Theory and Application. In: Mechanical Engineering Publications, (1989) 479–496.

[15] T. Coors, F. Pape, G. Poll, Enhancing Machine Elements by Residual Stresses, Optimized Surfaces and Tailored Forming. In: Proceedings, 21st International Colloquium Tribology, TAE, Esslingen, (2018) 277-278.

[16] F. Pape, T. Coors, Y. Wang, G. Poll, Fatigue life calculation of load-adapted hybrid angular contact ball bearings, Lecture Notes in Mechanical Engineering, Springer, doi: 10.1007/978-981-13-0411-8_36, (2018). https://doi.org/10.1007/978-981-13-0411-8_36

Residual Stresses 2018 – ECRS-10   Materials Research Forum LLC
Materials Research Proceedings 6 (2018) 221-226   doi: http://dx.doi.org/10.21741/9781945291890-35

Limitations and Recommendations for the Measurement of Residual Stresses in Welded Joints

Thomas Noël Nitschke-Pagel

Langer Kamp 8, 38106 Braunschweig, Germany

t.pagel@tu-braunschweig.de

Keywords: Residual Stresses, Welding Joints, X-Ray Diffraction

Abstract. Residual stresses in welded joints are often of extended interest in order to evaluate unexpected failures or distortions. Since the possibilities to calculate residual stresses in welds are still stronglyl limited the measurement techniques are still of great importance. Several measurement techniques with particular possibilities and limitations are available today where especially the different diffraction methods are used mostly. Material. weld type and the size of the components are important for the quality of the results obtained with different methods as well as the environment where the measurements have to be carried out. The paper shall give an overview of the results of a round robin test on the application of XRD on butt welded joints which has been carried out in cooperation of different experienced laboratories. The results show the high reliability of XRD-measurements in welds, if the measurements are performed under well defined boundary conditions. The experiences can be used as a recommendation about useful measurement conditions the expectable quality of the results.

Introduction

Residual stresses are mentioned very often if performance problems of welded construction are discussed [2]. Brittle fracture, low fatigue resistance, corrosion or unexpected distortion are related frequently to high tensile residual stresses. However knowledge about the magnitude of residual stresses is rarely precise but mostly assumptive or based on generalizing catalogues [1]. Many efforts are made since more than 30 years to calculate the welding residual stresses but in fact this possibility has not yet been established in practice because calculations are mostly limited to very small and simple weld geometries. Multipass welds or welded components with a high number of single weld seams cannot yet be handled with the numerical tools. Consequently in practice many simplifications are usual followed by a mostly poor reliability. A consideration of residual stresses in design rules reliable quantitative knowledge about the residual stresses in critical zones of a component are required and they should be determined non-destructive[1,2]. In practice XRD is applicable in several variations whereas uncertainty exists often when the customers are not experienced about the recommended conditions and the expectable accuracy of the results.

Background of the RR-test on welds

Several techniques are applicable to determine residual stresses in welds [3,4]. Mechanical techniques like the hole drilling use a locally restricted access to the material in order to determine the initial residual stresses while destructive sectioning techniques like the contour method allow the determination of the initial residual stresses in the entire cross section with preference to the longitudinal residual stress component. Non-destructive techniques like the micromagnetic method are available but the signals cannot be directly correlated to a strain or stress condition. The signals generally a calibration procedure which may complicate the measurements. The experiences of residual stress evidently show that the most reliable results can be expected by means of diffraction [5,6,7]. In addition to X-ray diffraction (XRD) high energy methods with particular advances and limitations are well established nowadays [8]. Limitations for the practical use are the restricted access to the facilities which makes them more useful in fundamental research. XRD is a widely used method with a lot of experiences in different materials and their conditions. Since a lot a different portable devices are available the method is preferable also in welding practice because it can be used under laboratory as well as under rough environmental conditions.

Special problems of XRD-application in welded joints

Residual stress determination in welded joints is usually complicated by different sources resulting from the microstructure in combination with a more or less complicate geometry.

Undercuts or sharp notch geometry at the weld toe and rough surface in the weld seam as well as bulging weld material lead to difficult adjustment conditions. In combination with coarse grain this ends in higher scattering range or incorrect results. Slag rests or condensed metal dust may cover the surface and act like beam absorber. Frequently the exposure time must be extended strongly but anyway the scatter range of the results will arise. Sometimes this problem can be solved by reducing the local resolution but frequently in such zones XRD-measurements are impossible like in the weld zone of Al- or Ni-based alloys or high alloyed steels.

A simple problem is generated by the shape of the weld. As Fig 1 illustrates the weld material may build a barrier for the incident or the reflecting beam which must end in an error. In butt welds the expected problems will be detachable as long as the height of the bulged weld material is not too high. In T-joints or welds with fillet welded longitudinal stiffeners the weld toe and the connected stiffener build a barrier which limites the measurement on one ψ-direction.

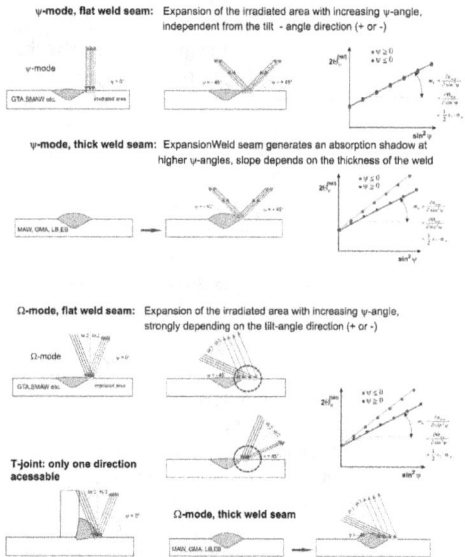

Fig 1. Relationship between weld shape and beam geometry with regard to the accessibility of the $\sin^2\psi$-method.

This is a serious problem because the peak determination in both ψ-tilt directions is a criterion for fulfilling the preconditions of the $\sin^2\psi$-method. Since in practice in welds with geometrical restrictions the adjustment of a precise ψ-startposition 0 at the measured point is not always possible as good as required. Here the movement in positive and negative ψ-direction enables the determination of a reliable average value. Limitation on one direction on the other hand will lead to an error depending on the discrepancy at the zero position and fin ally to a more or less greater uncertainty of the determined residual stress value. However it must be accepted that in welds with stiffeners movement in both directions is impossible. Then special attention must be payed for an adjustment of the sample as precise as possible.

Residual Stresses 2018 – ECRS-10   Materials Research Forum LLC
Materials Research Proceedings **6** (2018) 221-226   doi: http://dx.doi.org/10.21741/9781945291890-35

Round-Robin-Test on residual stress determination using XRD

Purpose of the Round-Robin-Test

An approach to establish residual stresses in welding design codes requires first an answqer on two basic questions. The first question is the possible and the required local resolution, the second is the accuracy.

The applicable resolution is mostly defined by the used devices and by the material condition. The technical range of applicable collimator diameters starts at approximately 0.3 mm and ends at some mm. For special applications also much smaller diameters can be used but this is usually not recommended for welded joints. The experiences of the practical application of XRD reveal that normally collimator sizes lower than 1 mm lead to higher scattering and large extensions of the required exposure times. Diameters higher than 3 mm are not useful because lateral surface residual stress gradients in the surrounding of a weld seam may be smoothened strongly. In practice diameters e.g. local resolutions of 1...2 mm represent a good compromise of an acceptable resolution and a limited time effort for exposure. Of course this agreement is useful for standard arc welds while in very narrow beam welds this limitation may be not acceptable (Fig.2).

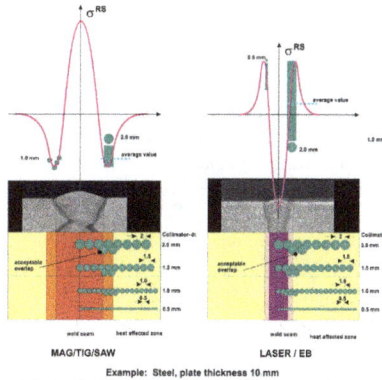

Fig 2. Influence of the collimator size on the local resolution of residual stresses and smoothing effect.

Standard rules for XRD-residual stress measurements are defined in [9], where recommendations for the measurement, the calculation procedure and useful material parameters as well as recommendations for the required calibration proceduresare are summarized. The careful consideration of the standard rules should guarantee that measured results are independent from the lab or the operator. This basic request was the initial idea of the presented RR-test where the quality of typical residual stress determination in standard welds performed by different experienced labs should be demonstrated.

Examined objects

The experiments were carried out using flat plates (200x150x6mm³) of a normalized fine grained normalized structural steel (german grade S460N) with a yield strength of 460 N/mm². The plates were planely ground and stress relief annealed before welding at 600°C/10min. A TIG-welded dummy seam was generated with a heat input of 10.92 kJ/cm As Fig 3 shows the TIG-process generates a very flat weld seam while the MAG-process leads to a strongly bulged weld seam with difficult geometrical measurement conditions. Due to the relatively low heat input the hardness values in the heat affected zone (HAZ) are rather similar.

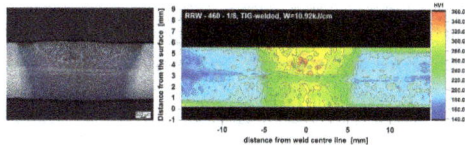

Fig 3. Weld geometry, structure and hardness distribution.

Results of the Residual stress measurements

Fig.4 (top) shows the representative distributions of the longitudinal and transverse residual stresses across the weld seam of a TIG-welded sample determined with a commonly used collimator-Ø of 1.0 mm. Here the typical symmetrical residual stress profiles are observed with some minor discrepancies in the weld material and the typical maximum and minimum peaks in the transition zone between the weld and the parent material. The average measurement error is below 10 MPa except in the weld seam where an error level of 50 MPa is reached. The error of the transverse residual stresses is necessarily higher. The reason is that due to the shift of the ψ-angle the geometry of the irradiated area changes with ψ and therefore the measured value is more influenced

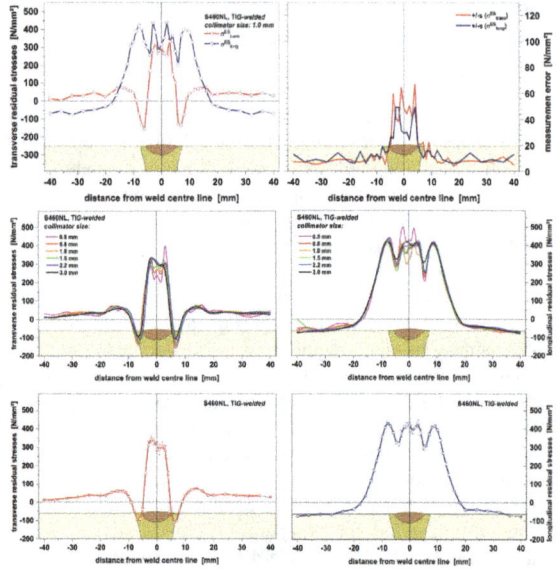

Fig 4. Comparison of the residual stress profiles measured with different local resolutions (collimator-Ø), average distributions and error-range.

by the stress gradient. In longitudinal direction the interval of the stress gradient is constant which leads to a lower error.

Influence of the local resolution on the quality of the measurement results

The influence of the local resolution on the measured residual stress profiles is represented in Fig.4 (middle) where the obtained profiles using collimators with diameters from 0.5 to 3.0 mm are compared. The measured results are obviously independent from the collimator size except the smallest diameters of 0.5 and 0.8 mm. Here a greater discrepancy in the weld material is present. However the deflection of the profiles with particular peaks and tales in the transition zone between the weld and the parent material is marginal in longitudinal and in transverse direction. The average values with the related error bars are shown in Fig.4 (bottom). The divergence in the weld seam is below ±30 N/mm². This is of great importance because it demonstrates that the residual stresses in arc welds can be measured with a good accuracy without the requirement of very high local resolution. Since the local resolution is very important for the exposure time and that is to say for the time effort for a single measurement this results offer the possibility to determine residual stresses in welds in an acceptable time span.

Results of the RR-tests

In Fig 5 (top) the residual stress distributions measured in TIG-welded plates by the participating labs results using uniform collimator sizes of 1.0 mm Ø are shown. As the resulting average profiles with the resulting scatterband reveal (Fig.5, bottom) the scatter range in the base material and in the heat affected zone is between ±25 N/mm² and ±40 N/mm². In the weld seam increasing deviations with an amount of ±100 N/mm² are present. Neverthe-less the agreement of the obtained residual stress profiles can be interpreted as

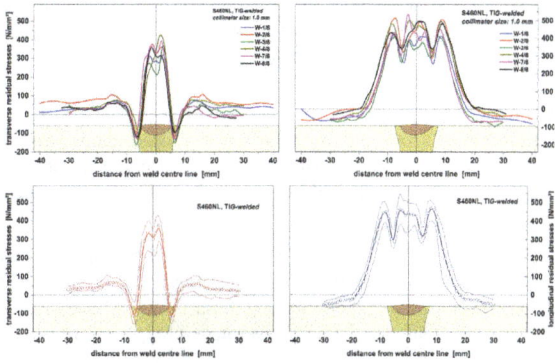

Fig 5. Comparison of the residual stress profiles obtained by the participating laboratories, average profiles and related error range.

excellent. Experiences with larger numbers of specimen of the same series, have shown that the variation of the residual stresses in welds may vary between ±20... ±80 N/mm² even if the specimen have been produced under comparative laboratory conditions. Considering this unpreventable variation the agreement of the obtained results is quite good. Larger deviations of the different measurements are connected with the integral width of the measured diffraction lines (Fig.6, left hand side). This is not really surprising because these values are much more influenced by the particular devices and assessment procedures (detector type, beam geometry, peak

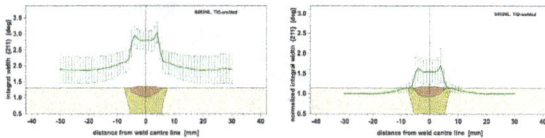

Fig 6 Comparison of the measured and normalized profiles of the integral width of the diffraction lines samples.

smoothing, etc) If the determined integral widths are normalized using the corresponding values of the base material this error is reduced significantly. Nevertheless in the weld material a greater discrepancy must be accepted.

Summary and conclusions

The determination of residual stresses by means of X-ray diffraction is the most suitable technique if confidential results are expected without any detraction of the investigated components even under rough or difficult environmental conditions. However the accuracy and the reliability of the obtained results depends strongly on the adherence of some standard [10] rules considering the special requirements of welds:

- *a sufficient number of ψ-angles is required to eliminate the disturbing effects of coarse grain and unsteady surface in the irradiated area in the surrounding of the weld seam.*

- *measurement in both ψ-direction is generally recommended. The consideration of both directions enables to compensate smaller adjustment errors.*

- *in welds with a bulged weld seam or shading stiffeners, where measurements cannot be performed in both ψ-directions a higher precision for the adjustment is strictly required.*

- *in a technical useful range the chosen resolutions do not affect strongly the accuracy of the measured residual stresses. In the weld seam the detrimental effect of the grain structure compensates the benefit of a higher resolution by scattering. Useful collimator sizes of 1...2 mm combine a sufficient local resolution with a lower error and an acceptable exposure time.*

Repeatable measurement results within a range of ±50 N/mm² or better can be expected independent from the operator or the used equipment. This perception is applicable for conventional arc welds with technically usual weld seam geometries. A proval of a general applicability for welds including small beam welds and comparable should be given separately.

Acknowledgement

The RR-test was carried out with help of the collaboration of René Fenzel (SLV Halle, Germany), Jens Gibmeier (KIT, Germany), Paul Lefèvre (SONATS, France), Fabien Lefebvre (CETIM, France) and Wolfgang Zinn (University of Kassel, Germany). The author wants to express his thank for the participation and helpful contributions.

References

[1] Barthelmey, J.Y.; Janosch,J.J.: Structural Integrity Assessment Procedure for Euopean Industry – SINTAP, Task 4 Compendium on Residual Stress Profiles. Brite-Euram-Sintap, BRPR-CT95-0024, 18.05.1999

[2] Bate, S.K.; Green,D.; Buttle,D.: A review of residual stress distributions in welded joints for the defect assessment of offshore structures. Offshore technology report,   OTH 482, HSE Books 1997

[3] Lu, J.: Handbook of Measurement of Residual Stress, Society for Experimental Mechanics Inc., American Institute of Physics, 1996

[4] Bahadur,A.; Kumar,B.R.; Kumar, A.S.; Sarkar,G.G.; Rao, S.: Development and comparison of residual stress measurement on welds by various methods. Materials Science and Technology, Volume 20, 2004 - Issue 2 , pp261-269

[5] I. C. Noyan, J. B. Cohen. Residual stress – measurement by diffraction and interpretation. Springer Series on Materials Research and Engineering, edited by B. Ilschner and N. J. Grant. Springer-Verlag, 1987.

[7] Eigenmann, B.; Macherauch, E.: Röntgenographische Untersuchung von Spannungszuständen in Werkstoffen. Materialwissenschaft und Werkstofftechnik Band 26 (1995). Teil I.; Heft 3; S. 148-160.; Teil II. Heft 4, S. 199-216; Teil III. Heft 9, S. 426-437

[8] Spiess, L.; Teichert, G.; Schwarzer, R.; Behnken, H.; Genzel, C.: Moderne Röntgenbeugung - Röntgendiffraktometrie für Materialwissenschaftler, Physiker und Chemiker: Springer Verlag 2009. https://doi.org/10.1007/978-3-8349-9434-9

[8] M.E. Fitzpatrick; A.T. Fry; P. Holdway; F.A. Kandil; J. Shackleton; L. Suominen: Determination of Residual Stresses by X-ray Diffraction – a national measurement good practice guide, Measurement Good Practice Guide No. 52, National Physical Laboratory Teddington, Middlesex, United Kingdom, 2005

[9] DIN EN 15303 Non-destructive testing – Test method for residual stress analysis by X-ray diffraction; German version EN 15305:2008

Residual Stresses 2018 – ECRS-10
Materials Research Proceedings 6 (2018) 227-232

Materials Research Forum LLC
doi: http://dx.doi.org/10.21741/9781945291890-36

Experimental and Computational Analysis of Residual Stress and Mechanical Hardening in Welded High-Alloy Steels

Nico Hempel[1,a,*], Thomas Nitschke-Pagel[1,b], Joana Rebelo-Kornmeier[2,c] and Klaus Dilger[1,d]

[1]TU Braunschweig, Institute of Joining and Welding, Langer Kamp 8, D-38106 Braunschweig, Germany

[2]Heinz Maier-Leibnitz Zentrum (MLZ), Technische Universität München, Lichtenbergstr. 1, D-85748 Garching, Germany

[a]n.hempel@tu-braunschweig.de, [b]t.pagel@tu-braunschweig.de, [c]Joana.Kornmeier@frm2.tum.de, [d]k.dilger@tu-braunschweig.de

Keywords: Welding, Residual Stress, Mechanical Hardening, Austenitic Steel, Neutron Diffraction, X-Ray Diffraction, Numerical Welding Simulation

Abstract. Due to the thermal cycle during welding, plastic deformation can occur in the heat-affected zone. After cooling, the yield stress can be locally increased due to the hardening effect and thus permits higher residual stresses in these areas. Therefore, a precise description of the hardening behavior in welding simulations is indispensable for reliably predicting residual stresses. Thus, this work is dedicated to the characterization of mechanical hardening in welded high-alloy steels and its effects on the residual stress state. Numerical welding simulations are performed using different hardening models and the outcomes in terms of both the residual stress and the hardening state are compared to experimental results gained by laboratory X-ray and neutron diffraction.

Introduction

Welding simulation is a powerful tool for predicting the residual stresses and distortion due to a welding process. However, the numerical outcome strongly depends on the choice of the hardening model and can differ significantly from experimental results [1]. Since the thermomechanical welding cycle induces plastic deformation and the residual stresses can be as high as the local yield stress, modeling the cyclic hardening behavior correctly is indispensable for predicting the residual stress state, which eventually may affect the structural performance of a welded component. Hardening models used for welding simulation usually take isotropic hardening, accounting for a successive growth of the yield surface, or kinematic hardening, represented by a translation of the yield surface, or both mechanisms into account. Numerous researchers have investigated the effect of the choice of the hardening model and were able to attain good agreement with experimental results, especially when sophisticated models like the Lemaitre-Chaboche model [2] are used, see e.g. [3] for an austenitic steel.

In this paper, the first results of an ongoing research project are shown which is aimed at clarifying which hardening model should be used for the welding simulation of different steels and which modeling effort is required. To this end, a bead-on-plate dummy weld without filler material is considered both experimentally and numerically. The welding residual stresses computed with different hardening models are compared to results of diffraction experiments. Additionally, characteristic quantities of hardening derived from the simulations are compared to the microstrain, which is a microstructural indicator of material hardening. Thus, this work will eventually help to make mechanically and microstructurally based recommendations for the choice of the hardening models in welding simulations.

Residual Stresses 2018 – ECRS-10   Materials Research Forum LLC
Materials Research Proceedings **6** (2018) 227-232   doi: http://dx.doi.org/10.21741/9781945291890-36

Experimental work

Sample preparation. Plates of the austenitic steel X2CrNi18-9 (AISI 304L) were used for the experiments. Tensile tests at room temperature revealed a yield stress of 255 MPa at 0.01 % plastic strain. The material was heat-treated at 1050 °C for 12 minutes and cooled in air. Bead-on-plate welding was then performed on workpieces of dimensions 150x200x10 mm³ using a mechanized tungsten inert gas (TIG) process without filler metal in flat position with a nominal heat input of 8.2 kJ/cm. The plates were placed on three ceramic balls during welding, guaranteeing accurately defined boundary conditions for the welding simulations, see also Fig. 1. Transient temperatures were recorded by NiCr-Ni thermocouples positioned at different distances from the weld.

Analysis of residual stress. The residual stresses and the mechanical hardening state in the welded samples were determined using X-ray diffraction (XRD) and neutron diffraction (ND). The measurements were taken at points along a line perpendicular to the welding direction at mid-length of the weld, see the measurement locations in Fig. 1. The residual stresses on the surface of the plates were determined by XRD using an Ω-diffractometer and a cylindrical collimator with a diameter of 2 mm. Interference lines of Cr-K_α radiation originating from the {220} lattice planes of austenite were recorded in a 2θ interval of about 121–137° using a 1D position sensitive detector. The sample was tilted for eleven angles ψ of 0°, 13°, 18°, 24°, 27°, 30°, 33°, 36°, 39°, 42° and 45°, allowing for an analysis of the shift of the interference line's center of gravity by the $\sin^2\psi$ method. Previously, the $K_{\alpha 2}$ doublet was eliminated using the Rachinger technique and the remaining $K_{\alpha 1}$ peak was smoothed with a Savitzky-Golay filter. ND measurements for residual stress analysis inside the plate were carried out at the STRESS-SPEC instrument [4] at the Heinz Maier-Leibnitz Zentrum (MLZ), Garching, Germany. The diffraction of neutrons with a wavelength of $\lambda \approx 1.67$ Å on the {311} lattice planes of austenite was analyzed. Strain mapping in transverse, longitudinal and normal directions was performed using a gauge volume of 1x1x1 mm³ at ten depth positions and 19 different distances from the weld centerline (WCL) in order to compute the multiaxial residual stress state. The stress-free lattice spacing was determined in a cylindrical sample of diameter 3 mm containing both molten and base material. Diffraction elastic constants $E^{\{220\}} = 256200$ N/mm², $v^{\{220\}} = 0.304$, $E^{\{311\}} = 186900$ N/mm² and $v^{\{311\}} = 0.325$ have been determined experimentally and were used for all analyses.

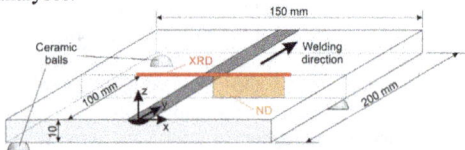

Figure 1: Experimental setup and locations of diffraction measurements

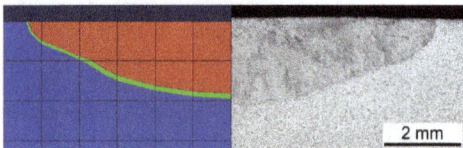

Figure 2: Simulated temperature distribution in cross-section (left) compared to macrograph (right) with T>$T_{liquidus}$ (red) and T>$T_{solidus}$ (green)

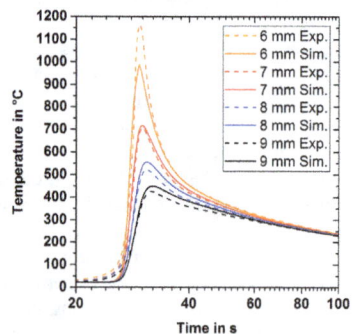

Figure 3: Transient temperatures at different distances from the WCL; experiment and simulation

Analysis of mechanical hardening. Mechanical hardening usually leads to an increase of the dislocation density and thus microstrain within the atomic lattice of a polycrystal. Therefore, the

Residual Stresses 2018 – ECRS-10   Materials Research Forum LLC
Materials Research Proceedings 6 (2018) 227-232   doi: http://dx.doi.org/10.21741/9781945291890-36

microstrain is an indicator of mechanical hardening, which can be determined using diffraction line profile analysis. To this end, the diffraction lines recorded in the normal direction of a welded sample as well as LaB_6 powder standard samples were utilized for both XRD and ND. Instrumental broadening was eliminated by Fourier deconvolution [5] and the remaining physical profiles were fitted using a pseudo-Voigt function. From its width parameters the microstrain was computed [6].

Numerical simulations
Welding simulations with the finite element method (FEM) were carried out using the software Abaqus. The FE mesh consisted of 37760 linear hexahedral elements with 43170 nodes and had a size of 1 x 2.5 x 1 mm³ in the weld and heat-affected zone. The thermal and mechanical boundary conditions of the welding experiments were replicated in the simulation. First, a thermal simulation was run using a circular surface heat source with constant distribution, whose parameters were fit to a macrograph, see Fig. 2, and temperature measurements, see Fig. 3, for an optimal representation of the actual TIG process. The transient temperatures were then input into the mechanical simulation where the residual stresses were computed. The temperature dependent physical properties were taken from a literature study by Voß [7] for the similar steel X5CrNi18-10. Since temperature-dependent stress-strain are not yet available for the X2CrNi18-9 steel investigated here, quasi-static flow curve data has also been taken from [7] and was scaled by fitting the flow curve at room temperature used by Voß to the one determined by tensile tests of X2CrNi18-9 at room temperature. In order to study the effect of the hardening model, the simulations were carried out with isotropic, kinematic and ideally plastic material behavior. For all cases, the combined isotropic-kinematic hardening law in Abaqus [2] was used, setting kinematic, isotropic or both hardening mechanisms to zero, respectively. For isotropic and kinematic hardening, the flow curves were modeled by the same asymptotic exponential functions, guaranteeing that the hardening behavior is quantitatively identical and differs only qualitatively. The annealing temperature, where the strain history is reset, is assumed to be 1000 °C, which is close to the solution heat treatment temperature of austenitic steels. Simulations with actual material data based on tensile and cyclic tests performed on X2CrNi18-9 will be shown in a later study.

Experimental results
Fig. 4 shows the longitudinal, transverse and von Mises residual stresses on the surface. It can be seen that the longitudinal residual stress component takes low compressive values far away from the weld, enters the tensile regime at about 20 mm from the WCL and reaches a maximum of about 550 MPa at about 8 mm from the WCL. The residual stress then decreases to about 300 MPa at the weld toe. Due to the coarse grain in the molten zone, the XRD measurements could not be analyzed there. The von Mises stress largely follows the longitudinal stress component, showing a maximum of about 500 MPa at about 8 mm from the WCL. In contrast to the longitudinal and von Mises residual stresses, the transverse component is not fully symmetric with respect to the WCL, which might be due to the fact that only tilt angles $\psi > 0°$ have been used and the negative side of the plate was sloped because of the welding distortion. Considering only the results on the positive side of the WCL, the transverse residual stress exhibits a maximum of about 240 MPa at 12 mm from the WCL and lower tensile values at the other positions. In general, it can be observed that the measurement uncertainty, resulting from the linear $\sin^2\psi$ fit, is rather high close to the weld toe, which is especially true for the transverse residual stress.

Due to space constraints, only the von Mises equivalent residual stress determined from the ND measurements is shown here, see Fig. 5. The highest equivalent residual stresses of up to 350 MPa are found around the molten zone that is indicated by the dashed line in the top left corner. While the band of the highest equivalent residual stresses is rather narrow near the top surface at

Residual Stresses 2018 – ECRS-10   Materials Research Forum LLC
Materials Research Proceedings **6** (2018) 227-232   doi: http://dx.doi.org/10.21741/9781945291890-36

about 8 mm from the WCL, it is wider below the molten zone and extends to the bottom of the plate.

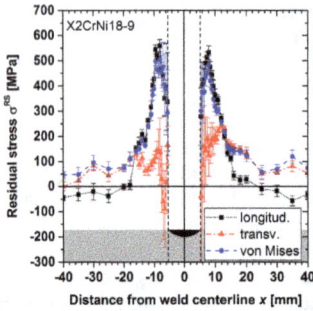

Figure 4: Longitudinal, transverse and von Mises residual stress on the plate surface

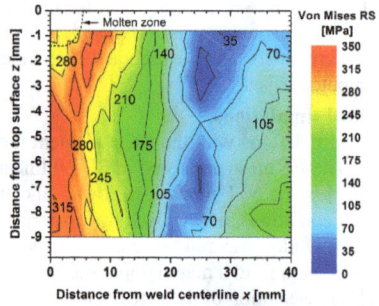

Figure 5: Von Mises residual stress in the cross-section of a welded plate

In Fig. 6 the microstrain as determined by line profile analysis of the XRD measurements is shown along a line on the surface of the plate. The distribution is almost symmetric with respect to the WCL, except for two outliers, and takes maximum values at 8 mm from the WCL. High gradients can be observed to both sides of the maxima. The microstrain takes relatively constant low values at distances larger than about 20 mm from the WCL.

The microstrain distribution derived from the ND measurements is depicted in Fig. 7. It can be seen that the molten zone is surrounded by an area with values that are higher than in the base material far away from the WCL. The maximum is found near the surface at about 7 mm from the WCL. It can also be observed that the values are of a similar magnitude than the results of the XRD measurements, see Fig. 6, but the relative difference between the minimum and maximum values is much lower for the ND results.

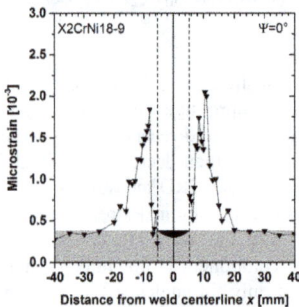

Figure 6: Microstrain distribution on the surface of a welded plate

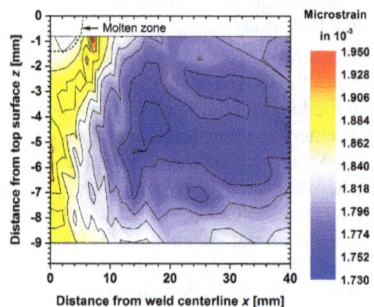

Figure 7: Microstrain distribution in the cross-section of a welded plate

Numerical results

A comparison of the numerical results obtained with different hardening models and the XRD measurements in terms of the von Mises equivalent residual stress is presented in Fig. 8. It can be seen that the three different hardening models yield qualitatively distinctly different results, especially within 20 mm of the WCL. Here, the isotropic model yields the highest residual stresses, taking maxima of 420 MPa at 7 mm from the WCL and showing slightly lower values in the molten zone, thus being closer to the experimental results than the other simulations. The

Residual Stresses 2018 – ECRS-10   Materials Research Forum LLC
Materials Research Proceedings 6 (2018) 227-232   doi: http://dx.doi.org/10.21741/9781945291890-36

results of the kinematic model exhibit a sharp bend at 20 mm from the WCL, remain on a level of about 200 MPa and increase to 330 MPa in the molten zone. The results of the ideally plastic model show the same bend, but remain at a constant plateau of 207 MPa.

In order to highlight the effect of the hardening model, the characteristic quantities for isotropic and kinematic hardening, i.e. the change of size of the isotropic yield surface $\Delta\sigma_y$ and the absolute value of the kinematic yield surface translation tensor $\|\alpha\|$ after cooling are presented in Fig. 9. It can be seen that isotropic hardening leads to a yield stress increase of up to 220 MPa in the heat-affected zone, whereas the values in the molten zone are about 130 MPa. Kinematic hardening, however, yields only a small residual effect on the yield surface in the heat-affected zone, but a much more pronounced effect in the molten zone, showing translations of the yield surface of about 12 MPa and 130 MPa, respectively.

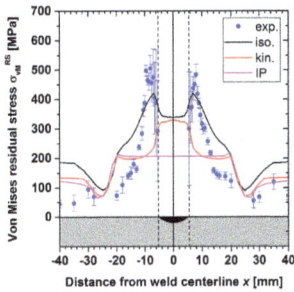

Figure 8: Von Mises residual stress determined by XRD (exp.) and computed with isotropic (iso.), kinematic (kin.) and no (IP) hardening

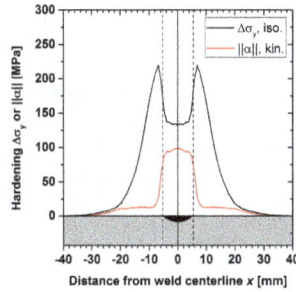

Figure 9: Change of size of isotropic yield surface $\Delta\sigma_y$ and absolute value of kinematic yield surface translation tensor $\|\alpha\|$

Discussion

Diffraction measurements revealed equivalent residual stresses that significantly exceed the initial yield stress of 255 MPa, see Figs. 4 and 5. For the welding geometry investigated here, the longitudinal stress component is dominant because of the self-constraint imposed by the plate, whereas the transverse residual stresses are relatively low due to free angular shrinkage.

The distributions of the microstrain, see Figs. 6 and 7, indicate significant mechanical hardening that must have occurred locally due to the thermomechanical welding cycle, permitting higher residual stresses. This is proven by the fact that the microstrain distributions coincide well with the areas with the highest residual stresses shown in Figs. 4 and 5.

The pronounced hardening effect explains why different hardening models yield distinctly different numerical results within 20 mm from the WCL. Here, the isotropic model gives an equivalent residual stress distribution that is qualitatively equal to the experimental one, see Fig. 8. Differences in the maximum residual stresses and the wider computed residual stress profile can be explained by the fact that external material data has been used which was only fitted to test results at room temperature. This cannot replace a full material characterization which is currently under progress and whose results will be used in future work.

The simulations with the kinematic model, however, underestimate the residual stresses in the heat-affected zone significantly. This is due to the fact that only a translation of the yield surface is considered. During heating, compressive stresses develop in the heat-affected zone due to the constrained thermal expansion, leading to plastic deformation and thus a translation of the yield surface in a certain direction. During cooling, however, the constrained shrinkage of the highly-heated areas results in tensile stresses and thus an earlier onset of plastic yield due to the previous shift of the yield surface. Thereby, the yield surface is shifted in almost opposite direction than before. In fact, this re-translation is not exactly opposite due to the non-proportional multiaxial stress state, but nevertheless leads to a rather small residual translation in the heat-affected zone, see $\|\alpha\|$ in Fig. 9.

Comparing the computational hardening effects, shown in Fig. 9, to the microstrain, which is an indicator of dislocations and thus mechanical hardening, it can be seen that only the results derived from the isotropic model are in qualitative agreement with the experimental findings, see Fig. 6. Thus, it can be inferred that the isotropic hardening model not only yields qualitatively correct residual stress distributions, but its choice is also legitimated by the microstructural mechanisms caused by the thermomechanical welding cycle.

Summary and future work

In this work, the effect of the choice of the hardening model in numerical welding simulations was analyzed studying a bead-on-plate weld on austenitic steel. It was found that the isotropic hardening model yields qualitatively correct residual stress distributions and that the continuum mechanical hardening effect agrees well with microstructural findings. This was neither the case for the kinematic hardening model nor for ideal plasticity.

However, these investigations could only yield qualitative results regarding the numerical simulations since material data from the literature has been used. Future work will therefore comprise the use of real material data, which also allows for the use of more sophisticated models of cyclic plasticity. Furthermore, the analysis of multilayer welds, i.e. of the effect of several thermomechanical cycles, and further microstructural analyses are planned.

Acknowledgement

This work was carried out in the research project Ni508/14-1 which was generously supported by the Deutsche Forschungsgemeinschaft (DFG). The authors would like to express their thanks for the support.

This work is based upon experiments performed at the Stress-Spec instrument operated by Helmholtz-Zentrum Geesthacht (HZG) at the Heinz Maier-Leibnitz Zentrum (MLZ), Garching, Germany. The authors gratefully acknowledge the financial support provided by HZG to perform the neutron scattering measurements at the MLZ.

References

[1] H. Wohlfahrt, Th. Nitschke-Pagel, K. Dilger, D. Siegele, M. Brand, J. Sakkiettibutra, T. Loose, Residual stress calculations and measurements – review and assessment of the IIW round robin results, Weld. World 56 (2012) 09/10, 120-140.

[2] J. Lemaitre, J.L. Chaboche, Mechanics of solid materials, Cambridge University Press, 1990.

[3] M.C. Smith, O. Muránsky, C. Austin, P. Bendeich, Q. Xiong, Optimised modelling of AISI 316L(N) material behaviour in the NeT TG4 international weld simulation and measurement benchmark, Int. J. Pres. Ves. Pip. (in press), https://doi.org/10.1016/j.ijpvp.2017.11.004.

[4] Heinz Maier-Leibnitz Zentrum, STRESS-SPEC: Materials science diffractometer, Journal of large-scale research facilities, 1, A6 (2015). http://dx.doi.org/10.17815/jlsrf-1-25.

[5] A.R. Stokes, A numerical Fourier-analysis method for the correction of widths and shapes of lines on X-ray powder photographs. Proc. Phys. Soc. 61 (1948) 382-391.

[6] T.H. de Keijser, J.I. Langford, E.J. Mittemeijer, A.B.P. Vogels, Use of the Voigt function in a single-line method for the analysis of X-ray diffraction line broadening, J. Appl. Cryst. 15 (1982) 308-314.

[7] O. Voß, Untersuchung relevanter Einflussgrößen auf die numerische Schweißsimulation, Dissertation, TU Braunschweig, 2001 (in German).

Residual Stresses 2018 – ECRS-10   Materials Research Forum LLC
Materials Research Proceedings 6 (2018) 233-238   doi: http://dx.doi.org/10.21741/9781945291890-37

Residual Stress Redistribution During Elastic Shakedown in Fillet Welded Plate

Jazeel R. Chukkan[1,2,a,*], Guiyi Wu[3,b], Michael E. Fitzpatrick[1,c], Xiang Zhang[1,d], and Joe Kelleher[4,e]

[1]Coventry University, Faculty of Engineering, Environment and Computing, Coventry, UK

[2]NSIRC, TWI Ltd, Granta Park, Great Abington, Cambridge, CB21 6AL, UK

[3]TWI Ltd, Granta Park, Great Abington, Cambridge, CB21 6AL, UK

[4]ISIS Pulsed Neutron & Muon Source, Harwell Campus, Didcot, Oxfordshire, UK

[a]chukkanj@uni.coventry.ac.uk, [b]guiyi.wu@twi.co.uk, [c]michael.fitzpatrick@coventry.ac.uk, [d]xiang.zhang@coventry.ac.uk, [e]joe.kelleher@stfc.ac.uk

Keywords: Residual Stress, Shakedown Limit Analysis, Elastic Shakedown, Neutron Diffraction, Stress Relaxation

Abstract. Welding residual stresses exist in various welded structures such as ships and offshore structures. According to the load levels during operation, the as-welded residual stresses can be relaxed or redistributed. The elastic shakedown phenomenon can be considered as one of the reasons for the stress relaxation or redistribution. This work studies the redistribution of welding residual stresses during different levels of shakedown in a fillet-welded plate manufactured in line with ship design and welding procedures. Fillet welding was performed on a ship structural steel, DH36. The fillet welds were subjected to different levels of shakedown under tensile cyclic load. Neutron diffraction was used to measure residual stresses in these plates in the as-welded state and after elastic shakedown. A mixed hardening model in line with the Chaboche model was determined for both weld and base material. A shakedown limit analysis based on plastic work dissipation was developed as the shakedown criterion to estimate the shakedown limit on the component. Further, the redistribution of residual stresses due to elastic shakedown was quantified through experimental measurements.

Introduction

Complex residual stresses are induced in load bearing members because of welding in offshore structures. For example, longitudinally stiffened plates which resist bending stresses consist of weld residual stresses due to fabrication [1]. Pre-existing residual stresses in load bearing members may be relieved or redistributed depending on the load levels during operation [2]. Elastic shakedown is one of the main phenomena contributing to this change in residual stress. Elastic shakedown is defined as a plastic deformation during the first few load cycles, followed by an elastic response which is associated with a limit called the shakedown limit [3]. The change in residual stresses is a result of plasticity induced during the first few cycles. The elastic shakedown limit in a component lies between the first yield and the plastic collapse load [4]. An applied load above the elastic shakedown limit will result in either plastic shakedown or ratcheting.

There are a number of research activities on the shakedown of residual stresses in welded structures [1,5-7]. However, the focus has been on the influence of shakedown on fatigue behaviour of the plate and hence on the relaxation of residual stresses at the surface of the plate. Paik et al. [6] conducted experiments to study the shakedown of residual stresses in butt-welded aluminium plates subjected to 3-point bending. Using hole drilling technique for residual stress measurements, they measured a 36% relaxation in the tensile longitudinal component of residual stresses after 3 load cycles with a magnitude equal to 88% of the yield strength. Lautrou *et al.* [5] studied a numerical model of a rectangular plate under plane stress conditions where initial residual stress was introduced using a non-uniform displacement at one side. The results show elastic behaviour within a few load cycles. Liangbi *et al.* [7] concluded that under constant amplitude load conditions, stress relaxation was limited to the first load cycle.

Fillet welds are more common in offshore applications and are considered more critical in fatigue due to the stress concentration at the weld toe. However, experimental research on the shakedown of residual stresses in fillet welds is very rare. Our recent study conducted on butt welded steel plates [8] found that the residual stresses, both transverse and longitudinal components, redistributed through the thickness depending on the relaxation at the top surface. This present paper is a continuation of that previous study conducted on butt-welded specimens [8]. The shakedown limit analysis was implemented in the fillet joints initially to determine the elastic shakedown limit. Absorbed plastic deformation energy at the end of each load cycle was used to determine the elastic shakedown limit [9]. Since linear or nonlinear isotropic and linear kinematic classics cannot represent realistic structural materials subjected to cyclic loadings, a mixed hardening model in line with Chaboche [10,11] was used for all numerical simulations. Experimental testing was used to study the effect of elastic shakedown in the redistribution of pre-existing residual stress fields in a fillet weld manufactured using DH36, a shipbuilding steel.

Methodology for estimating shakedown limit

Constitutive model: An elastic-plastic model with a combination of one isotropic hardening R and three kinematic hardenings α_1, α_2 and α_3, initially developed by Lemaitre and Chaboche [10] was implemented in the numerical model. Total strain, ε_{ij}^t is divided into elastic strains, ε_{ij}^e and plastic strains ε_{ij}^p. The elastic domain is described by a typical von Mises yield criterion $f=0$, where f is defined as:

$$f = J(\sigma - \alpha_1 - \alpha_2 - \alpha_3) - R - \sigma_y \qquad (1)$$

where σ is the Cauchy stress tensor, J is the von Mises equivalent stress and σ_y is the yield strength of the material. The isotropic hardening R and the kinematic hardening α_i are defined as below:

$$\dot{R} = b(Q - R)\,(p)^{\cdot} \qquad (2)$$

$$\dot{\alpha}_i = \frac{2}{3}C_i\dot{\varepsilon}_p - \gamma_i\dot{\alpha}_i\dot{p} \qquad (3)$$

where $b, Q, C_1, C_2, C_3, \gamma_1, \gamma_2$ and γ_3 are material parameters and $\dot{p}$ is the rate of accumulated plasticity. The material parameters obtained for DH36 base material and weld material in previous work [8] are used in this study which are given in Table 1.

Table 1: Chaboche constitutive parameters

DH36	E / GPa	ν	σ_y / MPa	Q / MPa	b	C_1 / MPa	γ_1	C_2 / MPa	γ_2	C_3 / MPa	γ_3
BM	200	0.3	350	−48.6	87.5	4360	16.4	38520	116	8000	40
WM	200	0.3	400	−102	14	8912	29.65	102300	400	8000	40

Shakedown limit analysis: The elastic shakedown limit of a material is between the first yield and the plastic collapse limit. Above the elastic shakedown limit, the material fails due to low-cycle fatigue in the presence of cyclic loading. In the previous work [8], it was shown that the plastic work dissipation at the end of each cycle can be considered as a shakedown criterion. On a component under cyclic loading with load cycles of nt to $(n+1)t$ time interval, the plastic strain increment after each load cycle can be expressed as:

$$\Delta\varepsilon_{ij}^{p} = \int_{nt}^{(n+1)t} \dot{\varepsilon}_{ij}^{p} \mathrm{d}t \tag{4}$$

Work done due to total strain can be decomposed into work done due to elastic strain, W^e and work done due to plastic strain, W^p. In the inital few load cycles, where plastic strains are active, the plastic work done is greater than zero. After a few initial load cycles, if the structure achieves elastic shakedown or the response is completely elastic, there will exist a time-independent residual stress field and the plastic work done will be equal to zero [8]. Since the structural response is elastic, it satisfies the lower bound theorem and hence it can be concluded that plastic work done at the end of individual load cycles, W^p defined in Eq .5 can be used as a shakedown criterion.

$$W^p = \sum_{n=1}^{N} \int_{V^e} \sigma_{ij} \Delta\varepsilon_{ij}^{p} \mathrm{d}V^e \tag{5}$$

where V^e is the volume of an element and N is the number of elements.

Numerical analysis

Weld model: A sequentially coupled thermo-mechanical analysis of the fillet welding was performed. An equivalent static heat source was used to introduce arc heat into the model. Filler deposition was simulated using a 'chunking' method as explained in R6 section III [12]. The model was initially developed as a T-joint with dimensions shown in Fig. 1a. The numerical fillet weld model was then cut to the dimension of a mechanical load model as shown in Fig.1b. Element activation/deactivation technique was implemented to simulate the cutting process.

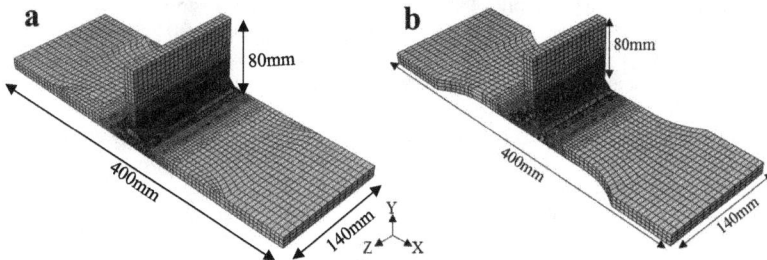

Fig. 1: a) Initial weld model, b) test specimen model following cutting

Shakedown limit: A half numerical model of the actual weld plate after cutting was developed to estimate the shakedown limit of the fillet weld under study. The procedure implemented and the detailed description of the procedure is explained in the previous work [8]. The elastic shakedown limit of the fillet weld geometry was determined using a step-by-step iterative procedure detailed in [8].

Experimental studies

Specimen manufacturing: Fillet weld was manufactured using Gas Metal Arc Welding process on DH36 steel plates. The dimensions of the base plate and the web were respectively $400 \times 140 \times 12.7$ mm³ and $140 \times 80 \times 12.7$ mm³, before cutting. The welding procedure specification was in accordance with the Lloyd's Register classification. Full restraint was implemented during welding using strongbacks to represent welding conditions in the ship and offshore structures. Thermocouples were used to monitor transient heat during welding. The welding set-up for the fillet weld is shown in Fig.2a. Thermocouple data were later used to

Residual Stresses 2018 – ECRS-10   Materials Research Forum LLC
Materials Research Proceedings **6** (2018) 233-238   doi: http://dx.doi.org/10.21741/9781945291890-37

validate the heat transfer model in the sequentially coupled welding simulation. Following welding, the weld plate was cut using EDM to prepare a mechanical test specimen.

Cyclic loading: Three tensile load cycles (R-ratio = 0) with 0.25 Hz frequency were applied on the plate. The maximum applied stress used on the plate was equivalent to achieving 68% of the yield strength of the parent material (350 MPa). The load was applied across the weld along the transverse direction of the specimen shown in Fig.2b.

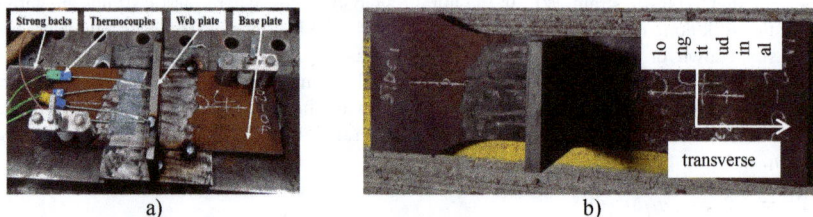

a)   b)

Fig. 2: a) Welding set-up b) Cyclic load specimen after EDM cutting

Neutron Diffraction: Neutron diffraction is a non-destructive technique which can measure strains through the thickness of a component. The ENGIN-X instrument at the UK's ISIS pulsed neutron source was used to determine residual stresses in the as-welded state and after different numbers of load cycles, at 30 locations in the plane shown in Fig.3.

Results and discussions

Elastic shakedown limit: The limit analysis FE model is subjected to increasing loads starting from 80% of yield strength of the parent material. The loading is gradually increased until a constant residual stress ceases to exist in the plate. The shakedown limit load factor λ estimated for the fillet weld is 1.2. This implies that in the absence of residual stress field any cyclic load within 1.2 times the yield strength applied in this set-up will achieve elastic-shakedown state.

As-welded residual stresses: Only the transverse residual stress distribution on the base plate is discussed in this paper for conciseness. Fig.4 compares the as-welded transverse residual stress component in the base plate predicted using finite element analysis and that measured using neutron diffraction. The stress values are taken from the plane across the weld at the mid-thickness. The comparison is drawn across the weld at 2.5mm below the top surface and 2.5mm above the bottom surface. High tensile transverse residual stresses are observed in both FE and experimental measurements at 2.5mm below the top surface of the base plate. The transverse residual stresses at 2.5mm above the bottom surface are significantly lower than those at 2.5mm below the top surface. The welding simulation captures the trend of residual stress distribution well. The minor difference between the predicted using welding simulation and experimental measurement may be due to experimental errors, including those arising from the EDM cutting, and material assumptions in the FE model. The maximum tensile transverse residual stress is almost identical for both numerical and experimental results, i.e., 250MPa. This value is equivalent to 71% of the yield strength of the base plate. It is demonstrated that the assumption of as-welded residual stress in structural integrity assessment guidance, BS 7910 is conservative.

Residual stress redistribution: Redistribution of transverse residual stresses following three tensile load cycles, measured using neutron diffraction is shown in Fig 5. The stresses at the top part of the base plate shown in Fig. 5a have a higher magnitude compared with the stress at the bottom part of the plate, shown in Fig. 5b. The redistribution of transverse stresses is minimum after the application of three load cycles. One of the reasons could be that the combination of pre-existing residual stress after EDM cuts and the applied load are not high enough to induce elastic shakedown in the transverse direction. Interestingly, it is found that after three loading cycles, the maximum value of tensile transverse residual stress at 2.5mm below the top surface is

Residual Stresses 2018 – ECRS-10   Materials Research Forum LLC
Materials Research Proceedings 6 (2018) 233-238   doi: http://dx.doi.org/10.21741/9781945291890-37

greater than that in the as-welded condition. However, the maximum tensile stress at 2.5mm above the bottom surface is similar in the as-welded condition and after one and three loading cycles. Based on this experimental evidence, the conservative level of the residual stress relaxation rule in BS 7910 may require re-investigation.

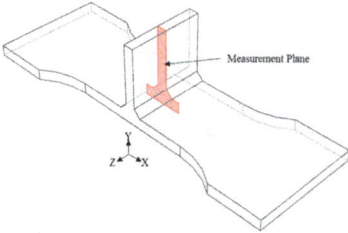

Fig. 3: Neutron diffraction measurement plane

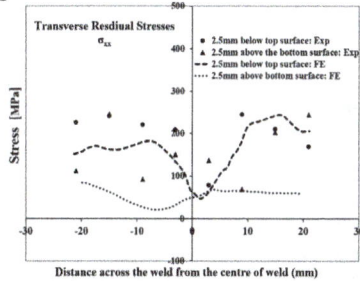

Fig. 4: RS redistribution across the weld at 2.5mm above and below the bottom and top surfaces respectively

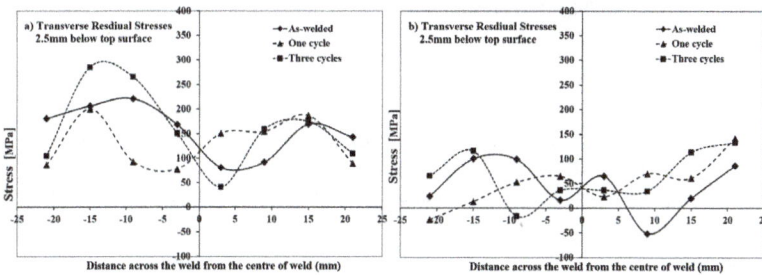

Fig. 5: Transverse residual stress redistribution by neutron measurement: a) 2.5mm below top surface, b) 2.5mm above the bottom surface

Effect of elastic shakedown: In the case of simple plates and weld joints like butt welds, if the combination of residual stresses along the load direction after a few cycles and the applied load is within the elastic shakedown limit, it can be said that the structure is stable and will achieve an elastic-shakedown state. However, in the case of fillet welds, since the residual stress component perpendicular to the loading direction is undergoing relaxation, both stress components parallel and perpendicular to the load direction should be considered to confirm elastic shakedown or steady state. If the applied load does not relax the residual stress components in the first few cycles, the structure is expected to follow cyclic plasticity or ratcheting over each load cycle.

It is noted that the relaxation or redistribution of residual stresses in fillet welds are very complex when compared to simple weld joints like butt welds [8]. This calls for further investigation of residual stresses in fillet welds under cyclic loading. The authors are currently investigating the effect of tensile load cycles on a plate with a fillet weld along the specimen's longer axis with a load application along with the longitudinal residual stress component.

Conclusions

The conclusions drawn from this study are:
- Shakedown limit of fillet welded geometry can be estimated based on a simplified method using plastic work done as a shakedown criterion.

- Experimental measurement of residual stress redistribution after three load cycles shows that there was only a minimal redistribution/relaxation in the transverse residual stress component even though the load applied was along this component.

Acknowledgements

This publication was made possible by the sponsorship and support of Coventry University and Lloyd's Register Foundation, a charitable foundation helping protect life and property by supporting engineering-related education, public engagement, and the application of research. The work was enabled through, and undertaken at, the National Structural Integrity Research Centre (NSIRC), a postgraduate engineering facility for industry-led research into structural integrity established and managed by TWI through a network of both national and international Universities.

References

[1] L. Gannon, N. Pegg, M. Smith, Y. Liu, Effect of residual stress shakedown on stiffened plate strength and behaviour, Ships and Offshore Structures. 8 (2013) 638-652. https://doi.org/10.1080/17445302.2012.664429

[2] J. Eckerlid, A. Ulfvarson, Redistribution of initial residual stresses in ship structural details and its effect on fatigue, Mar. Struct. 8 (1995) 385-406. https://doi.org/10.1016/0951-8339(94)00027-P

[3] M. Abdel-Karim, Shakedown of complex structures according to various hardening rules, Int. J. Pressure Vessels Piping. 82 (2005) 427-458. https://doi.org/10.1016/j.ijpvp.2005.01.007

[4] J. Bree, Elastic-plastic behaviour of thin tubes subjected to internal pressure and intermittent high-heat fluxes with application to fast-nuclear-reactor fuel elements, The Journal of Strain Analysis for Engineering Design. 2 (1967) 226-238. https://doi.org/10.1243/03093247V023226

[5] N. Lautrou, D. Thevenet, J. Cognard, A fatigue crack initiation approach for naval welded joints, 2 (2005) 1163-1170.

[6] J. Paik, O. Hughes, P. Rigo, Ultimate limit state design technology for aluminum multi-hull ship structures, Transactions-Society of Naval Architects and Marine Engineers. 113 (2006) 270-305.

[7] L. Li, T. Moan, B. Zhang, Residual stress shakedown in typical weld joints and its effect on fatigue of FPSOs, (2007) 193-201.

[8] J.R. Chukkan, G. Wu, M.E. Fitzpatrick, E. Eren, X. Zhang, J. Kelleher, Residual stress redistribution during elastic shake down in welded plates, 12th International Fatigue Congress (FATIGUE 2018), MATEC Web Conf.165 (2018) 21004.

[9] Y. Sun, S. Shen, X. Xia, Z. Xu, A numerical approach for predicting shakedown limit in ratcheting behavior of materials, Mater Des. 47 (2013) 106-114. https://doi.org/10.1016/j.matdes.2012.12.049

[10] J. Lemaitre, J. Chaboche, Mechanics of Solid Materials, Cambridge university press, 1994.

[11] R5, Assesment procedure for the high temperature response of structures, (2014).

[12] British Energy, R6 Revision 4: Assessment of the Integrity of Structures Containing Defects, Amendment 10, 2014.

Residual Stresses 2018 – ECRS-10   Materials Research Forum LLC
Materials Research Proceedings 6 (2018) 239-244   doi: http://dx.doi.org/10.21741/9781945291890-38

Modelling the Welding Process of an Orthotropic Steel Deck

Evy Van Puymbroeck[1,a,*], Nouman Iqbal[1,b] and Hans De Backer[1,c]

[1]Ghent University, Department of Civil Engineering EA15, Technologiepark 904, 9052 Ghent, Belgium

[a]Evy.VanPuymbroeck@UGent.be, [b]Nouman.Iqbal@UGent.be, [c]Hans.DeBacker@UGent.be

Keywords: Finite Element Modelling, Welding Simulation, Residual Welding Stresses, Orthotropic Bridge Deck

Abstract. A three-dimensional finite element welding simulation procedure is developed with the software Siemens NX and solver type SAMCEF in order to determine the residual stresses of a welded component of an orthotropic bridge deck. The welding process of a deck plate which is welded to a closed trapezoidal stiffener is simulated. A decoupled thermal-mechanical analysis is performed. During the thermal analysis, the temperatures introduced by the passage of the welding torch are calculated for different time steps. This temperature field is used during the thermal analysis to determine the residual welding stresses for the same time steps. The decoupled thermal-mechanical analysis gives the distribution of the residual stresses in the longitudinal direction. Only this direction is discussed since this is the direction that follows the welding path. On the deck plate there are tensile yield residual stresses near the weld region. In between the welded webs of the stiffener, there are compressive residual stresses. For the longitudinal stiffener, there are again tensile yield residual stresses near the weld which decrease at a greater distance and turn into compressive residual yield stresses. The finite element model results in a residual stress distribution introduced by the welding operation. This distribution can be used in future research to determine the effect of residual stresses on the fatigue life of this welded component in the orthotropic bridge deck.

Introduction

During the welding operations of bridge components, residual stresses are introduced due to local plastic deformation. The presence of residual stresses affect the final in-service performance of the weldment, such as fatigue and brittle fracture behavior [1]. Welding residual stresses may be beneficial or detrimental with respect to the load-induced stresses, depending on the magnitude, sign and distribution of the residual stresses. Tensile residual stresses, especially in the magnitude of the material's yield stress are detrimental because they increase crack growth and prevent crack closure. On the other hand, compressive residual stresses increase the crack closure, thus decreasing the crack growth rate. To increase the accuracy of fatigue life prediction of welded bridge components, it is essential to incorporate the effects of the residual welding stresses into the structural integrity and fatigue assessment [2]. Therefore, reliable residual stress prediction has to be developed. The advantage of a reliable finite element model is that it can be used for a wide range of welding conditions and bridge component geometries [3].

A three-dimensional finite element welding simulation is developed with the software Siemens NX in order to determine the residual stresses of a welded component of an orthotropic bridge deck. The welding process of a deck plate which is welded to a closed trapezoidal stiffener is simulated. A decoupled thermal-mechanical analysis is performed. During the thermal analysis, the temperatures introduced by the passage of the welding torch are calculated for different time steps. This temperature field is used during the thermal analysis to determine the residual welding stresses for the same time steps.

Residual Stresses 2018 – ECRS-10   Materials Research Forum LLC
Materials Research Proceedings **6** (2018) 239-244   doi: http://dx.doi.org/10.21741/9781945291890-38

Geometry

The welding process of the longitudinal stiffener-to-deck plate weld from an orthotropic bridge deck plate will be determined. The geometry of a large scale test specimen is used. It is a steel bridge deck with closed trapezoidal stiffeners constructed with constructional steel S235 with a yield strength of 235N/mm. The cross section of the longitudinal stiffener connected to the deck plate is the only interest for the welding simulation procedure. The closed trapezoidal longitudinal stiffeners have a height of 300mm and on top, they are 300mm wide while a width of 125mm is present at the lower soffit. The deck plate has a thickness of 15mm and the longitudinal stiffeners are 6mm thick. The dimensions of this connection is shown in Fig. 2. This cross section will be further considered for the welding simulation and a length of 250mm is assumed. This is the distance in between two tack welds.

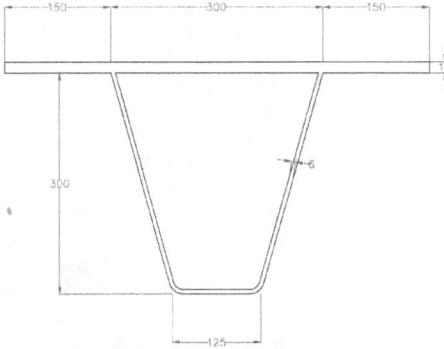

Fig. 2 Dimensions of longitudinal stiffener connected to deck plate (dimensions in mm)

Welding procedure

To construct the orthotropic bridge deck, the longitudinal stiffeners are welded to the deck plate with twin wire submerged arc welding. The orthotropic bridge deck is inverted to execute the welding operation. The stiffeners are kept on the right position and before they are entirely welded to the deck plate with the submerged arc welding, tack welds are provided to ensure their position on the deck plate and to provide distortion control of the stiffener. These tack welds are ground down before the welding is executed to minimize the tack weld profile. The length of the tack welds is equal to 25mm.

The welding can only be executed from the outside of the stiffener since the inside cannot be reached with a welding torch. Therefore, the welding process has to result in a weld melt-through towards the inside of the stiffener to obtain a good connection of the stiffener with the deck plate.

The parameters of the welding procedure for the simulated orthotropic bridge deck are given by the manufacturer. The diameter of the electrodes D is equal to 2mm while the extension of an electrode L is equal to 30mm. The welding is executed with a current I of 780A and a voltage of 29V. The advancing speed of the welding torches S is 950mm per minute or 15.83mm/s. DC+ submerged arc welding is used, this means that the current flows from the electrode to the base metal.

Finite element model

To determine the residual stress distribution caused by welding the longitudinal stiffener to the bridge deck plate, a decoupled thermal-mechanical analysis is performed with the software Siemens NX with solver SAMCEF. First a thermal analysis will be performed to calculate the temperature field introduced by the welding process. Subsequently a mechanical analysis is performed to determine the residual stresses. Solving the thermal and mechanical fields

Residual Stresses 2018 – ECRS-10   Materials Research Forum LLC
Materials Research Proceedings 6 (2018) 239-244   doi: http://dx.doi.org/10.21741/9781945291890-38

simultaneously is not recommended since it requires a large un-symmetric system of equations to be solved which is computationally demanding [4]. Therefore, the welding simulation is described by decoupled quasi-static thermo-elasticity equations. Thus the deformations are dependent on temperatures but temperatures are independent of deformation. The thermal field as a function of time is determined during the thermal analysis. This temperature field is used as input for the mechanical analysis [3].

A finite element mesh with three-dimensional 8-node brick elements is used in the thermal and mechanical analysis. The mesh comprises 27624 elements and 38870 nodes. Linear brick elements are the basic recommendation in plasticity which will arise due to the large temperatures of the steel during welding [4]. A true stress-plastic strain behavior is specified, this is natural strain which is conjugated to Cauchy stress and an isotropic hardening behavior is assumed.

The welding process will be modelled by subjecting the elements of the deck plate and the stiffener to a heat flux. The weld material itself will not be modelled but the width of the weld is taken into account by applying heat flux to the adjacent elements on the deck plate and stiffener.

Thermal analysis
In the analysis of the mechanical effects of welding, the fluid flow is not taken into account. This high temperature region is approximated by a given heat input in the thermal analysis. Thus, the actual physics that take place in the weld pool are simplified considerably and replaced by a heat input model [4]. During the thermal analysis, the temperature field is obtained by specifying the heat input.

The parameters necessary to describe the heat input to the weldment from the arc are essential to accurately compute the transient temperature field. The heat input model of Goldak [5], the so-called double ellipsoid heat source is used. This is a three-dimensional double ellipsoidal heat flux model used to examine the three-dimensional temperature, stress and strain fields. The heat flux can be determined based on the weld geometry [6].

For the thermal material behavior, the microstructure change is ignored and it is assumed that the material properties depend only on temperature [7]. The thermal conductivity and specific heat are specified in function of temperature. A constant mass density of the material is specified since the change in density is computed simultaneously with the temperatures when deformations are calculated. The mass density is 7872 kg/m^3.

Boundary conditions must be employed to account for surface heat losses. Natural convective heat transfer will be present and a convection coefficient of 15 W/m^2°C is specified. The emissivity in function of temperature is also considered. An initial temperature of 18°C is assumed for the whole structure.

Initially, the meshes of the bridge deck and the longitudinal stiffener are not connected to each other. Only after the passage of the welding torch, when a temperature of 1440°C is reached, the meshes are connected. This is realized by the finite element model with the TWELD command which defines the source welding temperature threshold that initiates the gluing and connecting of the meshes.

The heat input distribution is used to model the welding torch with a certain advancing speed. The heat flux for elements subjected to the welding process is calculated with the heat input distribution taking into account the dimensions of the elements. The heat flux is calculated for different time steps to simulate the passage of the welding torch with the considered welding speed.

The thermal analysis simulates the heat introduced in the orthotropic bridge deck by welding the longitudinal stiffener to the deck plate and calculates the temperatures of the steel bridge deck during the welding process. The applied heat flux is the largest in the connection of the

Residual Stresses 2018 – ECRS-10   Materials Research Forum LLC
Materials Research Proceedings **6** (2018) 239-244   doi: http://dx.doi.org/10.21741/9781945291890-38

deck plate with the stiffener where the welding torch is present and it gradually fades out when the distance to the connection increases. The welding torch also preheats a very small area in front of the torch where the heat source is going to pass. The melting temperature of the material is defined as 1440°C, at this temperature the melted material is located in the fusion zone. High temperatures are present at immediate vicinities of the fusion zone which defines the heat affected zone. The calculated temperatures for the front face of the welded connection for two time steps are displayed in Fig. 3.

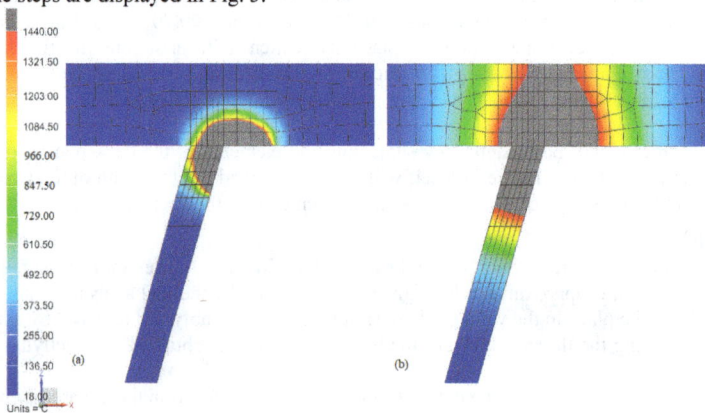

Fig. 3 Results from thermal analysis for (a) t=2.289s and (b) t=15.2055s for the front face

The fused zones and the weld penetration can clearly be recognized and is indicated by the grey zone which represents temperatures higher than 1440°C. This is the melting temperature which activates the command TWELD resulting in a connection of the meshes of the stiffener and the deck plate. It can be clearly seen that the temperature for the entire connection area is higher than this temperature resulting in a complete weld melt-trough. The zone with high temperatures increases when the time step is increased. This is due to the passage of the welding torch and the penetration of the weld into the material.

When the welding torch is passed by the entire length of the bridge component, the temperatures have to decrease due to the interaction of the model with the environment. The heat losses caused by natural convective heat transfer and radiation result in a decrease of the temperature for the bridge deck. This decrease in temperature can be recognized in the temperature results for later time steps.

Mechanical analysis

After the thermal analysis, the temperatures introduced during the welding of the orthotropic bridge deck are stored. The subsequent mechanical analysis introduces these temperatures as a time dependent load onto the model. A quasi-static mechanical analysis is sufficient since the inertia effects in the welding process is negligible due to the simplification of the weld pool region. Large deformation and rotation effects are the result of the large welding deformations. These large deformations are taken into account during the mechanical analysis which makes it more complex than the thermal analysis. Accounting for large deformations is realized by making use of the logarithmic natural strain definition [4].

It is important to have a correct description of the material behavior in order to have an accurate finite element model. The Young's modulus, thermal expansion and stress-strain relationship in function of temperature are defined. The shear modulus is equal to 80 GPa and

Residual Stresses 2018 – ECRS-10   Materials Research Forum LLC
Materials Research Proceedings **6** (2018) 239-244   doi: http://dx.doi.org/10.21741/9781945291890-38

the Poisson ratio is 0.3 in every direction of the material. These properties are the same for all temperatures.

Initially, the longitudinal stiffener and the deck plate are not connected. Before the welding process starts, the bridge deck is inverted and the stiffener is positioned on top of it. To ensure the exact location of the stiffener relative to the deck and to facilitate the welding process, two tack welds with a length of 25mm are provided at the beginning and ending of the considered cross section. The deck plate is positioned on top of a series of springs to simulate a rigid surface. The own weight is also assigned to the model by specifying the gravitational acceleration.

The boundary constraints must result in static equilibrium without introducing additional stresses. However, some additional constraints must be defined in order to prevent free body movement of the model. Therefore, symmetry points are defined and considered fixed in two directions in such a way that the deck plate and the stiffener cannot move in a horizontal direction during the welding process.

The load for the mechanical analysis consists out of the temperatures calculated during the thermal analysis and are implemented in the model as a time-dependent loading.

The residual stresses at the final time step, when the test piece is cooled down to environmental temperature after the welding, are of importance since these are the welding residual stresses that are present for the entire lifetime. To evaluate the residual stresses, the results in a zone at 25mm distance from the edges will be neglected to avoid the influence of the boundary conditions.

Residual stress results

The residual stresses for the orthotropic bridge deck will be discussed for the transverse direction only (perpendicular to the welding direction). A general overview of the residual stresses for the top of the deck is given in Fig. 4. The residual stress results for the middle section for the top and bottom of the deck plate are shown in Graph 1. The results for the longitudinal stiffener are not shown but will be discussed.

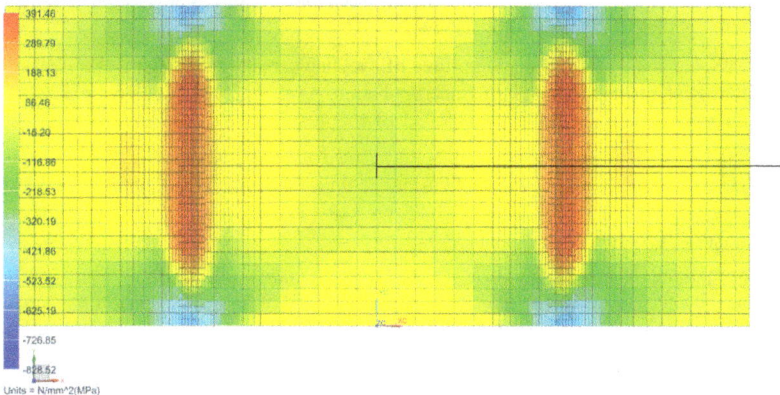

Fig. 4 Transverse residual stresses on top of the deck plate

There are tensile yield stresses present on top of the deck plate in the region near the weld seam which diminish at a greater distance from the welding line. On the bottom of the deck plate, the tensile yield stresses are at a distance of 20mm symmetrical around the weld toe, near the weld seam there are smaller tensile residual stresses present. The residual stress at the side of the test piece is close to zero while there is a compressive residual stress of about 45MPa in the middle. The longitudinal stiffener has a small tensile residual stress peak of 200MPa near the weld toe root while the residual stress at the weld toe is already close to zero. Along the stiffener there are small tensile residual stresses present and at the lower soffit, there is no longer residual stress present. These results agreed well with experimental data obtained with incremental hole-drilling.

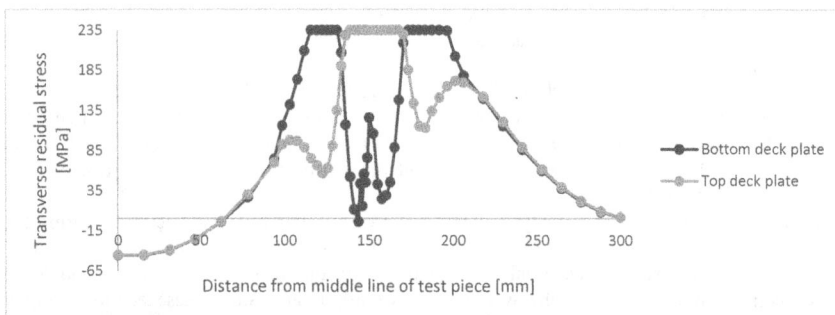

Graph. 1 Transverse residual stresses on for top and bottom of the deck plate

Summary

The residual welding stresses of the longitudinal stiffener-to-deck plate weld from an orthotropic bridge deck plate are determined. These residual stress results can be used in future research to increase the accuracy of fatigue life prediction of the welded orthotropic bridge component.

References

[1] S.W. Wen, P. Hilton and D.C.J. Farrugia, Finite element modelling of a submerged arc welding process, Journal of Materials Processing Technology, 2001, Vol. 119, pp. 203-209. https://doi.org/10.1016/S0924-0136(01)00945-1

[2] Z. Barsoum and A. Lundbäck, Simplified FE welding simulation of fillet welds – 3D effects on the formation residual stresses, Engineering Failure Analysis, 2009, Vol. 16, pp. 2281-2289. https://doi.org/10.1016/j.engfailanal.2009.03.018

[3] X. Shan, C.M. Davies, T. Wangsdan, N.P. O'Dowd and K.M. Nikbin, Thermo-mechanical modelling of a single-bead-on-plate weld using the finite element method, International Journal of Pressure Vessels and Piping, 2009, Vol. 86, pp. 110-121. https://doi.org/10.1016/j.ijpvp.2008.11.005

[4] L.-E. Lindgren, Numerical modelling of welding, *Computer methods in applied mechanics and engineering*, 2006, Vol. 195, pp. 6710-6736. https://doi.org/10.1016/j.cma.2005.08.018

[5] J.A. Goldak, A.P. Chakravarti and M. Bibby, A new finite element model for welding heat sources, *Metallurgical Transactions 15B*, 1984, pp. 299-305. https://doi.org/10.1007/BF02667333

[6] E.A. Bonifaz, Finite element analysis of heat flow in single-pass arc welds, Welding research supplement, 2000, pp. 121s-225s

[7] L.-E. Lindgren, Finite element modelling and simulation of welding part 2: Improved material modelling, *Journal of Thermal stresses*, 2001, Vol. 24, pp. 195-231. https://doi.org/10.1080/014957301300006380

Residual Stresses 2018 – ECRS-10
Materials Research Proceedings 6 (2018) 245-250

Materials Research Forum LLC
doi: http://dx.doi.org/10.21741/9781945291890-39

Strength Calculation of Stiffened Structures Taking Into Consideration Realistic Weld Imperfections

Christoph Stapelfeld[1,a,*], Benjamin Launert[1,b], Hartmut Pasternak [1,c], Nikolay Doynov[2,d], Vesselin Michailov[2,e]

[1] Chair of Steel and Timber Structures, Konrad-Wachsmann-Allee 2, Brandenburg University of Technology, Cottbus, Germany

[2] Chair of Joining and Welding Technology, Konrad-Wachsmann-Allee 17, Brandenburg University of Technology, Cottbus, Germany

[a]christoph.stapelfeld@b-tu.de, [b]benjamin.launert@b-tu.de, [c]hartmut.pasternak@b-tu.de, [d]nikolay.doynov@b-tu.de, [e]vesselin.michailov@b-tu.de

Keywords: Strength Calculation, Welding, Imperfections, Load Capacity, Stiffened Structures, Curved Structures, Hybrid Model

Abstract. The topic of this article is the application of an analytical numerical hybrid model for a realistic prediction of imperfections induced by welds. At the beginning, the analytical model, its physical basis as well as the physical interrelationships are explained. This is followed by the explanation of the coupling procedure between the analytical model and the numerical calculation. Afterwards, the coupled hybrid model is applied on the investigated stiffened curved structure for the determination of the weld imperfections. An ultimate load analysis gives information about the load carrying behavior under axial loading. The results are compared against the traditional approach using eigenmode-based imperfections. The comparison underlines the potential additional utilization of load bearing capacity by this new approach.

Introduction

The strength calculation of stiffened plates by the finite element method (FEM) has been part of the state of the art for a long time. Geometrical nonlinearities as well as the nonlinear material behavior are considered within the calculation. To simplify, both types of imperfections, geometrical and structural ones, are mostly combined in these strength calculations being considered as equivalent geometrical imperfections. Values for standard cases are included in EN 1993-1-5 in case of plated structures or slightly curved panels [1]. Because there are no specific rules for considering imperfections and its scaling in a load capacity calculation of stiffened curved panels they may be also assumed according to EN 1993-1-5 [2,3] as a first approach. However, it remains unclear to some extent how accurate these geometrical imperfections represent the actual residual stresses and deformations caused by welds, especially for more complex cases. The significance of numerical load capacity calculations could be increased enormously if these imperfections were known more exactly and could be considered directly during the computation.

Nowadays, the residual stresses and deformations can be determined by means of a thermo-mechanical FE simulation achieving quite realistic values. However, relevant structures and weld length are very large what leads to enormous calculation time and a huge demand of storage capacity [4]. Simplified numerical approaches are available and able to remedy this situation.

However, the application of these models partly demands more expertise than a conventional thermomechanical FE calculation [5] or the simplifications are so extensive that the weld imperfections calculated by the approach partially lose their validity [6]. In order to be able to take weld distortions and residual stresses directly into account in a load capacity calculation,

Residual Stresses 2018 – ECRS-10   Materials Research Forum LLC
Materials Research Proceedings 6 (2018) 245-250   doi: http://dx.doi.org/10.21741/9781945291890-39

fast but still sufficiently accurate procedures that, at the same time, are easy in their application are essential.

The Coupled Analytical Numerical Hybrid Model

The basic idea of the coupled analytical numerical hybrid model [7] is the linking of the major advantages of both, analytical and numerical procedures. On the one hand, the matchless very short calculation time of the analytical shrinkage force model and its simple application, and on the other hand the possibility to conduct a FE simulation to calculate stresses and distortions at any location of complex welded structures. According to this, all the determining factors on quality and quantity of weld imperfections are passed to an analytical calculation program, capturing the mathematical approach of the shrinkage force model. The output is a mechanical load and the point of action in longitudinal and transversal direction, equivalent to the heat effect of welding. The loads are then applied to the FE model of the weld structure and the distortions and stresses are calculated by a nonlinear elastic calculation. The influence of the weld sequence on the arising weld imperfections is captured by a back coupling. The numerically calculated stresses in the regarded weld caused from a previous weld are submitted to the analytical calculation. The results of the application of the hybrid model are subsequently superposed with additional fabrication tolerances followed by the load capacity calculation, Fig. 1.

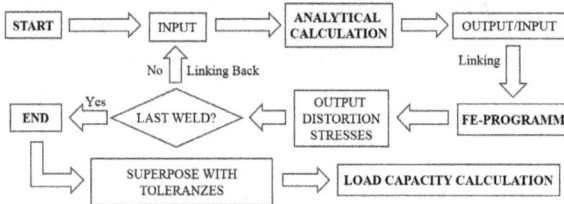

Figure 1: Scheme of the Load Capacity Calculation Taking Into Consideration Realistic Weld Imperfections.

Weld imperfections depend significantly on the maximum temperatures that every point perpendicular to the weld direction is exposed to and the stiffness of the structure. Equations for the calculation of the maximum temperatures were derived by Rykalin [8] constituting the basis of the shrinkage force model. For calculating a force alongside the weld, Okerblom [9] considered the border case of a line source in a rigid thin plate and integrated the thermal strains over the zone of plastic deformations:

$$F_x = v_x E = \int_{\varepsilon_F}^{2\varepsilon_F} 0,484 \frac{\alpha}{c\rho} \frac{q_s}{\varepsilon_{th}} E d\varepsilon_{th} = 0,355 \frac{\alpha}{c\rho} q_s E , \qquad (1)$$

with v_x, the shrinkage volume per unit length, ε_F, the yield point, α, the thermal expansion coefficient, c, the specific heat capacity, ρ, the density, q_s, the heat input per unit length, ε_{th}, the thermal strains and E, the Young's modulus. Kuzminov [10] enlarged the application area as well as increased the accuracy of the analytic approach significantly by including the second border case of a point source on a semi-infinite body by means of an influencing factor $K_{\chi\delta}$, mostly depending on the plate thicknesses and the heat exchange. Furthermore, the model considers the finite stiffness of the structure, K_k and the effect of existing stresses in the weld, K_σ:

$$F_x = 0,355 \frac{\alpha}{c\rho} q_s E K_{\chi\delta} K_k K_\sigma . \qquad (2)$$

Residual Stresses 2018 – ECRS-10   Materials Research Forum LLC
Materials Research Proceedings 6 (2018) 245-250   doi: http://dx.doi.org/10.21741/9781945291890-39

The longitudinal shrinkage force F_x is proportional to the width of the plastic zone:

$$b_{PZ} = \frac{F_x}{\varepsilon_m E \delta} \,. \tag{3}$$

Here, ε_m is the averaged yield strain and δ is the plate thickness. The transversal shrinkage caused by welds depends on further effects like heating through the thickness, stiffening cross beams, the effect of longitudinal strains, the degree of excessive heat as well as the effect of forced heat exchange [7,10]. Both, the longitudinal and the transversal shrinkage force have an appropriate point of action being equivalent to the centre of the plastic zone. They are significantly influenced by the material and its properties as well as the heat input per unit length and the plate thickness. Depending on the points of action, z_c, an equivalent linear strain distribution over the plate thickness can be calculated. Considering the point of origin in the centre of gravity of the plates cross section, the strain distribution ε_z follows as:

$$\varepsilon(z) = \varepsilon_m + \frac{12\varepsilon_m z_c}{\delta^2} z \,. \tag{4}$$

The calculated linear strain distributions over the plate thickness are now applied to the analytically calculated width of the plastic zone in the FE model of the welded structure. In the case of shell elements, each integration point is loaded with the strains according to eq. 4. The linking by means of stresses requires the consideration of the Poisson's ratio, eq. 5. The procedure is already validated and verified with Ansys®, LS Dyna®, Sysweld® and Abaqus®.

$$\sigma_{x,y} = \frac{\varepsilon_{x,y} E + \varepsilon_{y,x} E \nu}{1 - \nu^2} \,. \tag{5}$$

Previous comparisons of the results of the analytical numerical hybrid model with experimental data have already shown very good agreement [4,7]. The field of application covers different steel grades, several types of joints and the established welding techniques. Currently, the largest calculated structure was a deck section from ship building with 90 welds [7]. The CPU calculation time using a customary computer was approximately 30 min and it took about 30 h for preparing the CAD model and creating the FE mesh. Compared with a conventional thermomechanical simulation the calculation time saving using this new approach was more than 99 % [4,7].

The Finite Element Model of the Curved Stiffened Structure
For the application of the hybrid model and a subsequent load capacity calculation a panel similar to the panels of the Lyon Confluence footbridge in France, introduced in [2], was chosen, Fig 2. For defining the boundary conditions, a cylindrical coordinate system was created. The four edges of the panel are simply supported, $u_R = 0$. To restrain any movement or rotation of the structure, two nodes in the middle of the curved edges are fixed alongside θ and one node in the middle of the panel is fixed in the direction of the z coordinate. The steel grade is S355. Correspondingly, the Young's modulus is $E = 210$ GPa, the Poisson's ratio is $\nu = 0.3$ and the yield strength is $\sigma_F = 355$ MPa. For the GMNIA as well as the load capacity calculation with the consideration of realistic weld imperfections a slope of E/100 and an ultimate stress of 470 MPa was assigned. The structure is discretized with four-node shell elements (S4R in Abaqus®) with three integration points over the elements thickness. A quite fine mesh with an approximated element edge length of 30 mm is used.

Residual Stresses 2018 – ECRS-10   Materials Research Forum LLC
Materials Research Proceedings 6 (2018) 245-250   doi: http://dx.doi.org/10.21741/9781945291890-39

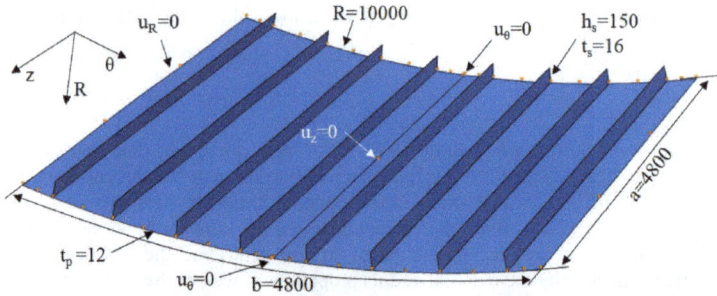

Figure 2: Geometrical Properties in mm and Boundary Conditions of the Curved Panel.

Considering the different plate thicknesses, the panel and the stiffeners are loaded with different shell edge loads targeting a homogeneous axial pressure, Fig 3(a). For the verification of the loading as well as the boundary conditions, the curvature of the plate was removed and a load capacity calculation was carried out. The numerically calculated plastic normal force was then compared with the analytical solution: $F = \sigma_F A = 27.26$ MN. For the determination of the lowest ultimate load, several load deformation calculations were executed considering different buckle modes as well as combinations of them. As a result, it turned out that the imperfection according to the first buckle mode, Fig 3(b), leads to the lowest load capacity.

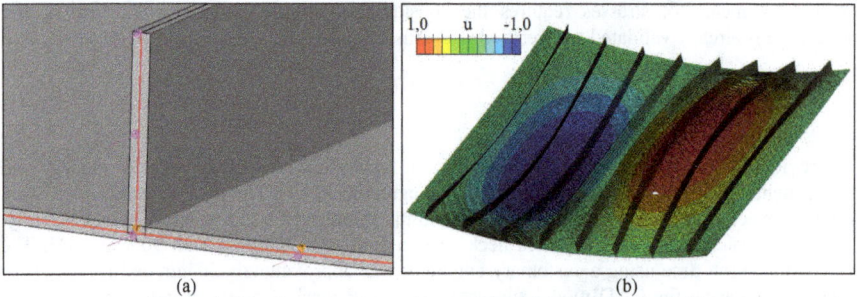

(a)   (b)

Figure 3: Loading Condition (a), and the First Buckle Mode (b).

Application of the Hybrid Model for Calculating Weld Imperfections and Results of a Subsequent Load Capacity Analysis

The requirements for the calculation of the mechanical loads with the analytical shrinkage force model are information about the type of joint and its dimensions, the material data as well as the welding technique and the welding parameters. Here, the initial assumption is made that the stiffeners are only welded one sided. The material data correspond to the material data of the steel S355J2. The weld technique is conventional MAG-welding with welding parameters targeting a fillet weld with a design throat thickness of 7 mm, which resulted in a heat input per unit length of $q_s = 2400$ J/mm. In this example case, the linking between the analytical model and the numerical approach was done by initial stresses, Table 1.

Table 1: Analytically Calculated Stresses for the Numerical Calculation of Weld Imperfections.

	Curved Panel			Stiffener		
Location of the Three Integration Points	Upper	Middle	Lower	Upper	Middle	Lower
Width of Plastic Zone b_{pz} [mm]		29.1			29.1	
Stresses in Weld Direction σ_z [MPa]	849.6	597.4	345.2	807.7	593.8	379.9
Stresses Transversal to the Weld σ [MPa]	σ_θ =1313	σ_θ =801	σ_θ =289.5	σ_R=1341	σ_R=789	σ_R=237

Residual Stresses 2018 – ECRS-10   Materials Research Forum LLC
Materials Research Proceedings **6** (2018) 245-250   doi: http://dx.doi.org/10.21741/9781945291890-39

For the proper loading of the FE model, two sections were created at each weld, representing the idealized zone of plastic deformations. The width of the sections is specified by the width of the analytically calculated plastic zone. When creating the sections, the position of the welds on the outside of the stiffeners is considered, fig 4(a). The initial stresses in longitudinal and transversal direction are defined as "initial state" in Abaqus®. Geometric nonlinear behavior is considered within the elastic calculation. The calculated deformation state indicates buckling of the curved panel in the positive z-direction and a tilt of the stiffeners towards the welds, Fig 4(b).

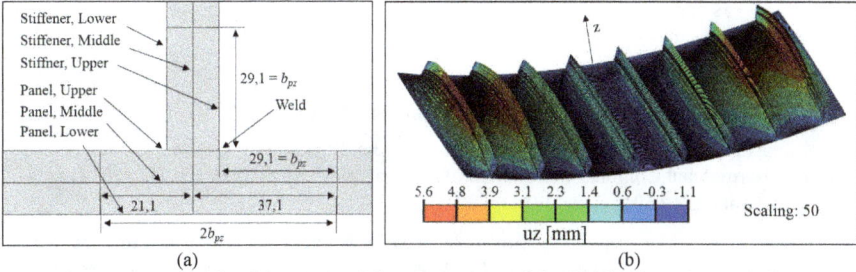

(a)   (b)

Figure 4: Assembly of the FE-Model in the Region of the Welds (a) and Out of Plane Deformations (b).

The ultimate load calculations were done assuming two kinds of imperfections. The first case is a combination of the geometrical and structural weld imperfections and manufacturing failures that were captured here by assuming an additional initial global buckle of b / 1000 = 4.8 mm. In the second case, all imperfections in the structure are caused only by the eight welds representing the case of an ideal model prior to welding. The comparison of the ultimate loads with the results of the calculation considering the critical first buckling mode as initial imperfection has shown increased load capacities of 16 % for case one and 31 % for the second case, Fig 5.

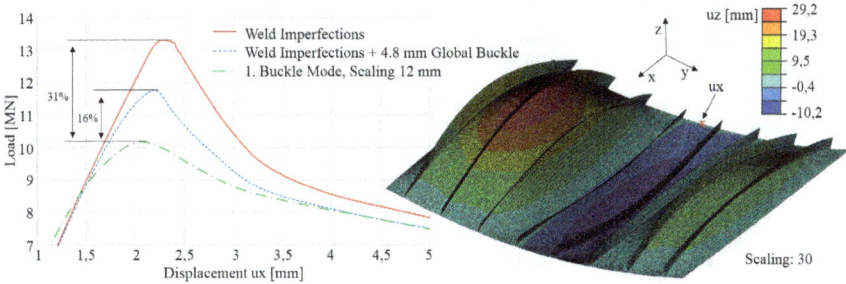

Figure 5: Load Displacement Curves and Corresponding Deformation State at Ultimate Load.

Summary

Numerous applications of the analytical numerical hybrid model for the calculation of weld imperfections indicate the significance of the results. In the case of consideration, the calculated longitudinal and transversal shrinkages as well as the out of plane deformations correlate well with known empirical values for one sided fillet welds. The use of calculated realistic weld imperfections instead of adequate imperfections in the load capacity calculation of the stiffened curved panel results in an increasing of the ultimate load in the range of 16 % till 31 %. However, a general statement on the effects of weld imperfections in load capacity calculations cannot be given. The results depend on the geometry, the steel grade, the quantity and positions of the weld as well as the weld parameters. A realistic consideration of these parameters is only possible by an accurate approach of the type that has been presented in this article. However, the approach in combination with loading calculations has to be validated by means of more experimentally determined load displacement curves. These works are ongoing at the moment.

Acknowledgements

These works are part of the IGF project No. 19173 BR of the German Research Association for Steel Application (FOSTA). This project is kindly funded by the German Federal Ministry of Economic Affairs and Energy (BMWi) by the AiF (German Federation of Industrial Research Associations) as part of the program for support of the Industrial Cooperative Research (IGF) on the basis of a decision by the German Bundestag.

References

[1] Eurocode 3: Bemessung und Konstruktion von Stahlbauten (EN 1993-1-5). 2017

[2] K. L. Tran, C. Douthe, K. Sab, J. Dallot, L. Devaine, Buckling of Stiffened Curved Panels Under Uniform Axial Compression, Journal of Constructional Steel Research 103 (2014) 140-147. https://doi.org/10.1016/j.jcsr.2014.07.004

[3] T. Manco, J. P. Martins, C. Rigueiro, L. S. da Silva, Numerical Analysis of Stiffened Curved Panels Under Compression, International Conference on Steel and Aluminium Structures, Hong Kong, China, 2016 1-12 (No. 28).

[4] V. Michailov, R. Ossenbrink, C. Stapelfeld, Anwendungsnahe Schweißsimulation komplexer Strukturen, DVS-Berichte, Band 282, DVS Media GmbH, Düsseldorf, 2010.

[5] Y. G. Duan, Y. Vincent, F. Boilot, J. B. Leblond, J. M. Bergheau, Prediction of welding residual distortions of large structures using a local/global approach, Journal of Mechanical Science and Technologie 21, 10 (2007) 1700-1706. https://doi.org/10.1007/BF03177397

[6] D. Thikomirov, B. Rietman, K. Kose, M. Makkink, Computing Welding Distortion: Comparison of Different Industrially Applicable Methods, SHEMET 11 (2008) 195-202.

[7] C. Stapelfeld, Vereinfachte Modelle zur Schweißverzugsberechnung, Dissertation, Shaker Verlag, Aachen, 2016.

[8] N. N. Rykalin, Berechnung von Wärmevorgängen beim Schweißen, VEB Verlag Technik, Berlin, 1957.

[9] N. O. Okerblom, Schweißspannungen in Metallkonstruktionen, VEB Carl Marhold Verlag, Halle (Saale), 1959.

[10] S. A. Kuzminov, Svarochnie deformazii sudovich korpusnich konstrukzii, Verlag Sudostroenie Leningrad, Leningrad, 1974.

Residual Stresses 2018 – ECRS-10   Materials Research Forum LLC
Materials Research Proceedings 6 (2018) 251-256   doi: http://dx.doi.org/10.21741/9781945291890-40

Contribution to Simplified Residual Stress Calculations of Multi-Layer Welds

Jakob Klassen[1,a,*], Thomas Nitschke-Pagel[1,b] and Klaus Dilger[1,c]

[1]Langer Kamp 8, 38106 Braunschweig, Germany

[a]j.klassen@tu-braunschweig.de, [b]t.pagel@tu-braunschweig.de, [c]k.dilger@tu-braunschweig.de

Keywords: Residual Stresses, Welding Simulation, X-Ray Diffraction, Neutron Diffraction

Abstract. Especially for larger structures destructive measuring techniques for residual stress determination can be neglected and non-destructive methods are subject to disadvantages such as the limited resolution. Furthermore, residual stress gradients in the direction of sheet thickness cannot be determined easily.

The growing use of FEM, on the other hand, gives an insight into residual stresses at each point of the computed model. However, FEM for residual stress calculation is subject to certain limits and error sources due to the fine discretization and the high degree of non-linearity. In particular, the calculation of multi-layer welds will reach high computation times. This known issue is often counteracted with radical simplification of the numerical model such as lumping or semi-transient method. The result inaccuracies of these methods are rarely quantified and published. To clarify which simplification strategies are applicable in numerical welding simulation reference models were produced and experimentally verified. On this basis, simplification approaches were investigated numerically and their effects on result quality were quantified.

It could be shown that most of the commonly used simplification approaches for the calculation of residual stresses and distortions are only partially permissible, if any. Each method has its limits and poses a risk to the user, if certain data for validation and verification are available only to a limited extent. This means that a strongly localized comparison between experiment and calculations is not necessarily a proof of correctness of the calculation approaches if a more refined experimental determination is dispensed with.

Introduction

Residual stresses are often suspected of having an influence on the service life of welded constructions, as those stresses are associated with crack formation and crack growth. In general, it is assumed that residual stresses superimpose with load stresses and thus can have a positive or negative effect depending on sign and value. Numerous works on these issues were published in the past. In any case, knowledge of the exact residual stress state is of great interest for the design of welded constructions. These exact stress states, however, are highly dependent on a multiplicity of factors. Therefore, a statement about the type and size of the respective residual stress distributions is not trivial. Thus, a compilation of representative residual stress distributions, for example in cataloged form, is desirable but not available for mentioned reasons.

There are numerous measuring techniques to obtain residual stresses. However, most techniques (e.g. diffraction methods, hole drilling etc.) are either expensive or limited in terms of accessibility, especially for through-thickness residual stress determination of thicker plates with possible residual stress gradients. Therefore, a reliable statement about the residual stress state in or near a critical location is often denied. These unavoidable experimental limitations may be overcome using numerical methods, in particular the finite element method (FEM). The FEM itself can be referred to as state of the art in calculation of residual stresses and distortion.

Residual Stresses 2018 – ECRS-10   Materials Research Forum LLC
Materials Research Proceedings 6 (2018) 251-256   doi: http://dx.doi.org/10.21741/9781945291890-40

Calculated results allow an insight in any desirable area of the computed model. However, volumetric models of multi-layer welds and larger structures result in large models and still unreasonable calculation times. Calculation engineers facing these problems are tempted to answer with a supposedly simple approach, namely simplifying dramatically. Common simplification approaches such as reduction of discretization, lumping or application of semi-transient approaches do lead to manageable computation times. However, results thus obtained and therefore their benefits are highly doubtful.

In the scope of the present work several (most common) simplification approaches were applied on a reference case. This reference was validated and verified through abundant experimental investigations by means of temperature and distortion measurements as well as residual stress determination using X-ray and neutron diffraction techniques.

Experimental investigations. Specimens' dimensions with a milled V-notch and welding parameters are given in Figure 1. For validation of numerical models temperature measurements were conducted. Type K thermocouples were used. Furthermore, distortions were obtained by means of laser trigonometry.

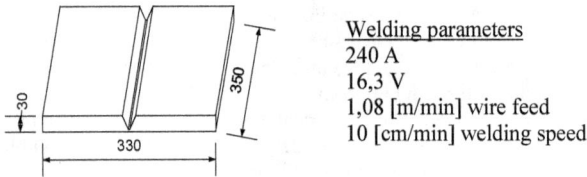

Welding parameters
240 A
16,3 V
1,08 [m/min] wire feed
10 [cm/min] welding speed

Figure 1: Specimen dimensions [mm] and welding parameters

Material properties. Experiments were carried out on a fine grained construction steel S355N with conspecific filler wire. This steel grade is commonly used in highly loaded constructions, e.g. steel bridges. Chemical composition and mechanical properties are given in Table 1.

Table 1: Chemical composition (by wt. [%]) and mechanical properties of used steel grade

	C	Si	Mn	P	S	Cr	Ni	Mo	Cu
S355N:	0,186	0,411	1,128	0,012	0,001	0,042	0,369	0,005	0,163
Tensile test:	R_{eH} = ~380 MPa;		R_m = ~540 MPa						

Residual stress measurement. Residual stresses (RS) are defined in reference to the welding direction. Thus, longitudinal RS are in welding direction and transverse RS perpendicular to the weld (Figure 2).

Near surface residual stresses were measured by means of X-ray diffraction (XRD). Diffraction lines ({211}-patterns) were obtained with CrK_α-radiation using a 2mm collimator. Afterwards residual stresses were calculated with the $\sin^2\psi$-method according to [1], [2]. Through thickness residual stress measurements by means of neutron diffraction (ND) were carried out at the E3 instrument at Helmholtz-Zentrum Berlin.

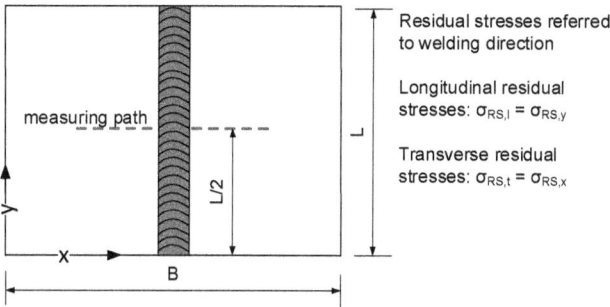

Figure 2: Orientation of measured and calculated residual stresses

Numerical investigations. Numerical welding simulations by means of three-dimensional fully transient structural finite element analyses were carried out with the SYSWELD code (Sysweld 2015.1, Version 17). Phase transformation was taken into account. Numerical models were calibrated by measured temperature fields and macrosections of each weld layer.

No clamping was applied, neither experimentally nor numerically. Therefore, deformations were inevitable. However, in numerical simulations the entire weld had to be modelled in advance ("chewing gum" or "quiet elements method" instead of "element birth method" - see [3], [4] and [5] for comparison) anticipating highly distorted elements of the final passes (according to [4]). Thus, elements' aspect ratio changes from bottom to the top layers, as shown in Figure 3. Corresponding mesh size information is given in Table 2. In further figures each FE mesh has been suppressed for clarity reasons.

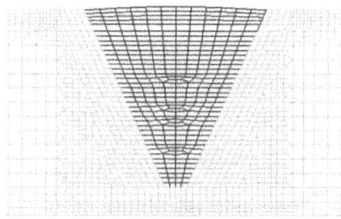

Figure 3: Finite element mesh cross section

Table 2: Details of FE mesh, V-model

Reference model-V	
# of hex elements	215308
# of nodes	224609
element size (x) [mm]	0,75-2,2
element size (y) [mm]	3,0
element size (z) [mm]	0,5-0,7

Simplification approaches. Some of most commonly applied simplification approaches are simplified geometries, applied thermal cycles as well as lumping several layers of multi-layer welds. However, there is hardly any scientific justification for some of these approaches. Within the framework of this paper these simplifications were implemented and compared with experimentally validated fully transient simulations. It is important to note that overall heat input in each model remained unchanged. This is a mandatory boundary condition, since calculation engineers only can rely on information obtained from welders or work shops.

Applied thermal cycle approach (TC). Weld build-up was adopted from V-model. Thermal cycles were experimentally obtained at certain points and assigned to numerical model. Thus, all elements of each pass were subject to certain thermal cycles (following measured values) at the same time.

Residual Stresses 2018 – ECRS-10   Materials Research Forum LLC
Materials Research Proceedings **6** (2018) 251-256   doi: http://dx.doi.org/10.21741/9781945291890-40

Lumping approach (L). FE meshes were adopted from V-model. In VL-model lumping was implemented such that all passes of each layer were summarized to one layer (compare Figure 4).

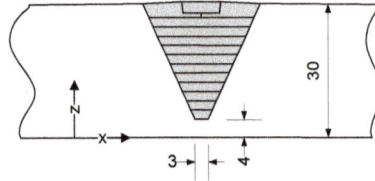

Figure 4: Weld build-up, VL-model [mm]

Results

The initial model (V) has been validated and verified with experimental data from temperature, distortion and residual stress measurements (comp. Figure 5). Thus, validation of the numerical model as well as its suitability for variations of further calculations is given (see [6] for further information).

It is important to note that points of residual stress measurements were linked by lines for clarity reasons only. Error bars are not plotted for the same reasons. However, errors' standard deviation of each measurement was calculated and was found to be less than ±10 MPa in average.

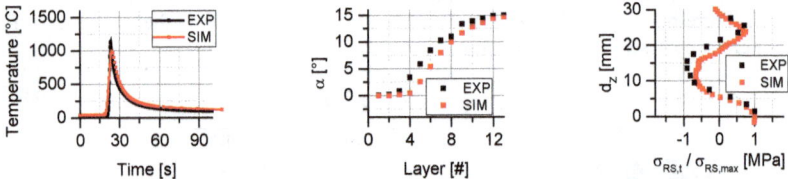

Figure 5: Examples for validation and verification of reference model-V

Applied thermal cycle approach. The results of a fully transient residual stress calculation with moving heat sources (model V) are compared with the results using the TC method (model TC). On the one hand, calculated longitudinal residual stresses at the bottom (not shown here) do not show significant differences. This was to be expected, since shrinking is equally hindered after first layer being welded. On the other hand, transverse residual stresses in the TC model are underestimated with an increasing number of layers. This effect is most likely to be explained with different conditions for hindered shrinkage in longitudinal direction when applying thermal cycle on the entire layer to be welded. Contrary to this, hindered shrinkage changes transiently with moving heat sources due to sequentially activation of filler wire elements. These influences of multi-layer thermal cycle approach on transverse residual stresses are even higher, as can be seen in Figure 6. Overall, transverse residual stresses are significantly smaller in magnitude and show even a change in sign.

Through-thickness residual stress distributions in the weld center are given in Figure 7. The assumptions stated above, in combination with tempering effects due to weld build-up, result in minor tensile residual stresses. Furthermore, the absence of plastic strains in transverse direction, which are mainly responsible for angular distortions, necessarily results in falsely predicted distortions.

Residual Stresses 2018 – ECRS-10
Materials Research Proceedings 6 (2018) 251-256

Materials Research Forum LLC
doi: http://dx.doi.org/10.21741/9781945291890-40

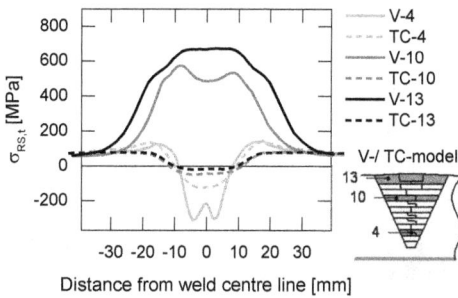

Figure 6: Comparison of calculated residual stresses (models V and TC)

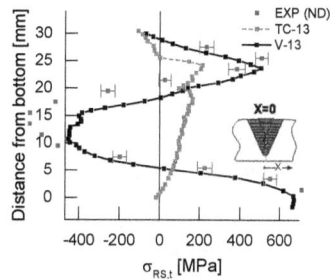

Figure 7: Comparison of calculated through-thickness residual stresses (models V and TC)

It can be noted that residual stress states of the transient calculation (which had been validated by measurements) is not reproduced in the TC approach, at all.

Lumping method. Lumping method was investigated on numerical basis only, for experimental validation is not possible. Calculated near surface residual stresses of one lumping approach were fairly comparable qualitatively. Distributions on each top side were comparable, which is due to equally modelled final passes. However, residual stresses in through-thickness direction show strong deviation in sign and value (see also Figure 8). Reasons for this are seen in increasing volumes of summarized passes which most likely are responsible for deeper penetrating annealing effects. Furthermore, these effects reduced plastic strains decreasing distortion.

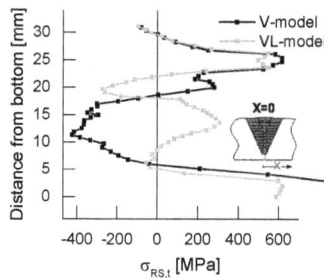

Figure 8: Comparison of calculated through-thickness residual stresses (models V and VL1)

It is important to note that calculated near surface residual stresses do not necessarily give information about the through-thickness distribution. Thus, obtained results might be misleading as was already found in a previous project [7].

Conclusions

In this paper simplification approaches have been presented that often can be found in literature. However, almost none of these publications do even try to validate their models. Instead, cherry picked experimental results are used for verification. Within the framework of this paper it could be shown that the investigated simplification approaches show less potential and more limitation

Residual Stresses 2018 – ECRS-10   Materials Research Forum LLC
Materials Research Proceedings **6** (2018) 251-256   doi: http://dx.doi.org/10.21741/9781945291890-40

in terms of result quality. For instance, it is well known that application-oriented simulation approaches need to keep energy input balanced. Using lumping, however, this cannot be obtained. The relationship between the introduced weld volume and its contact surfaces that are available for heat dissipation is now imbalanced. This mismatch results in a more profound annealing effect which again affects the residual stresses and deformations. Reducing energy input corresponding to those heat dissipating contact surfaces did not lead to any significant improvement in the calculation result.

Furthermore, some results appeared to be close to experiments. Although it is important to note that this most likely is by accident and, thus, might be misleading. In most cases numerical engineers will need to validate their models through distortion or X-ray measurements, if any. Though, it could be that apparently correct near surface residual stresses do not necessarily result in true through-thickness residual stress distributions.

References

[1] M. P. Macherauch E., „Das $sin^2\psi$-Verfahren der röntgenografischen Spannungsmessung," Zeitschrift für angewandte Physik, Bd. 13, 1961.

[2] C. Rohrbach, Handbuch für experimentelle Spannungsanalyse, Düsseldorf: VDI-Verlag GmbH, 1989.

[3] L.-E. Lindgren, H. Runnemalm und M. O. NässtrÖm, „Simulation of multipass welding of a thick plate," International Journal for Numerical Methods in Engineering, Nr. 44, pp. 1301-1316, 1999.

[4] L.-E. Lindgren und E. Hedblom, „Modelling of addition of filler material in large deformation analysis of multipass welding," Communications in Numerical Methods in Engineering, Nr. 17, pp. 647-657, 17 August 2001.

[5] D. W. Lobitz, J. D. McClure und R. E. Nickell, „Residual stresses and distortions in multi-pass welding," in Numerical Modeling of Manufacturing Processes, Winter Annual Meeting of the American Society of Mechanical Engineers, Atlanta, Giorgia, 1977.

[6] J. Klassen, „Beitrag zur vereinfachten Eigenspannungsberechnung von Mehrlagenschweißverbindungen," Shaker Verlag, Aachen, 2018.

[7] K. Dilger, W. Fricke und T. Nitschke-Pagel, „Entwicklung von Methodiken zur Bewertung von Eigenspannungen an Montagestößen bei Stahl-Großstrukturen," AiF-Schlussbericht, IGF-Vorhabennummer 17652N (in German), 2016

[8] F. Zhang, „Beitrag zum schweißbedingten Verzug unter Berücksichtigung seiner Wechselwirkung mit den Eigenspannungen," Institut für Schweißtechnik und Werkstofftechnologie, Braunschweig, 1998.

Stresses in Additive Manufacturing

Residual Stresses 2018 – ECRS-10
Materials Research Proceedings 6 (2018) 259-264

Materials Research Forum LLC
doi: http://dx.doi.org/10.21741/9781945291890-41

Residual Stresses in Selective Laser Melted Samples of a Nickel Based Superalloy

Arne Kromm[1,a,*], Sandra Cabeza[2,b], Tatiana Mishurova[1,c], Naresh Nadammal[1,d]
Tobias Thiede[1,e] and Giovanni Bruno[1,f]

[1]Bundesanstalt für Materialforschung und -prüfung (BAM), Unter den Eichen 87, 12205 Berlin, Germany

[2]Institut Max von Laue - Paul Langevin (ILL), 71 Avenue des Martyrs, 38000 Grenoble, France

[a]arne.kromm@bam.de, [b]cabeza@ill.fr, [c]tatiana.mishurova@bam.de,
[d]naresh.nadammal@bam.de, [e]tobias.thiede@bam.de, [f]giovanni.bruno@bam.de

Keywords: Additive Manufacturing, Selective Laser Melting, Residual Stresses

Abstract. Additive Manufacturing (AM) through the Selective Laser Melting (SLM) route offers ample scope for producing geometrically complex parts compared to the conventional subtractive manufacturing strategies. Nevertheless, the residual stresses which develop during the fabrication can limit application of the SLM components by reducing the load bearing capacity and by inducing unwanted distortion, depending on the boundary conditions specified during manufacturing. The present study aims at characterizing the residual stress states in the SLM parts using different diffraction methods. The material used is the nickel based superalloy Inconel 718. Microstructure as well as the surface and bulk residual stresses were characterized. For the residual stress analysis, X-ray, synchrotron and neutron diffraction methods were used. The measurements were performed at BAM, at the EDDI beamline of -BESSY II synchrotron- and the E3 line -BER II neutron reactor- of the Helmholtz-Zentrum für Materialien und Energie (HZB) Berlin. The results reveal significant differences in the residual stress states for the different characterization techniques employed, which indicates the dependence of the residual state on the penetration depth in the sample. For the surface residual stresses, longitudinal and transverse stress components from X-ray and synchrotron agree well and the obtained values were around the yield strength of the material. Furthermore, synchrotron mapping disclosed gradients along the width and length of the sample for the longitudinal and transverse stress components. On the other hand, lower residual stresses were found in the bulk of the material measured using neutron diffraction. The longitudinal component was tensile and decreased towards the boundary of the sample. In contrast, the normal component was nearly constant and compressive in nature. The transversal component was almost negligible. The results indicate that a stress re-distribution takes place during the deposition of the consecutive layers. Further investigations are planned to study the phenomenon in detail.

Introduction

Additive manufacturing (AM) offers the opportunity to produce geometrically complex parts compared to the traditional production technologies. An important AM technology for metals is selective laser melting (SLM), where a part is produced by melting a powder bed in layers [1]. However, residual stresses that arise during the process may limit the application of SLM parts by inducing unwanted distortion depending on the boundary conditions. Strategies for stress optimization must be developed to minimise distortion and with focus on other sensitive properties relying on residual stress. The material used in this study is the nickel based super Alloy 718 which has several applications in aerospace and chemical industry due its superior

Residual Stresses 2018 – ECRS-10   Materials Research Forum LLC
Materials Research Proceedings 6 (2018) 259-264   doi: http://dx.doi.org/10.21741/9781945291890-41

corrosion and heat resistance [2]. The SLM process is generally known to form a high amount of residual stresses due to the high temperature gradient present during laser melting. In principle the mechanisms of stress formation are similar to the fusion welding process. In the absence of solid state phase transformations tensile residual stresses are formed due to the hindered shrinkage of the already solidified material. The amount of stresses can be equal to the yield stress. After cooling to ambient temperature, a residual stress gradient between surface and core regions of the part is present. Its magnitude depends amongst others on the geometry and the stiffness of the whole part as well [3-8]. Adopting the scanning strategy during SLM can alter the level and distribution of the residual stresses [9-12]. Typically, the scanning is performed cyclic in sectors along the parts. This alters the heat flow and therefore the local temperature distribution. As a consequence, the stress formation is locally affected. However, the thermal-mechanical behaviour is complex and therefore to be evaluated by experimental investigation as a basis for modelling [5, 12-14]. This enables deliberate adjustment of process parameters to control the residual stresses and associated distortion during fabrication. In order to evaluate localised residual stress distributions in SLM parts different regions have to be investigated. Beside the surface and the bulk, the sub-surface area is of particular interest. Diffraction methods enable for non-destructive measurement of spatial resolved stress distributions. X-ray and neutron diffraction are appropriate to cover surface and bulk. On the other hand, the intermediate area (sub-surface) is accessible by the application of high energy synchrotron diffraction [15]. This study aims at the characterization of residual stresses in SLM parts by using different diffraction measurement techniques.

Experimental
Alloy 718 was processed and provided by SIEMENS AG, Power and Gas, Berlin, Germany. The specimens were produced on an EOS M290 machine using the standard EOS parameter set for Alloy 718. The processing parameters, like track width, scan speed and power input, are confidential but were kept constant during fabrication. The deposition of each layer during the SLM process was identical with the hatching and scanning along the length and the width of the specimens, respectively. The residual stresses were determined using three complimentary techniques. The surface and the sub-surface was characterised by laboratory X-ray as well as high energy synchrotron diffraction (instrument EDDI at Bessy HZB, Berlin) using the $\sin^2\psi$-method [16]. Due to high energy, up to 150 keV, penetration depths of up to 100 μm are achievable at EDDI depending on the lattice plane evaluated [17]. The sample bulk was measured by neutron diffraction (instrument E3 at HZB Berlin) [18]. Table 1 gives important parameters for each measurement setup. The {311} diffraction line of nickel was used for stress evaluation in case of X-ray diffraction and for neutron diffraction. Synchrotron measurements were conducted in energy dispersive (EDXRD) mode using a white beam [17]. Therefore, up to four diffraction lines (see Table 1) could be utilised. Their mean value (weighted by their multiplicity) was taken for the stress evaluation.

 The SLM specimens were cuboids with a length of 100 mm and a width of 20 mm. The longitudinal edges were shaped round. Each measurement was conducted in a quarter of the sample along two lines as indicated in Fig. 1. Symmetry of the residuals stresses with respect to the centre of the sample was assumed. One measuring line was placed along the centre of the sample, while the other one was in parallel along the edge. For X-ray and synchrotron diffraction the measurement was conducted in the surface according to the penetration depth of the radiation up to 60 μm. In case of neutron diffraction, the measuring lines were placed at a distance of 2 mm below the sample top surface. The microstructural characterisation included optical microscopy and Electron Back-Scattered Diffraction (EBSD). For this purpose, a small volume representing the transverse cross section of the sample was prepared by electrical discharge machining (EDM), see Fig. 1. Additionally, powder taken from this area served as stress free reference for neutron diffraction.

Residual Stresses 2018 – ECRS-10   Materials Research Forum LLC
Materials Research Proceedings 6 (2018) 259-264   doi: http://dx.doi.org/10.21741/9781945291890-41

Table 1. Measuring and evaluation parameters for residual stress analyses

	X-ray diffraction	Synchrotron diffraction	Neutron diffraction
Radiation	Mn Kα	White beam E = 50 keV-150 keV	$\lambda = 1.476$ Å
Diffraction line	Ni{311}	Ni{111}, Ni{200}, Ni{220}, Ni{311}	Ni{311}
2Θ angle	156°	10°	85.3°
Gauge volume	ø 2 mm collimator	1×1 mm² (primary beam)	4 × 4 × 2 mm³
Stress components	long, trans	long, trans	long, trans, norm
Exposure time	5 s	300 s	900 s
DEC	$s_1(hkl)$ and $\frac{1}{2}s_2(hkl)$ calculated from the single crystal constants using the Eshelby/Kröner-model		

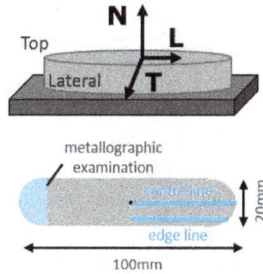

Figure. 1. Schematic of the sample with measuring lines located in the centre and along the edge as well as the area considered for metallographic examination

Results

Microstructure. The parts produced by SLM process showed typical features characteristic to multi-pass welds but on a micro scale. As can be seen from Fig. 2 the bulk of the sample consisted of small overlapping runs. The heat flow was opposite to the building direction which leads to a columnar growth of the grains. Strong rotated cube texture specific of the heat dissipation along the specimen thickness was observed along the building direction, which is partly attributed to the shorter hatch length utilized for the SLM.

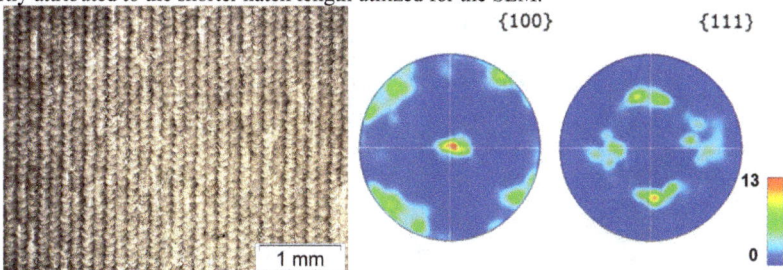

Figure 2. Cross section of the bulk by light optical microscopy (left), {100} and {111} pole figures obtained by EBSD showing rotated cube texture (right)

Residual Stresses 2018 – ECRS-10   Materials Research Forum LLC
Materials Research Proceedings **6** (2018) 259-264   doi: http://dx.doi.org/10.21741/9781945291890-41

Residual stresses. The residual stresses obtained by X-ray diffraction are shown in Fig. 3 (left). In longitudinal as well as transverse direction the stresses were in tension. No gradient was present along the sample centreline. While the stresses in longitudinal direction showed values between 600 MPa and 850 MPa the stresses in transverse direction were significantly higher. Values around 1000 MPa indicate stresses comparable to the yield point of the wrought alloy. No differences were observed for the sample centre and the edge line. The stress distribution is uniform along the top surface.

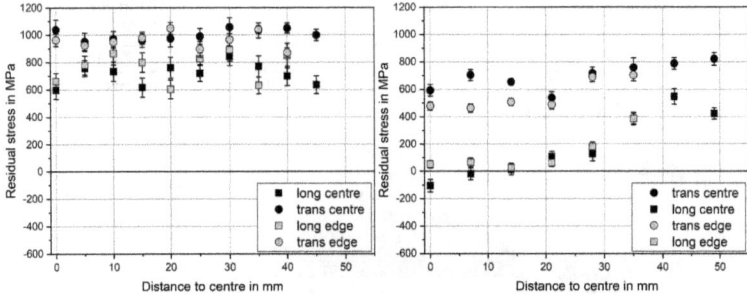

Figure 3. Longitudinal and transverse residual stresses in the top-surface obtained by X-ray diffraction at the sample centre and edge line (left) and in the sub-surface obtained by synchrotron diffraction at the sample centre and edge line (right)

With increasing penetration depth, the residual stresses changed their magnitude and also distribution as shown in Fig. 3 (right). In longitudinal direction the residual stress level was slightly shifted in parallel to values between 400 MPa and 800 MPa. Only a slight gradient was present along the measuring line indicating lower stresses in the sample centre. This gradient was more pronounced in transverse direction were the stresses in the centre were significantly lowered even to compressive values of -100 MPa. Far from the centre of the sample the stresses remained in tension showing approximately 400 MPa. As already indicated by X-ray diffraction the measured centre line and edge line of the sample are comparable showing the same stress gradients.

Different stress characteristics are to be found in the bulk of the sample 2 mm below the top surface (see Fig. 4). Lower tensile longitudinal stresses were present along the sample length which turned into compression (-250 MPa) in the border region. Transverse stresses were balanced around zero at the measured centre line.

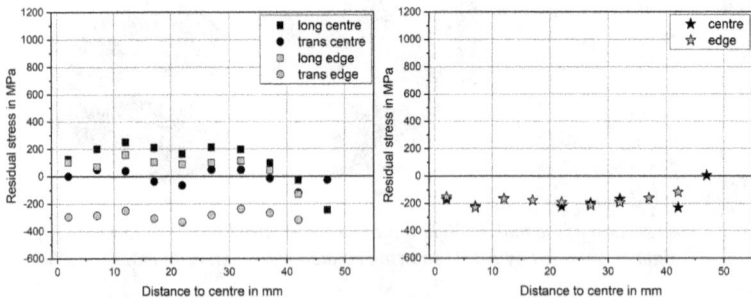

Figure 4. Longitudinal and transverse residual stresses (left) and normal residual stresses (right) in the bulk obtained by neutron diffraction at the sample centre and edge line

Considering the measured edge line, the transverse stresses were shifted in parallel into compression up to -300 MPa. On the other hand, longitudinal stresses remained in tension at a similar level compared to the centre line.

The normal stress components in the sample bulk are shown in Fig. 4 (right). The distribution and also the level of the stresses is similar in the centre and the edge line. The stresses are constant around -200 MPa.

Taking into account all the stresses determined by the different measuring techniques applied it becomes clear that a significant stress redistribution took place during deposition of the single layers during the SLM process. The highest tensile stresses were always formed in the very last deposited layer. Immediately after reheating by deposition of the following layers the stresses were lowered especially transversely in the sample centre region. With ongoing deposition of layers and increasing distance to the top surface the stresses were then shifted to lower tensile or even compressive values due to the balance of forces.

Summary

Three different diffraction measuring techniques featuring different penetration depths were capable to determine localised residual stress profiles in samples of nickel-based superalloy 718 manufactured by SLM. Using X-ray diffraction, longitudinal and transverse stress components in the top surface showed high tensile values up to the yield limit. In the intermediate zone below the surface, synchrotron diffraction disclosed a pronounced stress gradient along the length of the sample for both stress components. Particularly, the transverse residual stress relaxed very quickly towards the centre of the sample. Residual stresses analysed by neutron diffraction in the bulk showed considerably lowered stresses proving the stress re-distribution during deposition of the SLM layers. Even compressive stresses were found in transverse and predominantly normal direction in the sample.

References

[1] L.E. Murr, S.M. Gaytan, D.A. Ramirez, E. Martinez, J. Hernandez, K.N. Amato, P.W. Shindo, F.R. Medina, R.B. Wicker, Metal Fabrication by Additive Manufacturing Using Laser and Electron Beam Melting Technologies, J Mater Sci Technol 28(1) (2012) 1-14. https://doi.org/10.1016/S1005-0302(12)60016-4

[2] K.N. Amato, S.M. Gaytan, L.E. Murr, E. Martinez, P.W. Shindo, J. Hernandez, S. Collins, F. Medina, Microstructures and mechanical behavior of Inconel 718 fabricated by selective laser melting, Acta Mater 60(5) (2012) 2229-2239. https://doi.org/10.1016/j.actamat.2011.12.032

[3] P. Mercelis, J.P. Kruth, Residual stresses in selective laser sintering and selective laser melting, Rapid Prototyping J 12(5) (2006) 254-265. https://doi.org/10.1108/13552540610707013

[4] C.E. Protasov, V.A. Safronov, D.V. Kotoban, A.V. Gusarov, Experimental study of residual stresses in metal parts obtained by selective laser melting, Laser Assisted Net Shape Engineering 9 International Conference on Photonic Technologies Proceedings of the Lane 2016 83 (2016) 825-832.

[5] L. Van Belle, G. Vansteenkiste, J.C. Boyer, Investigation of residual stresses induced during the selective laser melting process, Key Eng Mater 554-557 (2013) 1828-1834. https://doi.org/10.4028/www.scientific.net/KEM.554-557.1828

[6] M.S. Abdul Aziz, T. Furumoto, K. Kuriyama, S. Takago, S. Abe, A. Hosokawa, T. Ueda, Residual Stress and Deformation of Consolidated Structure Obtained by Layered Manufacturing Process, J Adv Mech Des Syst 7(2) (2013) 244-256. https://doi.org/10.1299/jamdsm.7.244

[7] T. Mishurova, S. Cabeza, T. Thiede, N. Nadammal, A. Kromm, M. Klaus, C. Genzel, C. Haberland, G. Bruno, The Influence of the Support Structure on Residual Stress and Distortion in SLM Inconel 718 Parts, Metallurgical and Materials Transactions A (2018).

[8] N. Nadammal, A. Kromm, R. Saliwan-Neumann, L. Farahbod, C. Haberland, P. Portella, Influence of Support Configurations on the Characteristics of Selective Laser-Melted Inconel 718, Jom-Us 70(3) (2018) 343-348. https://doi.org/10.1007/s11837-017-2703-1

[9] B. Chong, S. Shrestha, Y.K. Chou, Stress and Deformation Evaluations of Scanning Strategy Effect in Selective Laser Melting, Proceedings of the Asme 11th International Manufacturing Science and Engineering Conference, 2016, Vol 3 (2016). https://doi.org/10.1115/MSEC2016-8819

[10] Y. Liu, Y.Q. Yang, D. Wang, A study on the residual stress during selective laser melting (SLM) of metallic powder, Int J Adv Manuf Tech 87(1-4) (2016) 647-656. https://doi.org/10.1007/s00170-016-8466-y

[11] Y.J. Lu, S.Q. Wu, Y.L. Gan, T.T. Huang, C.G. Yang, J.J. Lin, J.X. Lin, Study on the microstructure, mechanical property and residual stress of SLM Inconel-718 alloy manufactured by differing island scanning strategy, Opt Laser Technol 75 (2015) 197-206. https://doi.org/10.1016/j.optlastec.2015.07.009

[12] N. Nadammal, S. Cabeza, T. Mishurova, T. Thiede, A. Kromm, C. Seyfert, L. Farahbod, C. Haberland, J.A. Schneider, P.D. Portella, G. Bruno, Effect of hatch length on the development of microstructure, texture and residual stresses in selective laser melted superalloy Inconel 718, Mater Design 134 (2017) 139-150. https://doi.org/10.1016/j.matdes.2017.08.049

[13] M. Shiomi, K. Osakada, K. Nakamura, T. Yamashita, F. Abe, Residual stress within metallic model made by selective laser melting process, Cirp Ann-Manuf Techn 53(1) (2004) 195-198. https://doi.org/10.1016/S0007-8506(07)60677-5

[14] J.P. Kruth, J. Deckers, E. Yasa, R. Wauthle, Assessing and comparing influencing factors of residual stresses in selective laser melting using a novel analysis method, P I Mech Eng B-J Eng 226(B6) (2012) 980-991.

[15] C. Genzel, C. Stock, W. Reimers, Application of energy-dispersive diffraction to the analysis of multiaxial residual stress fields in the intermediate zone between surface and volume, Mat Sci Eng a-Struct 372(1-2) (2004) 28-43. https://doi.org/10.1016/j.msea.2003.09.073

[16] E. Macherauch, P. Müller, Das sin²ψ - Verfahren der röntgenografischen Spannungsmessung, Zeitschrift für angewandte Physik 13 (1961) 305-312.

[17] C. Genzel, I.A. Denks, J. Gibmeler, M. Klaus, G. Wagener, The materials science synchrotron beamline EDDI for energy-dispersive diffraction analysis, Nucl Instrum Meth A 578(1) (2007) 23-33. https://doi.org/10.1016/j.nima.2007.05.209

[18] T. Poeste, R.C. Wimpory, R. Schneider, The new and upgraded neutron instruments for material science at HMI - current activities in cooperation with industry, Residual Stresses Vii 524-525 (2006) 223-228. https://doi.org/10.4028/0-87849-414-6.223

Residual Stresses 2018 – ECRS-10   Materials Research Forum LLC
Materials Research Proceedings **6** (2018) 265-270   doi: http://dx.doi.org/10.21741/9781945291890-42

Evaluation of In-Process Laser Heat Treatment on the Stress Conditions in Laser Metal Deposited Stellite® 21

Grant L. Payne[1,a], Ioannis Violatos[1,b], Stephen Fitzpatrick[1,c,*],
David Easton[1,d] and Joshua Walker[1,e,*]

[1]Advanced Forming Research Centre, 85 Inchinnan Drive, Glasgow, Scotland (UK), PA 9LJ

[a]grant.payne@strath.ac.uk, [b]ioannis.violatos@strath.ac.uk, [c]s.fitzpatrick@strath.ac.uk,
[d]david.easton@strath.ac.uk, [e]joshua.walker@strath.ac.uk

Keywords: Additive Manufacturing, Laser Metal Deposition, Residual Stress, XRD, Contour Method, Micro-Hardness, Energy Density, Micro-Structural Analysis

Abstract. Laser Metal Deposition with Powder (LMD-*p*) has been investigated as a means of Remanufacturing high value components, such as tooling, dies and moulds. However, the LMD-*p* process is known to develop high levels of residual stresses within the builds, which may have an effect on the mechanical performance of the components. Heat treatment is a common method for stress relieving, however, large components or those undergoing ReManufacturing may not be suitable for conventional stress relieving heat treatments processes, such as those using a furnace. Therefore, localised and dynamic heat treatment using the laser installed on the LMD-*p* apparatus has been investigated as means of providing stress relieving heat treatment. As such, research to understand the generation and distribution of stresses has been undertaken in conjunction with micro-structural analysis to provide a robust evaluation. A combination of Contour Method, XRD, Micro-Hardness and SEM imaging was used for analysis. Preliminary assessments have largely shown positive results as the specimen with in-process heat treatment has exhibited low and relatively uniform stress fields.

Introduction

ReManufacturing tools and dies is increasing in popularity owing to its time efficiency, waste reduction and cost savings with respect to machine/production down time [1-4]. LMD-*p* can selectively deposit material in regions of wear and/or fatigue to extend a product's life or increase its performance, however, a limitation of the process is thermal shock to the substrate and high residual stresses created within the builds. Traditional heat treatment processes such as annealing rely on placing the entire component within a furnace to undergo stress relieving heat treatment.

Industrial interest and government led policy have led to increased research activity in the field of additive manufacturing. In one such project, in a baseline study Stellite® 21 was deposited on AISI H13 Steel with a standard set of parameters, this build strategy exhibited high stresses [5]. A similar project reported the high stresses in LMD-*p* Stellite® 21 to have been reduced through conventional furnace based heat treatment [6].

Experimental Method

Cylindrical billets of AISI H13 Tool Steel (diameter 21mm x 26mm height) were cut from wrought blocks using EDM and turning, ensuring that the recast layer was removed before deposition. Custom LMD-*p* apparatus was used to deposit the Stellite® 21 using a *Trumpf TruDisk 2002 1030nm Laser* at Laser Additive Solutions Ltd, Doncaster using the process parameters defined in Table 1.

Residual Stresses 2018 – ECRS-10   Materials Research Forum LLC
Materials Research Proceedings 6 (2018) 265-270   doi: http://dx.doi.org/10.21741/9781945291890-42

Localised heat treatment was achieved by defocusing the laser beam (increasing the Laser Stand Off distance), lowering the Laser Power and running successive, overlapping passes across the substrate's top surface. The material was preheated to ~450°C and the temperature was monitored using a *LAND ARC Thermal Camera* before deposition began. This heating program was then repeated at 6.5mm intervals, with a final heat treatment at build completion as detailed in figure 1.

Table 1. LMD-p Process Parameters

Laser Power	Laser Stand Off	Feed Rate	Spot Size	Z Build Height	Energy Density (Eq. 1)
Deposition: 575W	12 mm	8 mm/s	1.5 mm	0.3	47.92 J/mm³
Heat Treatment: 250W	112 mm	10 mm/s	10 mm	mm/layer	2.50 J/mm³

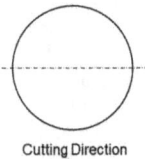

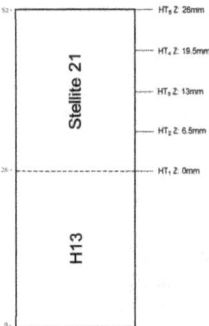

Cutting Direction

Figure 1. Specimen Schematics

Table 2. Chemical Composition of Stellite® 21, Manufacturer Specification in wt.-%

Co	Cr	Mo	C	Ni	Other Elements Present
Base	26 – 29	4.5 – 6.0	0.20 – 0.35	2.0 –3.0	Fe, Si, Mn

Table 3. Chemical Composition of AISI H13 Tool Steel, Manufacturer Specification in wt.-%

Fe	C	Mn	Si	Cr	Mo	V
Base	0.4	0.4	1.0	5.2	1.4	1.0

Table 4. Mechanical Properties of AISI H13 Steel & Stellite® 21 [1]

	Young's Modulus [MPa]	Poisson's Ratio
AISI H13	210000	0.30
Stellite 21	215000	0.21

$$Energy\ Density \left(\frac{J}{mm^3}\right) = \frac{Laser\ Power\ (W)}{Feed\ Rate\left(\frac{mm}{s}\right) * Spot\ Size\ (mm^2)} \qquad \text{Eq. 1}$$

Zavala-Arrdendo developed an equation Eq. 1 in which it is apparent that energy density is a key process driver that is governed by three of the most imperative process parameters [7]. It determines to the amount of energy transferred to a given area (Spot Size) at any given moment.

Each parameter can be independently modified to reach a desired energy density, however, they must be carefully balanced in order to achieve suitable material properties and build speed. Increasing Laser Power to increase energy density will raise thermal gradients and affect cooling rates. If the Feed Rate and or Laser Spot Size are lowered to increase the energy density then the build time will significantly increase, the latter is due to the track width becoming narrower. If large amounts of energy are rapidly transferred to the substrate this will cause: thermal shock, distortion and high levels of oxidisation, therefore, energy must be transferred, incrementally, to avoid the aforementioned defects.

Residual Stresses 2018 – ECRS-10   Materials Research Forum LLC

Materials Research Proceedings **6** (2018) 265-270   doi: http://dx.doi.org/10.21741/9781945291890-42

Micro-Structural Analysis

The characterisation of the micro-structure was carried out on a *FEI Quanta 250 FEG SEM* at the AFRC. As previously described, the specimens were mechanically ground and polished, but not etched as the commonly used *Kallings no. 2* etchant has been observed to adversely affect the H13. A suitable etching method to enhance micro-structure visibility has not yet been found for H13 / Stellite® 21.

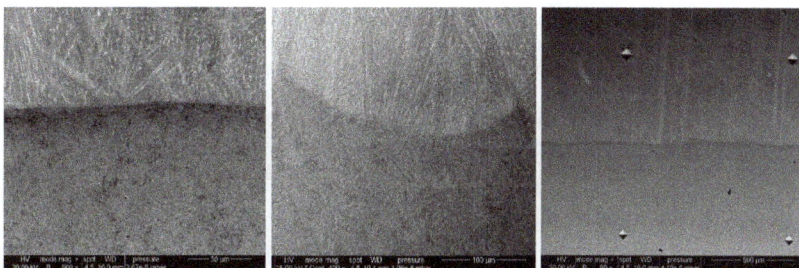

Figure 2. (a) H13 / Stellite® Interface (b) Stellite Weld Bead (c) Interface Showing Hardness Indentations

Micro-structural analysis revealed columnar grain growth perpendicular to the interface. Similar structures were observed at the build edges. Grain structure within the central bulk of the build was mostly equiaxed, this is most likely a product of slower cooling rates. No significant defects or cracking was observed at the heat-treatment z-levels. Small amounts of porosity (trapped carrier gas) and build defects were observed throughout the build, the magnitude of which did not raise concerns. The H13 revealed a HAZ ranging from ~500μm to ~750μm shown as a lighter coloured region on the images. However, the thermally softened region was not visible by SEM analysis but was revealed in the Micro-Hardness examination (Figure 3).

Micro-Hardness Analysis

The specimen was analysed on a *Zwick ZHV1 – Micro Vickers Hardness Tester* using a 1mm x 1mm matrix and HV1 indentations. Points were carefully aligned with the interface, with a row falling ~500μm either side of the interface as shown in Figure 2 (c).

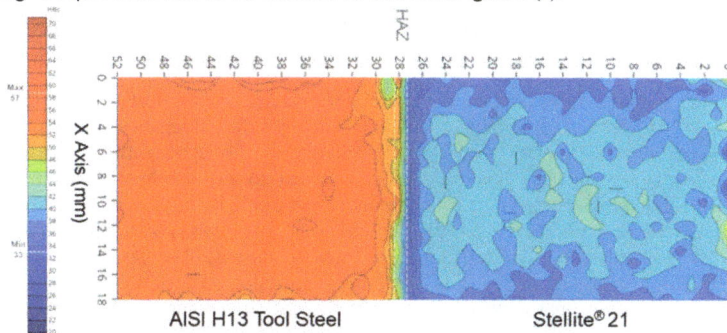

Figure 3. Micro-Hardness Contour Map

Measurements were converted as per ASTM E140 -12be1 [8] from Vickers to Rockwell and a contour map (Figure 3) was created using Plotly online software. The average value for H13 was

54.0 HR_C, 39.76 HR_C for Stellite 21 and 41.05 HR_C in the HAZ. A decline in hardness is observed in the H13 Steel at the interface, this extends beyond the HAZ (~500µm - 750µm) to ~3mm into the substrate. The change is fairly minor 54.0 HR_C - 49.35 HR_C indicating that the preheating has had a limited effect on the H13.

Residual Stress Measurement - Contour Method
The Contour Method calculates the out of plane residual stress by precisely cutting a specimen into two pieces, measuring the resulting deformation and relaxation due to residual stress redistribution and lastly running a Finite Element simulation {Prime, 2001 #1;Prime, 2013 #12}.

The component to be measured is sectioned along the plane of interest using EDM wire erosion. The two cut faces are then measured to determine the surface topology to which a surface contour can be analytically determined. This can be incorporated into a FE simulation to determine the pre-existing residual stresses prior to EDM sectioning. This makes Contour Method a useful tool for assessing complex residual stress fields such as those found in weld interfaces or ALM parts.

The specimen was clamped and cut longitudinally with a 0.25mm wire across the diameter using EDM to create to equal halves, parallel and perpendicular to the LMD-p build as noted in Figure 1. Cut surfaces were measured using a CMM at the AFRC with a 0.2mm x 0.2mm measurement matrix. A *Mitutoyo Crystal Apex C CMM* was used with a 1mm diameter Ruby attached to a *Renishaw® PH10T* probe. The measured surface topography was prepared using MATLAB® for a Finite Element simulation in Abaqus™. Using the outline from the CMM a geometry was recreated in Abaqus™ and partitioned into two regions with their respective elastic-mechanical properties Table 4.

Figure 4. Contour Residual Stress Map

Figure 4 shows a compressive residual stress field originating at the interface and extending 3mm into the H13 material. This correlates with the thermally softened region (Figure 3). A zone of tensile stress adjacent to the build finish in Stellite is displayed with an equilibrated compressive stress region. This is most likely attributed to the accelerated cooling rates experienced. In ReManufacturing some of this tensile zone would be machined away, however, this would not remove these stresses and it is crucial to note that work must be undertaken to analyse stress redistribution and relaxation in such conditions. The specimen exhibited comparatively lower stresses to the baseline process/operation [5]. EDM cutting artefacts introduced errors in the residual stress analysis (grey areas shown in Figure 4).

Residual Stress Measurement – X-Ray Diffraction Method
The specimen was examined using a PROTO LXRD diffractometer at the AFRC with using the parameters detailed in Table 5 and the best practice standards set out by the National Physics Laboratory [10]. Round apertures with diameters of 0.15mm and 1mm were used with

Residual Stresses 2018 – ECRS-10
Materials Research Forum LLC
Materials Research Proceedings **6** (2018) 265-270
doi: http://dx.doi.org/10.21741/9781945291890-42

measurements beginning at the H13/Stellite® interface in the centre of the build running from (10.5, 26) to (10.5, 21.25). Measurements were concentrated at the interface, to capture the transition and magnitude of the axial stresses in the HAZ. Points were then positioned at 0.5mm intervals to capture bulk trends.

Table 5. XRD Parameters

Material	Radiation Source	Tube Voltage	Tube Current	Wavelength	Bragg Angle	{hkl} Plane
Ferrite BCC	Chromium Kα	30 KV	25 mA	2.291 Å	150.41°	211

A Pre-Build H13 specimen without LMD-*p* was measured to provide guideline stress conditions for the substrate. Care was taken to mechanically polish both specimens in order to remove the recast layer caused by EDM sectioning – they were then electro-polished for 45 seconds to remove any mechanically induced stresses from polishing.

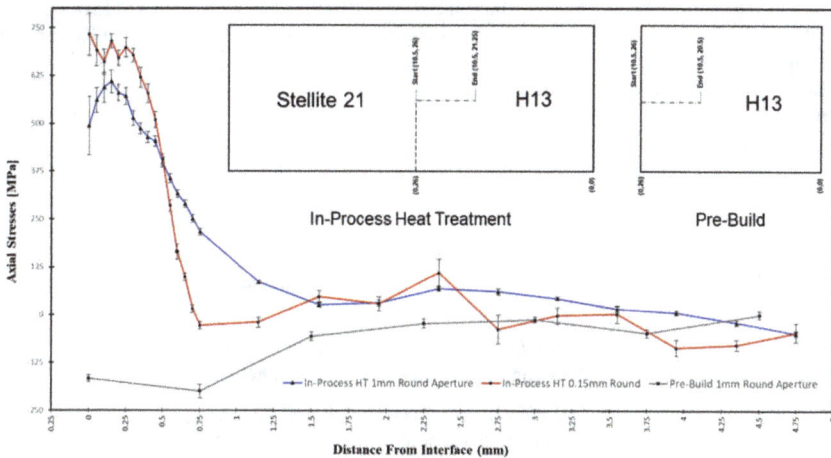

Figure 5. Axial Stresses in Relation to Distance from Interface and Measurement Location Schematic

Elasticity theory for isotropic solids and Hooke's law under plane stress conditions have been combined with the strains in terms of inter-planar spacing under different tilt angles (ψ) obtained from Bragg's Law for diffraction, to produce the stress acting along a specific direction, Eq. 2.

$$\sigma_\varphi = \frac{E}{(1+v)sin^2\psi}\left(\frac{d_\psi - d_n}{d_n}\right)$$ Eq.2

where E is the Young's modulus, v is the Poisson's ratio, d_ψ inter-planar spacing of planes at a tilt angle ψ to the surface and d_n is inter-planar spacing of planes normal to the surface [10].

In the HAZ, a trend from high tension (~700MPa) to a low compression stress state is observed. Immediately following this (25μm to 125μm) in the 0.15mm aperture series there is a dip to ~50MPa of compressive stresses, this is most likely a localised equilibrium effect caused by the tensile stresses in the HAZ. This effect is not fully observed in the 1mm aperture series due to an averaging effect in the diffraction area, this is a limitation of using a larger aperture. Therefore, future work could use this combination in order to observe localised trends and to validate global trends.

Residual Stresses 2018 – ECRS-10   Materials Research Forum LLC
Materials Research Proceedings 6 (2018) 265-270   doi: http://dx.doi.org/10.21741/9781945291890-42

Summary

Preliminary investigation broadly showed positive results as the specimen with in-process heat treatment has exhibited lower and more uniform residual stress fields with respect to the baseline LMD-*p* specimen [5]. However, thermal softening and axial stresses at the H13 interface appeared greater in the heat-treated specimen. This partially contradicts the results from the Contour Method data, thus warranting further examination of the residual stresses. Crystallographic analysis using EBSD and EDX could be undertaken to examine the possibility of micro-structural alterations influencing these readings. Further mechanical testing and finite element analysis will provide a more comprehensive understanding of the effect of in-process heat treatment on parts ReManufactured using LMD-*p*.

Acknowledgments

I would like to thank the EPSRC (EP/I015698/1), HVM Catapult and the AFRC for their financial assistance without which this research would not have been possible. The following people should be acknowledged for their technical assistance and expert advice: Liza Hall, Ryan O'Neil, Giribaskar Sivaswamy, Himanshu Lalvani, Shanmukha Moturu, Kornelia Kondziolka & Abigail Ellison Plot.ly should be thanked for their software. The Materials & Residual Stress team and Machining group at the AFRC should be acknowledged for their contribution in this research project.

References

[1] Toumi, M., et al. Residual Stress Analysis of Stellite-coated Forging Tool Steel Using Synchrotron Radiation Diffraction. in Materials Science Forum. 2012. Trans Tech Publ.

[2] Birol, Y., Thermal fatigue testing of Inconel 617 and Stellite 6 alloys as potential tooling materials for thixoforming of steels. Materials Science and Engineering: A, 2010. **527**(7-8): p. 1938-1945. https://doi.org/10.1016/j.msea.2009.11.021

[3] Mudge, R.P. and N.R. Wald, *Laser engineered net shaping advances additive manufacturing and repair.* Welding Journal-New York-, 2007. **86**(1): p. 44.

[4] Payne, G., et al., *Remanufacturing H13 Steel Moulds and Dies Using Laser Metal deposition.* Advances in Manufacturing Technology XXX; IOS Press: Amsterdam, The Netherlands, 2016: p. 93.

[5] Cullen, C., et al., *AFRC CATP 764 Suitability of LMD for Oil and Gas Cladding and Die Remanufacture.* 2018, University of Strathclyde: Advanced Forming Research Centre.

[6] Cullen, C., et al., *AFRC CATP 463 Additive Die ReManufacturing.* 2017, University of Strathclyde: Advanced Forming Research Centre.

[7] Zavala-Arredondo, M., et al., Laser diode area melting for high speed additive manufacturing of metallic components. Materials & Design, 2017. **117**: p. 305-315. https://doi.org/10.1016/j.matdes.2016.12.095

[8] Designation, A., *Standard Hardness Conversion Tables for Metals Relationship Among Brinell Hardness.* Vickers Hardness, Rockwell Hardness, Superficial Hardness, Knoop Hardness, Scleroscope Hardness, and Leeb Hardness, ASTM international E140-12b, 2013. **1**.

[9] Prime, M.B., *Cross-sectional mapping of residual stresses by measuring the surface contour after a cut.* Journal of engineering materials and technology, 2001. **123**(2): p. 162-168. https://doi.org/10.1115/1.1345526

[10] Fitzpatrick, M., et al., Determination of residual stresses by X-ray diffraction. 2005.

Residual Stresses 2018 – ECRS-10
Materials Research Proceedings 6 (2018) 271-276

Materials Research Forum LLC
doi: http://dx.doi.org/10.21741/9781945291890-43

Residual Stresses Analysis in AISI 316L Processed by Selective Laser Melting (SLM) Treated by Mechanical Post-Processing Treatments

Q. Portella[1,a,*], M. Chemkhi[1,2,b], D. Retraint[1,c],

[1] Charles Delaunay Institute, LASMIS, CNRS, University of Technology of Troyes, 10000 Troyes, France

[2] EPF Graduate School of Engineering, 2 Rue Fernand Sastre, Troyes, France

[a]quentin.portella@utt.fr, [b]mahdi.chemkhi@epf.fr, [c]delphine.retraint@utt.fr

Keywords: Residual Stresses, 316L, Selective Laser Melting, SMAT, Roughness, Micro-Hardness

Abstract. Selective Laser Melting (SLM) is a metal additive manufacturing process widely used in industry for its extraordinary versatility and minimal waste of material. Mechanical properties of SLM parts strongly depend on the process parameters such as power, scanning speed, hatch space and scanning strategy. Depending on these latter parameters, the SLM parts can be porous or fully dense. However, the high thermal gradients which are characteristic of this process induce complex distributions of residual stresses and defects such as micro-cracks. In the case where these internal stresses have a negative effect on the physical and mechanical integrity of SLM parts, the well-known solution proposed to reduce them is a post-heat treatment. Incidentally, superficial compressive residual stresses can also be generated by mechanical treatments and can improve the fatigue performance of the treated parts. The aim of this work is to examine the effect of Surface Mechanical Attrition Treatment (SMAT) on the residual stresses present in AISI 316L parts processed by SLM. The X-ray diffraction (XRD) results show that this mechanical treatment is a promising method to avoid the negative effect of tensile residual stresses resulting from the SLM process and can introduce a beneficial superficial compressive residual stress state. Moreover, work hardening and surface roughness were evaluated in all the untreated and SMATed samples.

Introduction

Residual stresses are associated to any fabrication process and can actually be harmful for the samples, leading to lower mechanical properties or even plastic deformation [1]. Manufacturers usually prevent the negative effect of these stresses by using post-heat treatments that fix the desired microstructure and mechanical properties depending on the thermal stability of parts. Another way to reduce these residual stresses is the use of a post-mechanical treatment, such as SMAT (Surface Mechanical Attrition Treatment) [2]. During this treatment, the sample is impacted by a multitude of balls which moves in multiple directions contrary to shot peening where the shot direction is generally normal to the surface sample or presents a specific angle [3]. Consequences of this mechanical post-treatment are multiple, depending on the intensity of the treatment. It can deeply affect the microstructure of the treated part, generate a superficial nanocrystalline layer [4], increase the micro-hardness [5], reduce the roughness [6], or even enhance the nitrogen diffusion [7].

In this paper, the studied material is a 316L austenitic stainless steel (SS) which was produced by a metallic additive manufacturing process, Selective Laser Melting (SLM). These new methods of manufacturing pave the way to new structures, such as porous ones [8], with many advantages, like less waste of matter or faster manufacturing but can also introduce defects like

Residual Stresses 2018 – ECRS-10   Materials Research Forum LLC
Materials Research Proceedings 6 (2018) 271-276   doi: http://dx.doi.org/10.21741/9781945291890-43

cracks, internal porosity, poor surface state and residual stresses [1]. Parts made by SLM usually exhibit tensile residual stresses [8]. Several authors who have studied the effect of SMAT on conventional (cast, wrought) 316L SS showed that SMAT can introduce superficial compressive residual stresses [10]. To the best of the authors' knowledge, there is no previous study investigating the effect of SMAT on the mechanical properties and thermal residual stresses of additive manufactured parts. In this work, the SMAT effect on SLM 316L SS is studied in terms of residual stresses, roughness, micro-hardness, phase analysis and tensile behaviour.

Materials and methods

The 316L powder used to produce the SLM samples was obtained by nitrogen atomisation process. The particles are spherical with a size ranging between 10 and 45 μm. The chemical analysis of the 316L powder is presented in Table 1.

Table 1 : Chemical analysis of the 316L powder

Element	Fe	C	Si	Mn	P	S	Cr	Ni	Mo	N
wt%	Balance	0.01	0.53	1.12	0.03	0.013	16.61	10.18	2.15	0.08

The 316L samples were produced by the Platinium 3D Technology Platform. The geometry and dimensions of the samples are described in Figure 1.

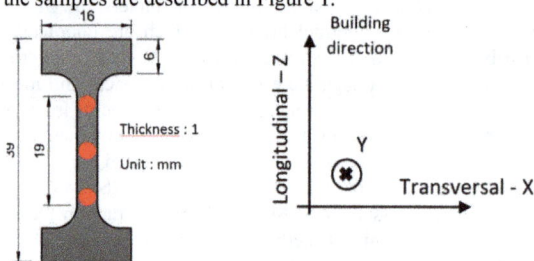

Fig. 1 : Sample geometry according to ISO 6892-1 [11]

Samples were made under Ar atmosphere with a residual oxygen fraction inside the chamber equal to 0.02%. The building platform was heated at 150°C during the entire process. The samples were built vertically, according to Fig. 1., The vertical position of the samples was chosen to ensure the same surface state between the two sides of each sample. The layer thickness was set to 50 μm and the hatch space to 80 μm.

The Vickers micro-hardness measurements were performed using a FM300e equipment from the Future-Tech Company, with a load of 25g during 10s. The measurements were performed in the cross-section perpendicular to the bilding direction. Each hardness experimental value is an average of 7 indentations along the transversal axis performed at the same distance from the surface.

Roughness measurements were performed using a Surtronic 3+ equipment from the Taylor-Hobson company. The evaluation length for Ra parameter was performed in accordance with the standard NF EN ISO 4287 [12]. Before performing the measurements, two tests were carried out on a reference specimen. After these validation tests, the obtained results are an average of 8 measurements. Residual stresses were measured at the surface of each sample, at three different points in the longitudinal and transverse directions (see Fig.1) in order to check the homogeneity of the superficial residual stress state. The X-ray diffraction $\sin^2\psi$ method was used with Cr Kα wavelength.Tensile tests were performed at room temperature using a 5KN Microtest Extended Tensile tester equipment from Deben UK Ltd. with a motor speed of 1.25mm/min.The SMAT parameters that were used are summarised in Table 2.

Residual Stresses 2018 – ECRS-10   Materials Research Forum LLC
Materials Research Proceedings **6** (2018) 271-276   doi: http://dx.doi.org/10.21741/9781945291890-43

Table 2 : SMAT treatment

Configuration of treatment	SMAT Very High (SVH)
Duration of treatment	20 min
Nature/diameter of shot	100Cr6/3mm
Amplitude	27 µm

Results and Discussion

Surface roughness: To evaluate the surface roughness, the most universally used roughness parameter for general quality control is the arithmetic average height Ra [13]. As-fabricated parts have a roughness average of 11 ± 1 µm on each side of the samples. This value of roughness is similar to what was found by other authors who worked on 316L SLM, like [14], [15] with a measured Ra of ~12 µm. This similarity in roughness measurement is due to the nominal size of powder initially used. The general parameters for fabrication (power, scan speed, thickness, hatch space) are indeed different but the size of the used powder is the same (10-45µm). The surface of the as-fabricated samples is a compound of partially melted powder, which is at the origin of this roughness value.

As shown in Figure 2, the SMAT drastically decreases the surface roughness up to 95% so that the surface roughness reaches a value of 1.2 µm.

SMAT exists among other methods designed to reduce surface roughness. Yasa and Kruth used laser re-melting to lower the surface roughness from 12 µm to 1.5 µm [15].

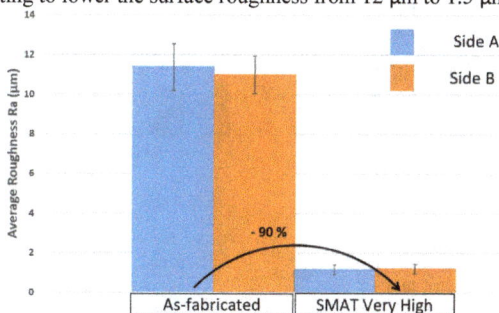

Fig. 2 : Evolution of the surface roughness Ra

Micro-hardness: In order to have some indication on the micro-hardness of parts produced by additive manufacturing, previous studies reveal a value of micro-hardness of 316L SS SLM between 210 HV_{1000} and 240 HV_{1000}: [16] speaks about 235 HV, [17] reports values between 210 and 240 HV_{1000}, [18] 216 HV_{1000}, [19] 228 HV, [11] 225 HV_{1000}. In our case, the as-fabricated sample shows a micro-hardness around 240 $HV_{0.025}$ near the surface and 205 $HV_{0.025}$ in the bulk which is in accordance with literature. As shown in Fig. 3, SMAT has increased the micro-hardness by more than 50% in the bulk of the sample. The strain hardening effect has affected the entire width of the sample. For the as-fabricated one, the evolution of the micro-hardness shows an increase from the centre to the surface, a trend which is much less present after the SVH treatment.

Residual stresses: The as-fabricated sample shows tensile residual stresses up to 30-40 MPa in the transversal direction and 100 MPa in the longitudinal direction (see Fig. 1). In the work of [9] the evaluated residual stresses are approximatively equal to 200 MPa with an evolution from 100 to 150 MPa in the sample's thickness. Fig. 4 shows that SMAT generates superficial compressive residual stresses ranging from -170 MPa to -280 MPa. The differences recorded between the three different points of measurement is rather small. Moreover, it has to be underlined that, these measurements do not give any information on the maximum compressive residual stress which is reached after SMAT; on the one hand, Roland et al. [10] indicate that the highest compressive residual stress value after SMAT is reached at the surface of a 316L steel sample, but on the other hand, Gallitelli et al. [20] show that the maximum compressive residual stresses is reached below the surface for a titanium sample subjected to SMAT. Residual stress profile measurements are thus under progress using X-ray diffraction coupled with electropolishing steps.

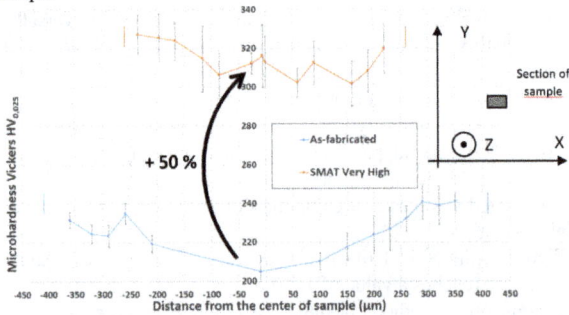

Fig. 3 : Evolution of micro-hardness profiles according to the transversal direction for different sample conditions

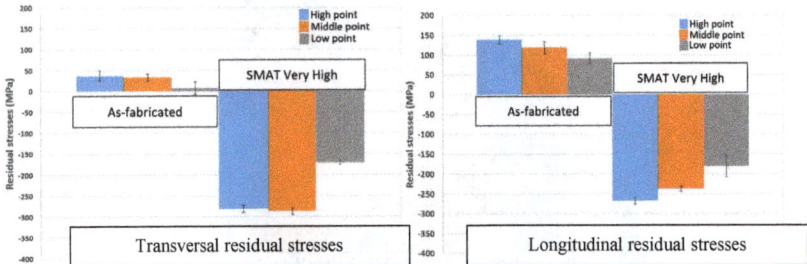

Fig. 4 : Superficial residual stresses evaluated at three different points (see Fig. 1) on the untreated and SMATed surface samples

Tensile tests: Fig. 5 represents the tensile curves obtained for the untreated and SMATed samples. It shows that SMAT modifies the mechanical behaviour drastically, with a high reduction of elongation and an increase in yield strength and tensile strength. Indeed, as-fabricated samples exhibit a uniform elongation of 31.19 ± 2.33 % with mechanical properties of 557 ± 15 MPa for Yield Strength (YS) and 859 ± 11 MPa for Ultimate Tensile Test (UTS). SVH treatment increases YS by 53 % reaching a value of 852 ± 38 MPa and 7 % for UTS setting the value at 918 ± 6 MPa. The biggest modification is for the Uniform Elongation (UE) which decreases by 71.7 %, with a final value of 7.78 ± 0.6 %. These modifications induced by SMAT on the YS, UTS and UE strongly depend on the treatment intensity: Roland et al. [5] showed the same tendency for conventional 316 L SS before and after SMAT; the more intense the treatment is, the more affected the mechanical properties are [21].

Conclusion

This study shows that 316L SS SLM subjected to SMAT can have its properties strongly modified (micro-hardness, surface roughness, residual stresses) in order to improve its mechanical behaviour. Other SMAT treatments with different configurations (lower intensity)

are currently under way in order to evaluate their effects on this stainless steel properties. Residual stresses should be examined at different depths below the untreated and SMATed surfaces to identify their evolution. SMAT proved to be especially promising to enhance several mechanical properties such as yield and ultimate strengths, nevertheless, it also decreases the ductility (elongation). Besides, additional experiments are under progress in order to understand the effect of SMAT on the strengthening of SLM parts.

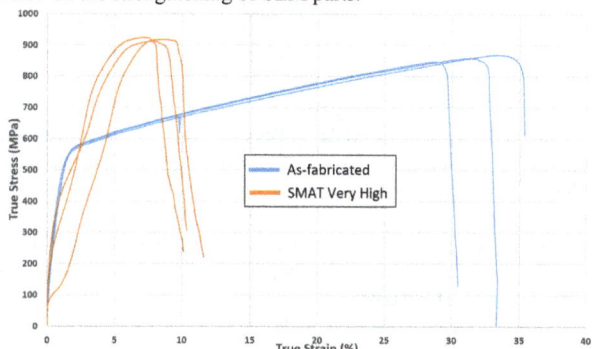

Fig. 5 : Tensile tests

Acknowledgements
The authors wish to thank the Aube Department Council (France) and Europe (FEDER) for financial support. The PLATINIUM3D Technology Platform is also acknowledged for providing the samples.

References
[1] W. J. Sames, F. A. List, S. Pannala, R. R. Dehoff, and S. S. Babu, "The metallurgy and processing science of metal additive manufacturing," *Int. Mater. Rev.*, vol. 61, no. 5, pp. 315–360, Jul. 2016. https://doi.org/10.1080/09506608.2015.1116649
[2] K. Lu and J. Lu, "Nanostructured surface layer on metallic materials induced by surface mechanical attrition treatment," *Mater. Sci. Eng. A*, vol. 375–377, pp. 38–45, Jul. 2004. https://doi.org/10.1016/j.msea.2003.10.261
[3] B. Arifvianto, Suyitno, M. Mahardika, P. Dewo, P. T. Iswanto, and U. A. Salim, "Effect of surface mechanical attrition treatment (SMAT) on microhardness, surface roughness and wettability of AISI 316L," *Mater. Chem. Phys.*, vol. 125, no. 3, pp. 418–426, Feb. 2011. https://doi.org/10.1016/j.matchemphys.2010.10.038
[4] Y. Samih, B. Beausir, B. Bolle, and T. Grosdidier, "In-depth quantitative analysis of the microstructures produced by Surface Mechanical Attrition Treatment (SMAT)," *Mater. Charact.*, vol. 83, pp. 129–138, Sep. 2013. https://doi.org/10.1016/j.matchar.2013.06.006
[5] T. Roland, D. Retraint, K. Lu, and J. Lu, "Enhanced mechanical behavior of a nanocrystallised stainless steel and its thermal stability," *Mater. Sci. Eng. A*, vol. 445–446, pp. 281–288, Feb. 2007. https://doi.org/10.1016/j.msea.2006.09.041
[6] B. Arifvianto, Suyitno, and M. Mahardika, "Effects of surface mechanical attrition treatment (SMAT) on a rough surface of AISI 316L stainless steel," *Appl. Surf. Sci.*, vol. 258, no. 10, pp. 4538–4543, Mar. 2012. https://doi.org/10.1016/j.apsusc.2012.01.021
[7] M. Chemkhi, D. Retraint, A. Roos, and C. Demangel, "Role and effect of mechanical polishing on the enhancement of the duplex mechanical attrition/plasma nitriding treatment of AISI 316L steel," *Surf. Coat. Technol.*, vol. 325, pp. 454–461, Sep. 2017. https://doi.org/10.1016/j.surfcoat.2017.06.052

[8] J. Čapek et al., "Highly porous, low elastic modulus 316L stainless steel scaffold prepared by selective laser melting," *Mater. Sci. Eng. C*, vol. 69, pp. 631–639, Dec. 2016. https://doi.org/10.1016/j.msec.2016.07.027

[9] T. Simson, A. Emmel, A. Dwars, and J. Böhm, "Residual stress measurements on AISI 316L samples manufactured by selective laser melting," *Addit. Manuf.*, vol. 17, pp. 183–189, Oct. 2017. https://doi.org/10.1016/j.addma.2017.07.007

[10] T. Roland, D. Retraint, K. Lu, and J. Lu, "Fatigue life improvement through surface nanostructuring of stainless steel by means of surface mechanical attrition treatment," *Scr. Mater.*, vol. 54, no. 11, pp. 1949–1954, Jun. 2006. https://doi.org/10.1016/j.scriptamat.2006.01.049

[11] "ISO 6892-1:2009(en), Metallic materials — Tensile testing — Part 1: Method of test at room temperature."

[12] "ISO 4287:1997 - Geometrical Product Specifications (GPS) -- Surface texture: Profile method -- Terms, definitions and surface texture parameters."

[13] E. S. Gadelmawla, M. M. Koura, T. M. A. Maksoud, I. M. Elewa, and H. H. Soliman, "Roughness parameters," *J. Mater. Process. Technol.*, vol. 123, no. 1, pp. 133–145, Apr. 2002. https://doi.org/10.1016/S0924-0136(02)00060-2

[14] J. A. Cherry, H. M. Davies, S. Mehmood, N. P. Lavery, S. G. R. Brown, and J. Sienz, "Investigation into the effect of process parameters on microstructural and physical properties of 316L stainless steel parts by selective laser melting," *Int. J. Adv. Manuf. Technol.*, vol. 76, no. 5–8, pp. 869–879, Feb. 2015. https://doi.org/10.1007/s00170-014-6297-2

[15] E. Yasa and J.-P. Kruth, "Microstructural investigation of Selective Laser Melting 316L stainless steel parts exposed to laser re-melting," *Procedia Eng.*, vol. 19, pp. 389–395, Jan. 2011. https://doi.org/10.1016/j.proeng.2011.11.130

[16] I. Tolosa, F. Garciandía, F. Zubiri, F. Zapirain, and A. Esnaola, "Study of mechanical properties of AISI 316 stainless steel processed by 'selective laser melting', following different manufacturing strategies," *Int. J. Adv. Manuf. Technol.*, vol. 51, no. 5–8, pp. 639–647, Nov. 2010. https://doi.org/10.1007/s00170-010-2631-5

[17] E. Liverani, S. Toschi, L. Ceschini, and A. Fortunato, "Effect of selective laser melting (SLM) process parameters on microstructure and mechanical properties of 316L austenitic stainless steel," *J. Mater. Process. Technol.*, vol. 249, pp. 255–263, Nov. 2017. https://doi.org/10.1016/j.jmatprotec.2017.05.042

[18] Z. Sun, X. Tan, S. B. Tor, and W. Y. Yeong, "Selective laser melting of stainless steel 316L with low porosity and high build rates," *Mater. Des.*, vol. 104, pp. 197–204, Aug. 2016. https://doi.org/10.1016/j.matdes.2016.05.035

[19] A. A. Deev, P. A. Kuznetcov, and S. N. Petrov, "Anisotropy of Mechanical Properties and its Correlation with the Structure of the Stainless Steel 316L Produced by the SLM Method," *Phys. Procedia*, vol. 83, pp. 789–796, Jan. 2016. https://doi.org/10.1016/j.phpro.2016.08.081

[20] D. Gallitelli, D. Retraint, and E. Rouhaud, "Comparison between Conventional Shot Peening (SP) and Surface Mechanical Attrition Treatment (SMAT) on a Titanium Alloy," *Adv. Mater. Res.*, vol. 996, pp. 964–968, Aug. 2014. https://doi.org/10.4028/www.scientific.net/AMR.996.964

[21] T. Roland, D. Retraint, K. Lu, and J. Lu, "Generation of Nanostructures on 316L Stainless Steel and Its Effect on Mechanical Behavior," *Materials Science Forum*, 2005. https://doi.org/10.4028/0-87849-969-5.625

Residual Stresses 2018 – ECRS-10
Materials Research Proceedings 6 (2018) 277-282

Materials Research Forum LLC
doi: http://dx.doi.org/10.21741/9781945291890-44

Analytical Model for Distortion Prediction in Wire + Arc Additive Manufacturing

Jan Roman Hönnige[a,*], Paul Colegrove[b] and Stewart Williams[c]

Welding Engineering and Laser Processing Centre (WELPC), Cranfield University, Cranfield, Bedfordshire, MK43 0AL, UK

[a]j.honnige@cranfield.ac.uk, [b]p.colegrove@cranfield.ac.uk, [c]s.williams@cranfield.ac.uk

Keywords: Residual Stress, Bending Distortion, Neutron Diffraction, Geometry Factor, Material and Process Factor, Critical Wall Height

Abstract. An analytical model was developed to predict bending distortion of the base-plate caused by residual stresses in additively manufactured metal deposits. This avoids time-consuming numerical simulations for a fast estimation of the expected distortion. Distortion is the product of the geometry factor K, which is determined by the cross-section of substrate and deposit, and the material and process factor S, which is the quotient of residual stress and the Young's Modulus. A critical wall height can be calculated for which the structure distorts the most. This critical height is typically less than 2.5 times the thickness of the substrate. Higher walls increase the stiffness of the cross-section and reduce the distortion with increasing height.

Introduction

Wire + Arc Additive Manufacturing (WAAM) can be used to manufacture large-scale near-net-shape preforms for structural components with medium complexity [1]. Typical materials are structural metals, such as steel, titanium, Inconel or aluminium alloys. Residual Stresses are amongst the largest challenges in WAAM. They can cause serious distortion [2] and premature failure during the deposition or in service of the component [3]. It was widely reported using different experimental approaches and numerical simulations that stresses can reach values between 50% and 100% of the materials yield strength [2–7]. Thermal stress relieving can be used to eliminate these stresses [8]. However, unclamping the part after the deposition causes distortion before the part can be heat treated. The magnitude of the distortion depends on the residual stresses and therefore on the thermo-mechanical history of the deposition, as well as the material characteristics and the geometry. The repetitive deposition procedure requires extensive computing in numerical approaches to capture the full thermal history, required for stress calculation [6]. Simplified models can be accurate and more efficient [9], but parametric investigations would still require the simulation of each individual case. This makes it time consuming to compare cross-sectional variations. Analytical approaches, as they exist for butt-welding [10] or cladding processes [11], can estimate distortion fast and effortless. Colegrove et al. [2] proposed the analogy of the behaviour of residual stresses in WAAM and bending distortion according to cantilever beam theory, which will be elaborated and validated in this paper.

Methodology

Analytical Model for Distortion Prediction. Previous investigations allowed fundamental assumptions for this analytical approach. Uniform residual stresses are produced along the wall height during deposition, as shown in the schematic in Figure 1 (a). The balancing compressive stresses in the substrate are assumed constant as well with a non-continuous transition in the interface. Unclamping causes distortion and redistribution of the stresses in a way that these tensile stresses drop linearly towards the top of the wall (Figure 1 (b)) [2]–[7]. This stress development and redistribution is furthermore assumed to be constant along the length L of the wall, disregarding any edge effects. The distortion $w(x)$ is therefore a function of the length L and the constant curvature κ (the inverse of the curvature radius ρ), as illustrated in Figure 1 (c). The analytical model assumes a deposition of a straight linear wall. Wall width WW and wall height WH, as well as substrate thickness t and width b are geometrical variables. Figure 1 (d) and Eq. (1) and (2) show how these variables can be used to find the centroid of the substrate and deposit. The centroid of the overall area z_0 is also the location of the neutral axis of the T-section and can be calculated using Eq. (3) to determine the moment of inertia I_{yy} with Eq. (4).

$$z_s = \frac{1}{2}t \tag{1}$$

$$z_d = \frac{1}{2}WH + t \tag{2}$$

$$z_0 = \frac{\sum z_{s,i}A_i}{\sum A_i} = \frac{(z_d \times WW \times WH) + (z_s \times b \times t)}{WW \times WH + bt} \tag{3}$$

$$I_{yy} = \frac{1}{12}(bt^3 + WW \times WH^3) + bt \times (z_0 - z_s)^2 + WW \times WH \times (z_d - z_0)^2 \tag{4}$$

$$F_d = F_s = F = -\iint \sigma_{xx,d}(y,z) * dA(WW, WH) \tag{5}$$

$$F = \sigma_{xx} * WW * WH \tag{6}$$

$$M = F_d(z_d - z_0) + F_s(z_0 - z_s) = F(z_d - z_s) = F(\frac{1}{2}t + \frac{1}{2}WH) \tag{7}$$

$$\kappa\left[\frac{1}{m}\right] = \frac{1}{\rho} = \frac{M}{EI_{yy}} = \overbrace{\frac{\widetilde{\sigma_{xx}}}{E}}^{material\ \&\ process\ factor\ S} \times \overbrace{\frac{WW \times WH \times (t + WH)}{2\,I_{yy}}}^{geometry\ factor\ K} = S\left[\frac{MPa}{GPa}\right] \times K\left[\frac{1}{m}\right] \tag{8}$$

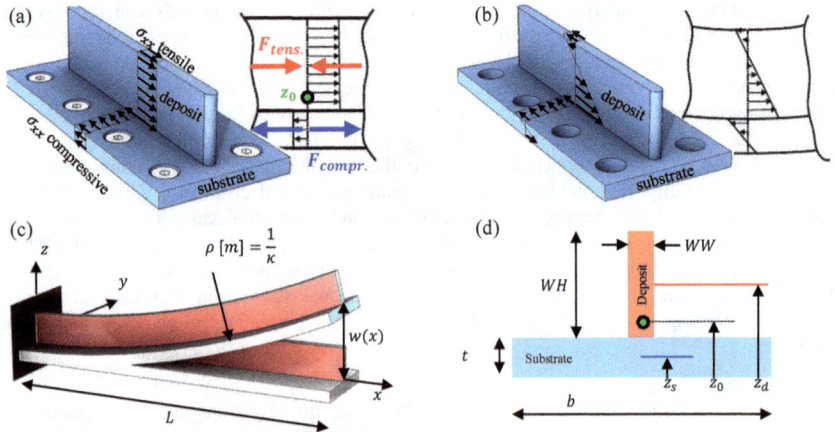

Figure 1: (a) unbalanced stress field development during the WAAM deposition; (b) redistributed and self-balanced stress field after distortion; (c) cantilever beam with constant curvature ρ and cross-section along the length L and the distortion w(x); (d) variables in cross-section.

Residual Stresses 2018 – ECRS-10   Materials Research Forum LLC
Materials Research Proceedings 6 (2018) 277-282   doi: http://dx.doi.org/10.21741/9781945291890-44

The internal forces F_d caused by the residual stresses σ_{xx} in the deposit can be calculated with Eq. (5) and act at the centre gravity of this integrated volume. The compressive stress in the substrate F_s must be equal to F_d. Since the stresses were assumed to be constant across the rectangular cross-section, the internal force F_d acts at z_d and F_s acts at z_s and the magnitude simplifies as per Eq. (6). The location of the concentrated forces F_d and F_s result in a bending moment around the neutral axis, which is calculated with Eq. (7).

The curvature κ can now be calculated using Eq. (8), in which the quotient of the Young's modulus and the residual stress define the process and material factor S, while the geometry factor K is entirely dependent on the four cross-sectional variables. The value of the tensile residual stress σ_{xx} needs to be available for distortion prediction and depends on material characteristics and the thermal history. It can be determined experimentally [2]–[7] or analytically [11].

Stress redistribution. The redistributed residual stress field after distortion (Figure 1 (b)) is the sum of the initial stress field before unclamping (Figure 1 (a)) and the stress caused by the internal bending moment, according to Eq. (9), where z = 0 is located in the neutral axis z_0. Both, the bending moment M and the I_{yy} are constant along the length. The equilibrium residual stress field is therefore the as-clamped stress field, after pivoting around the neutral axis. The stress gradient, which drops linearly towards the top of the wall can be calculated using Eq. (10).

$$\sigma_{xx,res}(z) = \sigma_{xx} + \frac{M}{I_{yy}} \times z \tag{9}$$

$$\frac{\Delta \sigma_{xx}}{\Delta z} = \frac{M}{I_{yy}} \tag{10}$$

Validation of Prediction for Distortion and Stress Redistribution. To validate the distortion prediction three geometrically identical walls were built with WAAM using Ti−6Al−4V, In718 and the aluminium alloy AA2319. The dimensions were L =250, WW = 6, WH = 22, t = 7 and b = 60 (Figure 1; units in mm). Neutron diffraction strain measurements were performed for stress calculation using the ENGIN-X instrument at ISIS in Didcot for the Ti−6Al−4V [4] and aluminium [7] deposits and the SALSA beam line at ILL in Grenoble for the Inconel deposit[1]. For the latter an incoming wavelength of 1.62 Å was used to diffract the {311} crystallographic plane with an angle of 2θ = 96.2°. The scan was performed along the vertical centreline in the middle of the deposit. All three specimen were scanned before and after unclamping. The constant longitudinal stress value in the clamped condition was used to predict the distortion and the associated stress gradients after redistribution to compare them with the experimental results.

Results

The geometry factor K of any profile can be plotted numerically, as shown in Figure 2 (a). It is here plotted as a dimensionless function of practical ratios of WH/t and b/WW, allowing the determination for any arbitrary combination of realistic wall geometries on any substrate. The cross-section in the present experiments has a value K = 68.57 [1/m] (X in Figure 2 (a)). The critical wall height is defined as the value which gives the maximum value of t*K in Figure 2 (a). This is plotted as a function of b/WW in Figure 2 (b). Walls that exceed the critical height result in increased stiffness of the cross-section and therefore reduce the bending distortion.

The determined residual stress from the neutron diffraction experiment are shown in Figure 3, from which the constant residual stress values for each alloy can be taken. The dashed lines are not trend lines, but the predicted stress redistribution using the analytical model. For this

---

[1] The instrument and typical methodology are described in Pirling et al. [12]

Residual Stresses 2018 – ECRS-10   Materials Research Forum LLC
Materials Research Proceedings **6** (2018) 277-282   doi: http://dx.doi.org/10.21741/9781945291890-44

geometry, the neutral axis z_0 almost coincides with the interface (z = 6.97 mm), which is why the both stress plots, before and after unclamping meet in the interface.

The calculated results of the analytical model in Table 1 are compared with the measured stress gradients and the actual distortion of the specimen. The distortion of 200 mm long specimen was calculated using the constant curvature.

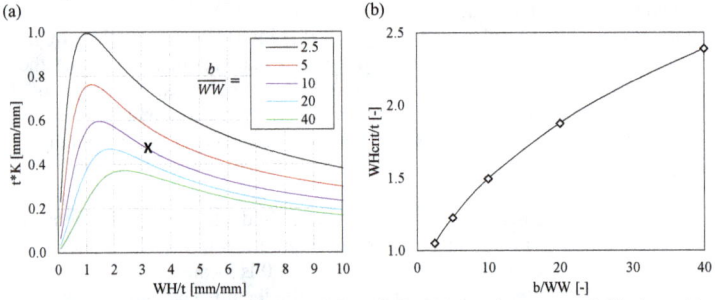

Figure 2: (a) Plot of the geometry factor K for all dimensional ratios and (b) the critical wall height

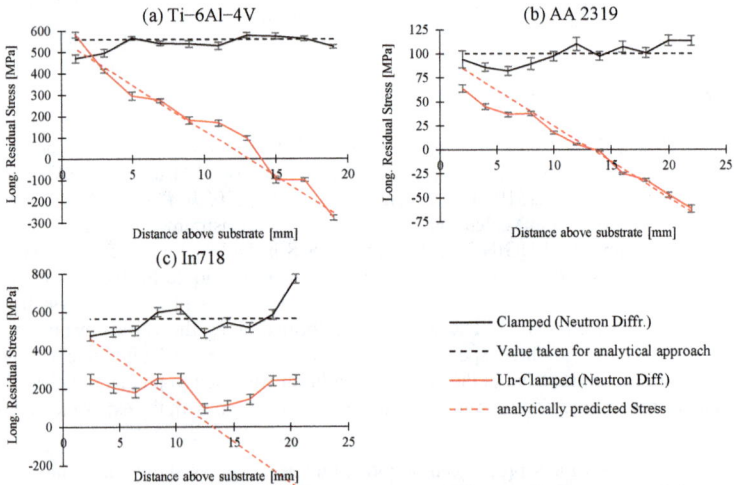

Figure 3: Measured Longitudinal Residual Stress using Neutron Diffraction and that predicted from the analytical model using identical wall geometries of (a) Ti–6Al–4V [4] (b) Aluminium AA 2319 [7] and (c) Inconel In718.

Table 1: Experimentally and analytically determined distortion characteristics for K = 68.57 [1/m]

	σ_{xx} [MPa]	S [MPa/GPa]	κ [1/m]	ρ [m]	w (L = 200mm)		$\Delta\sigma/\Delta z$ [MPa/mm]	
	exper.	exper.	analyt.	analyt.	analyt.	exper.	analyt.	exper.
Ti-6Al-4V	565	5	0.348	2.873	6.92	7.1	42.8	40
In718	560	2.8	0.199	5.02	3.98	5.64	41.6	11
AA 2319	130	1.75	0.092	10.84	1.85	1.75	7.58	7

Residual Stresses 2018 – ECRS-10   Materials Research Forum LLC
Materials Research Proceedings **6** (2018) 277-282   doi: http://dx.doi.org/10.21741/9781945291890-44

Discussion

The analytical model is in good agreement with the experimental results of titanium and aluminium deposition (Table 1). The Inconel results do not agree, which may be due to experimental errors, such as insufficient clamping and slip of the substrate during the deposition. As a result bending distortion may have occurred during the deposition already, leading to more distortion than predicted and less redistribution during unclamping. However, the good agreement with the other two alloys gives confidence that the analytical model can be used to provide a good first estimation of distortion and stress redistribution. A high geometry factor K and a high material and process factor S produce large distortion. Titanium alloys are therefore particularly susceptible due to high residual stresses (~600 MPa) combined with a relatively low elastic modulus (113 GPa) [4]. Inconel and Steel are less susceptible due to their stiffness above 200 GPa [2], while having comparable stresses. Aluminium is the least susceptible material due to very low residual stresses (~100 MPa) that over-compensate the low stiffness (72 GPa) [7]. The analytical model requires the input of these values. If no residual stress data is available, then assuming yield strength as residual stress is justifiable and represents the worst case scenario. The value can otherwise be estimated analytically based on thermal material properties [11].

Another interesting observation is the existence of a critical wall height WH_{crit} (Figure 2 (b)), which depends only on geometry and indicates the wall height that results in the greatest distortion. This value appears to be smaller than 2.5 times the thickness of the substrate for realistic geometries. Exceeding this wall height would result in reduced distortion. The figurative reason is that the added height increases the stiffness of the cross-section more than the bending moment increases. This results in shortening of the substrate, rather than in bending distortion. For example, building a 70 mm high wall (instead of 22 mm) would reduce the geometry factor to $K = 32.86$ and the final distortion of the titanium wall from 6.92 mm to 3.26 mm. It should be noted that the critical wall height is independent of material and process.

For the purpose of comparing this approach with experimental results, simple geometries like a rectangular cross-section are easier to display, allowing the dimensionless and universally valid plot in Figure 2. However, more complex deposits can be predicted as well (e.g. several walls parallel to each other, non-rectangular cross-sections, cross-sectional variations along length, asymmetric double sided deposits, 2D deposits, stress gradients).

For more complex geometries, the integral in eq. (5) would have to be solved, assuming that the force in the deposit acts in the center of the integrated volume and the moment of inertia of the new cross-section has to be calculated accordingly.

Conclusion

The analytical model is a very effective tool for estimating the distortion and stress redistribution in additively manufactured structures on a substrate. The accuracy is reasonably good, considering the assumptions and boundary conditions. The main input is the as-deposited residual stress, which can be found experimentally. This value can otherwise be estimated analytically [11] or simply be assumed to be as high as the materials yield strength.

It was furthermore found that for any possible geometry combination there is a critical wall height, which would result in the greatest possible distortion for a particular AM process and material. This critical wall height is typically smaller than 2.5 times the thickness of the substrate. It is therefore also possible to determine the most-critical substrate geometry for a deposit in terms of bending distortion.

Acknowledgement

The findings were generated with the financial support of the WAAMMat program.

Residual Stresses 2018 – ECRS-10
Materials Research Forum LLC
Materials Research Proceedings 6 (2018) 277-282
doi: http://dx.doi.org/10.21741/9781945291890-44

References

[1] S. W. Williams, F. Martina, a. C. Addison, J. Ding, G. Pardal, and P. Colegrove, "Wire+Arc Additive Manufacturing," *Mater. Sci. Technol.*, vol. 32, no. 7, pp. 641–647, 2015.

[2] P. A. Colegrove, H. E. Coules, J. Fairman, F. Martina, T. Kashoob, H. Mamash, and L. D. Cozzolino, "Microstructure and residual stress improvement in wire and arc additively manufactured parts through high-pressure rolling," *J. Mater. Process. Technol.*, vol. 213, pp. 1782–1791, 2013.

[3] J. Zhang, X. Wang, S. Paddea, and X. Zhang, "Fatigue crack propagation behaviour in wire + arc additive manufactured Ti-6Al-4V: Effects of microstructure and residual stress," *Mater. Des. Des.*, vol. 90, pp. 551–561, 2016.

[4] J. R. Hönnige, S. Williams, M. J. Roy, P. Colegrove, and S. Ganguly, "Residual Stress Characterization and Control in the Additive Manufacture of Large Scale Metal Structures," in *10th International Conference on Residual Stresses*, 2016.

[5] F. Martina, M. J. Roy, B. A. Szost, S. Terzi, P. A. Colegrove, S. W. Williams, P. J. Withers, J. Meyer, and M. Hofmann, "Residual stress of as-deposited and rolled wire+arc additive manufacturing Ti–6Al–4V components," *Mater. Sci. Technol.*, vol. 0836, no. April, 2016.

[6] J. Ding, P. Colegrove, J. Mehnen, S. Ganguly, P. M. S. Almeida, F. Wang, and S. Williams, "Thermo-mechanical analysis of Wire and Arc Additive Layer Manufacturing process on large multi-layer parts," *Comput. Mater. Sci.*, vol. 50, no. 12, pp. 3315–3322, 2011.

[7] J. R. Hönnige, P. A. Colegrove, S. Ganguly, E. Eimer, S. Kabra, and S. W. Williams, "Control of Residual Stress and Distortion in Aluminium Wire + Arc Additive Manufacture with Rolling," *Addit. Manuf.*, vol. 22, pp. 775–783, 2018.

[8] J. R. Hönnige, P. A. Colegrove, B. Ahmad, M. E. Fitzpatrick, S. Ganguly, T. L. Lee, and S. W. Williams, "Residual stress and texture control in Ti-6Al-4V wire+arc additively manufactured intersections by stress relief and rolling," *Mater. Des.*, vol. 150, pp. 193–205, 2018.

[9] J. Ding, P. Colegrove, J. Mehnen, S. Williams, F. Wang, and P. S. Almeida, "A computationally efficient finite element model of wire and arc additive manufacture," *Int. J. Adv. Manuf. Technol.*, vol. 70, pp. 227–236, 2014.

[10] J. Winczek, R. Parkitny, K. Makles, and M. Kukuryk, "Analytical model of stress field in submerged arc welding butt joint with thorough penetration," in *MATEC Web of Conferences - MMS 2017*, 2018, vol. 157, pp. 1–11.

[11] N. Tamanna, R. Crouch, I. R. Kabir, and S. Naher, "An analytical model to predict and minimize the residual stress of laser cladding process," *Appl. Phys. A Mater. Sci. Process.*, vol. 124, no. 2, pp. 1–5, 2018.

[12] T. Pirling, G. Bruno, and P. J. Withers, "SALSA-A new instrument for strain imaging in engineering materials and components," *Mater. Sci. Eng. A*, vol. 437, pp. 139–144, 2006.

Residual Stresses 2018 – ECRS-10
Materials Research Proceedings 6 (2018) 283-288

Materials Research Forum LLC
doi: http://dx.doi.org/10.21741/9781945291890-45

An Analytical Method for Predicting Residual Stress Distribution in Selective Laser Melted/Sintered Alloys

Dibakor Boruah[1,2,a,*], Xiang Zhang[2,b] and Matthew Doré[3,c]

[1]National Structural Integrity Research Centre (NSIRC), TWI Ltd, Cambridge, UK

[2]Faculty of Engineering, Environment and Computing, Coventry University, Coventry, UK

[3]Integrity Management Group, TWI Ltd, Cambridge, UK

[a]boruahd@uni.coventry.ac.uk, [b]xiang.zhang@coventry.ac.uk, [c]matthew.dore@twi.co.uk

Keywords: Residual Stress, Modelling, Selective Laser Melting, Additive Manufacturing

Abstract. Residual stresses that build up during selective laser melting or sintering (SLM/SLS) process can influence the dimensional accuracy, mechanical properties and in-service performance of SLM/SLS parts. Therefore, it is crucial to understand, predict and effectively control residual stresses in a part. The present study aims at developing an analytical model to predict the through-thickness distribution of residual stresses in an SLM part-substrate system. The proposed model demonstrates how residual stresses built up in the substrate and previously deposited layers are related to the stress induced by a newly deposited layer, based on the stress and moment equilibrium requirements. The model has been validated by published experimental measurements and verified with existing analytical/numerical models. The outcomes of the study suggest that the proposed analytical model can be used for quick estimation of residual stress distribution and the order of magnitude.

Nomenclature

a_n , b_n	Constants (related to material, process parameter and part geometry)
h, Δh	Substrate height, layer height, respectively (mm)
k	Residual stress in a newly deposited layer (MPa)
m	An individual deposited layer, m = 1, 2, 3, 4, … … …. $(n-1)$
n	Number of deposited layers $(1, 2, 3, 4, … … … …. n)$
y	Distance from substrate bottom surface (mm)
$\Delta\sigma_n(y)$	Stress increment in the substrate due to deposition of the nth layer (MPa)
$\Delta\sigma_{TS(nL)}(y)$	Total(T) stress increment in substrate(S) due to deposition of 'n' layers(nL) (MPa)
$\Delta\sigma_{T(Lm)_{(Ln)}}(y)$	Total(T) stress increment in m^{th} layer(Lm) due to depositing the nth layer(Ln) (MPa)

Introduction

Selective laser melting (SLM) and selective laser sintering (SLS) are two commonly employed additive manufacturing processes, belong to the laser powder bed fusion (LPBF) technology, offering great advantages and opportunities compared to traditional subtractive techniques. They are primarily used to build complex geometry, lightweight and customized functional parts directly from CAD data by consolidating successive layers of powder by using a high power-density laser to melt and fuse metallic powders together. The main difference between SLM and SLS is the binding mechanism between the powder particles. In SLS, powder particles are partially molten and requires post treatment to improve part's density and mechanical properties. In SLM, powder particles are fully molten. Since the difference between SLM and SLS is

Residual Stresses 2018 – ECRS-10   Materials Research Forum LLC
Materials Research Proceedings 6 (2018) 283-288   doi: http://dx.doi.org/10.21741/9781945291890-45

somewhat ambiguous, the stress-inducing mechanisms in LPBF are usually described for the case of SLM [1,2].

SLM is known for introducing significant amounts of residual stresses owing to the large thermal gradients inherently present in the process [1]. During the SLM process, the material experiences large localised heat fluctuations in a short period of time. This causes a high-temperature gradient and the resulting residual stress can cause part warpage, crack formation both during laser processing and after cutting parts from the substrate. Process induced residual stresses can affect the functional performance of the parts by undesired strength reduction, which limits the applicability of the process [1,3–5].

Although the experimental methods for measuring residual stresses possess various advantages, modelling of residual stresses is an alternative in many cases as it is quicker, inexpensive and has no restriction on specimen size, surface finish, etc. For successful manufacturing of an SLM part, effective evaluation of residual stress is very important. Therefore a simple analytical model is required. In this study, the original idea of Shiomi, et al. [5] is extended (through changes in assumptions) to calculate the through-thickness distribution and magnitude of residual stresses in an SLM part-substrate assembly due to the deposition of 'n' number of SLM layers. The mathematical formulation by Shiomi is limited to stress increment in the substrate $(\Delta\sigma_1(y))$ due to the deposition of only one layer of SLM. Moreover, the proposed model provides a better explanation of layer by layer building up of residual stresses (with mathematical representations) due to the progressive deposition of additive layers (starting from a single layer to 'n' SLM layers), taking into account of the force and moment equilibrium.

Formulation of the proposed analytical model

Mechanism of residual stress development. The mechanism of residual stress development in SLM can be distinguished into two descriptive models: (i) Temperature gradient mechanism (TGM) model, resulting from the large thermal gradients that occur around the laser spot owing to the rapid heating of the upper surface (substrate or a previously deposited layer) by the laser beam. Since the thermal expansion of the heated top layer is restricted by the underlying colder material, yielding a compressive stress-strain condition in the irradiated zone [1,3]. (ii) The cool-down phase model, which represents a cooling stage of the irradiated zone after the laser beam leaves that area, as a result, the material tends to shrink. The shrinkage is partially inhibited by the underlying material layers or substrate, yielding a residual tensile stress condition in the newly added layer and compressive stress in below [1,3].

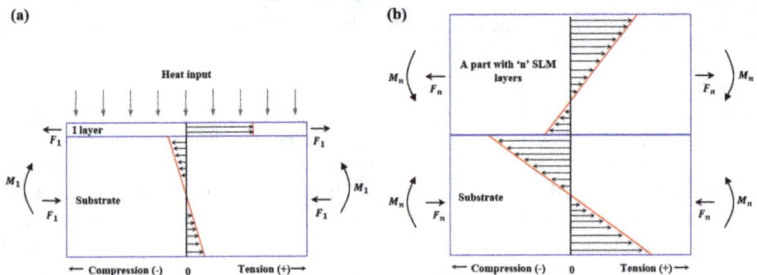

Fig. 1: Conceptual model showing the mechanism of building up of residual stresses during SLM process: (a) stress distribution after deposition of the first layer, (b) stress distribution after deposition of 'n' layers

Residual Stresses 2018 – ECRS-10   Materials Research Forum LLC
Materials Research Proceedings **6** (2018) 283-288   doi: http://dx.doi.org/10.21741/9781945291890-45

A conceptual model has been developed to interpret the mechanism of building up of residual stresses in the SLM process as illustrated in Fig 1. Fig. 1 (a) shows the generation of an upward bending moment (M_1) in the SLM part-substrate assembly by a pair of equal and opposite forces to balance the process induced tensile residual stress (k) while depositing 1st layer. This process gets more complicated as more layers build up, and the part height becomes significant as compared to the substrate. Each successive SLM layer induces the same amount of misfit strain each time on the building part of changing height. Therefore, the final stress distribution in a multilayer SLM part-substrate assembly is significantly different from the starting system with a single layer (Fig. 1a), which can be determined by a succession of force and moment balance calculations. Fig. 1 (b) represents the final state of residual stress distribution in an SLM part-substrate assembly after deposition of 'n' SLM layers, having an upward moment in the substrate and a downward moment in the part to maintain the force and moment equilibrium condition. As a result, residual stresses at the free surface of the newly deposited material and substrate are tensile in nature, at the interface residual stresses are compressive in nature [1,3].

Assumptions. Based on the known distribution of residual stresses in [1,3–5] and the aforementioned conceptual model, the following assumptions are made:

1) The part is being built at room temperature on top of a substrate of height 'h'.
2) The substrate is free from residual stresses prior to material deposition, no external forces are applied to the part-substrate assembly, and the general beam theory is valid.
3) The substrate and deposited layers have the same length and same width.
4) Residual stress of a newly deposited layer has a value 'k', which is tensile in nature and constant throughout the layer height (Δh).
5) The height of each newly deposited layer (i.e. Δh) and the resulting stress induced by each newly deposited layer is the same.
6) The increment of residual stress '$\Delta\sigma$' owing to deposition of a new layer distributes linearly in the substrate following a linear equation.

Formulations. Fig. 2 shows the building up of residual stresses in the substrate and the SLM part due to the deposition of the first to n^{th} layer. A summary of the key equations of the proposed analytical method is presented in Eqs. 1-5.

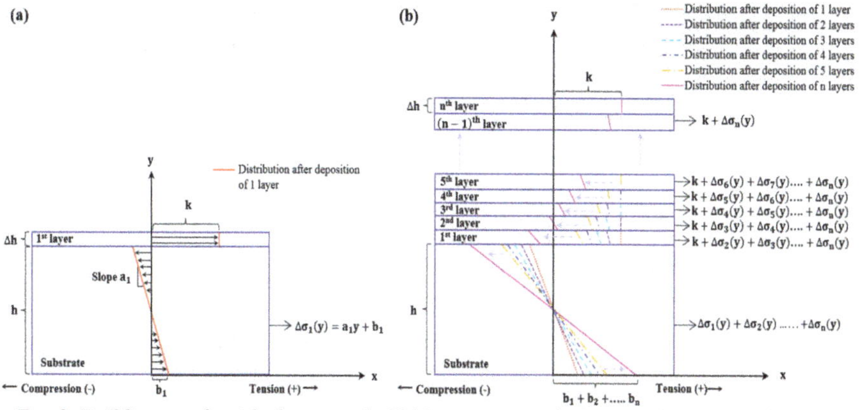

Fig. 2: Building up of residual stresses by SLM process: residual stress distribution due to the stress induced by deposition of (a) the first layer, (b) 'n' layers additively.

Assume that stress increment distributes linearly in the substrate due to the addition of $1^{st}, 2^{nd}, \ldots\ldots\ldots n^{th}$ layer respectively as follows:

$$\Delta\sigma_1(y) = a_1 y + b_1, \quad \Delta\sigma_2(y) = a_2 y + b_2, \quad \dots\dots\dots\dots, \Delta\sigma_n(y) = a_n y + b_n \tag{1}$$

After solving the force and moment equilibrium equations, the total stress increment in the substrate due to the deposition of $1, 2, \dots\dots n$ layers, respectively, can be expressed by Eq. 2:

$$\Delta\sigma_{TS(1L)}(y) = -6k\Delta hy\left(\frac{h+\Delta h}{h^3}\right) + k\Delta h\left(\frac{2h+3\Delta h}{h^2}\right)$$

$$\Delta\sigma_{TS(2L)}(y) = -6ky\Delta h\left\{\frac{h+\Delta h}{h^3} + \frac{h+2\Delta h}{(h+\Delta h)^3}\right\} + k\Delta h\left\{\frac{2h+3\Delta h}{h^2} + \frac{2h+5\Delta h}{(h+\Delta h)^2}\right\}$$

$$\dots\dots\dots\dots\dots\dots\dots\dots\dots\dots\dots\dots\dots\dots\dots\dots\dots\dots\dots\dots\dots\dots\dots\dots\dots\dots$$

$$\Delta\sigma_{TS(nL)}(y) = -6ky\Delta h\left\{\frac{h+\Delta h}{h^3} + \frac{h+2\Delta h}{(h+\Delta h)^3} + \frac{h+3\Delta h}{(h+2\Delta h)^3}\dots\dots + \frac{h+n\Delta h}{\{h+(n-1)\Delta h\}^3}\right\} + k\Delta h\left\{\frac{2h+3\Delta h}{h^2} + \right.$$

$$\left.\frac{2h+5\Delta h}{(h+\Delta h)^2} + \frac{2h+7\Delta h}{(h+2\Delta h)^2}\dots\dots + \frac{2h+(2n+1)\Delta h}{\{h+(n-1)\Delta h\}^2}\right\} \tag{2}$$

Therefore, the total increment of residual stresses in the substrate due to the deposition of 'n' layers of SLM can be expressed by a generalised Eq. 3:

$$\Delta\sigma_{TS(nL)}(y) = -6ky\Delta h\sum_{n=1}^{n}\left\{\frac{h+n\Delta h}{\{h+(n-1)\Delta h\}^3}\right\} + k\Delta h\sum_{n=1}^{n}\left\{\frac{2h+(2n+1)\Delta h}{\{h+(n-1)\Delta h\}^2}\right\} \tag{3}$$

The total stress increment in the $1^{st}, 2^{nd}, \dots\dots\dots (n-1)^{th}$ layers respectively due to the deposition of the nth layer of SLM can be expressed by Eq. 4:

$$\Delta\sigma_{TL1(Ln)}(y) = k - 6ky\Delta h\left\{\frac{h+2\Delta h}{(h+\Delta h)^3} + \frac{h+3\Delta h}{(h+2\Delta h)^3}\dots\dots + \frac{h+n\Delta h}{\{h+(n-1)\Delta h\}^3}\right\} + k\Delta h\left\{\frac{2h+5\Delta h}{(h+\Delta h)^2} + \right.$$

$$\left.\frac{2h+7\Delta h}{(h+2\Delta h)^2}\dots\dots + \frac{2h+(2n+1)\Delta h}{\{h+(n-1)\Delta h\}^2}\right\}$$

$$\Delta\sigma_{TL2(Ln)}(y) = k - 6ky\Delta h\left\{\frac{h+3\Delta h}{(h+2\Delta h)^3} + \frac{t+4\Delta t}{(t+3\Delta t)^3}\dots\dots + \frac{h+n\Delta h}{\{h+(n-1)\Delta h\}^3}\right\} + k\Delta h\left\{\frac{2h+7\Delta h}{(h+2\Delta h)^2} + \right.$$

$$\left.\frac{2h+9\Delta h}{(h+3\Delta h)^2}\dots\dots + \frac{2h+(2n+1)\Delta h}{\{h+(n-1)\Delta h\}^2}\right\}$$

$$\dots\dots\dots\dots\dots\dots\dots\dots\dots\dots\dots\dots\dots\dots\dots\dots\dots\dots\dots\dots\dots\dots\dots\dots\dots\dots$$

$$\Delta\sigma_{T\{L(n-1)\}(Ln)}(y) = k - 6ky\Delta h\left\{\frac{(h+n\Delta h)}{\{h+(n-1)\Delta h\}^3}\right\} + k\Delta h\left\{\frac{2h+(2n+1)\Delta h}{\{(h+(n-1)\Delta h)\}^2}\right\} \tag{4}$$

Eq. 4 can be written in a generalised form as Eq. 5 for the stress in the mth layer:

$$\Delta\sigma_{T(Lm)(Ln)}(y) = k - 6ky\Delta h\sum_{m+1}^{n}\left\{\frac{h+n\Delta h}{\{h+(n-1)\Delta h\}^3}\right\} + k\Delta h\sum_{m+1}^{n}\left\{\frac{2h+(2n+1)\Delta h}{\{h+(n-1)\Delta h\}^2}\right\} \tag{5}$$

Eqs. 3 and 5 can be used to calculate the distribution of residual stresses in an SLM part-substrate assembly if the residual stress of a newly deposited layer (k) is known. In SLM, the value of 'k' can be considered as the yield strength of the material [1,4,5]. However, it is recommended to calibrate the 'k' by experimentally measuring the near-surface residual stress of the SLM part.

Parametric study

A parametric study has been performed in terms of the influence of number of deposited layers, layer height and substrate height on residual stress distribution. For all cases, residual stress in the new SLM layer (k) is assumed as 300 MPa (yield strength of steel) [1,5]. Residual stress distribution due to the variation in number of deposited layers is shown in Fig. 3 (a). Four different cases were studied with number of layers being 20, 40, 60, and 80 respectively, with layer height 0.1 mm. For all cases, the substrate height was 10 mm. Residual stress distribution due to the variation of layer height with a substrate of 10 mm height is shown in Fig. 3 (b). The SLM part height was 8 mm with four different layer heights: 0.05 mm (160 layers), 0.1 mm (80 layers), 0.2 mm (40 layers), and 0.32 mm (25 layers) respectively. Fig. 3 (c) demonstrates residual stress distribution variations due to the different substrate heights: 9 mm, 10 mm, 11 mm and 12 mm, respectively, keeping same individual layer height (0.1 mm) and total part height (8 mm) for all.

Residual Stresses 2018 – ECRS-10   Materials Research Forum LLC
Materials Research Proceedings 6 (2018) 283-288   doi: http://dx.doi.org/10.21741/9781945291890-45

The parametric study has revealed that: (i) the larger the number of deposited layers, the higher the resulting residual stresses, (ii) variation of individual layer height has no significant effect on stress distribution, and (iii) the lower the substrate height, the higher the resulting stress in it.

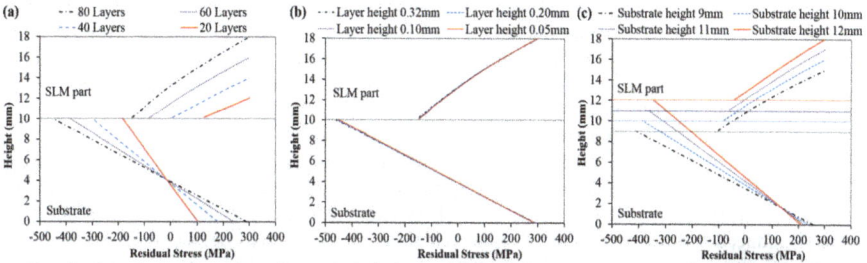

Fig. 3: Residual stress distributions with the variation of (a) number of layers, (b) individual layer height, (c) substrate height by keeping same individual layer height and total part height.

Validation and verification of the proposed analytical model

The proposed analytical model has been validated with experimental measurements in [5], and verified with predictions based on [1,4,5]. Validation and verification examples are presented as three cases of different combinations of part-substrate assemblies. Table 1 represents the parameters used in the models. For all cases, residual stress value in the newly deposited layer was considered as the yield strength of the material. Fig. 4 shows calculated residual stress distributions in this study and comparison with published values.

Table 1: Details of published results used for the validation/verification

SLM part/substrate	Stress evaluation method	k (MPa)	h (mm)	Δh (μm)	n
Chrome molybdenum steel (JIS SCM440) on stainless steel [5]	Layer removal method and analytical modelling	300	8	100	60
Stainless steel 316L on stainless steel [1]	Analytical modelling	300	20	50	50
Steel on steel [4]	Finite element modelling	410	1	150	1

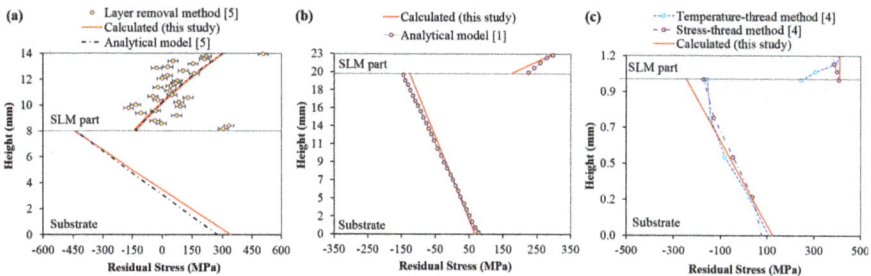

Fig. 4: Comparison of calculated residual stress distributions with experimental measurements and predicted values from references: (a) Chrome molybdenum steel (JIS SCM440) on stainless steel [5], (b) Stainless steel 316L on stainless steel [1], (c) Steel on steel [4]

Discussion

A good agreement was achieved by comparison with literature experimental measurements, numerical [4] and analytical [1,5] models. Predictions by the proposed analytical model and existing models [1,4,5] are very close. However, the mathematical representation by Shoimi [5] is limited to a single SLM layer. Also, they have neglected higher order terms for layer height (Δh^2), which will apparently induce error to the final stress distribution. Likewise, Mercelis's

287

analytical model [1] doesn't provide a final mathematical expression for calculating the through-thickness residual stresses distribution for a part-substrate assembly having 'n' number of SLM layers. Moreover, conventional numerical analysis for several single scans with a fine mesh model is very complicated and requires large computational time [4]. Therefore, numerical analyses are limited to a few SLM layers. Conversely, the proposed analytical model is much simpler compared to the existing modelling techniques and can be used for quick prediction of through-thickness residual stress distribution in an SML part-substrate assembly with 'n' number of layers.

Limitations of the proposed method are: (i) It does not explain the phenomenon of relaxation and redistribution of residual stress once the part is removed from the base plate. However, formulation for stress redistribution after baseplate removal can be found in [1], or by performing FEA; (ii) There is a discontinuity in the stress value on either side of the interface due to the differences in the elastic modulus of the two materials (SLM and substrate), although process induced residual strains are continuous at the interface. (iii) The parametric study using the proposed model is limited to the geometrical variables of an SLM part-substrate assembly. Further investigation is required to link 'k' to the SLM process parameters (laser power, scan speed, scan strategy, etc.), so that it can be used as an evaluation tool for SLM process parameter design.

Conclusions

An analytical model is presented for predicting residual stress distribution in SLM parts. The model is based on the force and moment equilibrium of induced stresses by progressive deposition of material layers. The model has been validated with experimental measurements, verified with predictive models from the literature. Based on the study, the following conclusions can be drawn:

1) To calculate the through-thickness distribution of residual stresses in an SLM part-substrate system, the proposed model requires only four parameters: the layer height (Δh), substrate height (h), number of deposited layers (n), and residual stress in newly deposited layer (k).
2) Compared to other analytical and numerical methods, this approach is simpler and can give a quick estimation of through-thickness residual stress distribution and magnitude.
3) With good calibration of stress value in a newly deposited layer (k), this method can be used to predict residual stress profile in an SLM part-substrate system with much less cost and time.

Acknowledgement

This publication was made possible by the sponsorship and support of the Lloyd's Register Foundation, which is a charitable organization that helps to protect life and property by supporting engineering-related education, public engagement and the application of research.

References

[1]   P. Mercelis, et al., Residual stresses in selective laser sintering and selective laser melting, Rapid Prototyp. J. 12 (2006) 254–265. https://doi.org/10.1108/13552540610707013

[2]   J.P. Kruth, et al., Binding mechanisms in selective laser sintering and selective laser melting, Rapid Prototyp. J. 11 (2005) 26–36. https://doi.org/10.1108/13552540510573365

[3]   J.P. Kruth, et al., Assessing and comparing influencing factors of residual stresses in selective laser melting using a novel analysis method, Proc. Inst. Mech. Eng. Part B J. Eng. Manuf. 226 (2012) 980–991. https://doi.org/10.1177/0954405412437085

[4]   C. Li, et al., Prediction of residual stress and part distortion in selective laser melting, Procedia CIRP. 45 (2016) 171–174. https://doi.org/10.1016/j.procir.2016.02.058

[5]   M. Shiomi, et al., Residual stress within metallic model made by selective laser melting process, CIRP Ann. - Manuf. Technol. 53 (2004) 195–198. https://doi.org/10.1016/S0007-8506(07)60677-5

Residual Stresses 2018 – ECRS-10   Materials Research Forum LLC
Materials Research Proceedings 6 (2018) 289-294   doi: http://dx.doi.org/10.21741/9781945291890-46

Determination of the X-Ray Elastic Constants of the Ti-6Al-4V Processed by Powder Bed-Laser Beam Melting

Nathan Dumontet[1,a,*], Guillaume Geandier[2,b], Florian Galliano[3,c],
Bernard Viguier[1,d] and Benoit Malard[1,e]

[1] CIRIMAT, Université de Toulouse, CNRS, INP- ENSIACET 4 allée Emile Monso - BP44362, 31030 Toulouse Cedex 4 – France

[2] Institut Jean Lamour, UMR 7198 CNRS - Université de Lorraine, Campus ARTEM 2 allée André Guinier BP 50840 54011 Nancy Cedex France

[3] MBDA France, Site de Bourges Aeroport, Route d'Issoudun, 18030 Bourges Cedex France

[a]nathan.dumontet@ensiacet.fr, [b]guillaume.geandier@univ-lorraine.fr,
[c]florian.galliano@mbda-systems.com, [d]bernard.viguier@ensiacet.fr, [e]benoit.malard@ensiacet.fr

Keywords: X-Ray Elastic Constants, Laser Beam Melting, Martensitic Phase, Ti64

Abstract. The determination of residual stresses (RS) through X-Ray Diffraction (XRD) necessitates the knowledge of the X-Ray Elastic Constants (XEC) S_1 and $\frac{1}{2} S_2$ for the specific phase in presence. In the case of Ti-6Al-4V additive manufacturing (AM) elaboration method such as Powder Bed Laser Beam Melting (PB-LBM) results in a microstructure fully constituted α' martensitic phase. The properties of this α' martensite are largely unknown. The purpose of the present paper is to determine the XEC of the α' martensitic phase. This is performed using an original methodology involving 2D X-Ray detection. The results shows significant differences with XEC values available for the α phase from Titanium.

Introduction

AM is a process developed to build parts with less building steps than conventional manufacturing routes. There exists a lot of different processes for AM, among them the LBM consists in melting a thin metallic layer of powder using a localized laser beam and building layer by layer a part in three dimensions. Because of the cooling rate induced by the laser, in the range of 10^7-10^8 K/s [1,2], α' martensitic phase forms in the Ti-6Al-4V. Elaboration process such as PB-LBM also induces localized high thermal gradient within the part [3,4]. Mismatches in dilatation caused by the thermal gradient generates RS [5] resulting in a stress state were the bulk is in a compressive state and last layers in tensile state [6,7]. Since this stress state can be a limit for part production in AM, RS determination is an important concern. Such determination is often performed by XRD technique using the $\sin^2\psi$ method [8]. This technique allows to relate the variation of a given lattice plane spacing to local stress components. This requires the knowledge of the XEC S_1 and $\frac{1}{2}S_2$ of the given lattice plane for the phase under consideration. The XEC are connected to Young modulus and Poisson ratio for the considered lattice plane through the relations:

$$S_1^{\{hkl\}} = -\frac{\nu_{\{hkl\}}}{E_{\{hkl\}}}, \text{ and } \frac{1}{2}S_2^{\{hkl\}} = \frac{1+\nu_{\{hkl\}}}{E_{\{hkl\}}}. \tag{1}$$

The present study concerns the Ti-6Al-4V alloy which is widely used in industry and which is the most used titanium alloy for AM. After processing via LBM process the microstructure of the alloy is mainly constituted by the α' martensitic phase, as shown in Fig. 1. This α' martensitic phase presents the same hexagonal structure than the α phase but different composition since it incorporates β-forming elements such as vanadium. In order to measure the RS in LBM parts of Ti-6Al-4V alloys one needs to know the values of XEC for this specific α' martensitic phase. However, to our knowledge the only XEC values available in the literature for Ti alloys refer to either pure Ti α phase [9] or one lattice plane of the α phase of a wrought Ti-6Al-4V alloy [10]. In this paper, a methodology is proposed and discussed to determine XEC values of the α' martensitic phase, by performing synchrotron X-ray diffraction on tensile specimens elaborated by PB-LBM process. The values are compared to the literature data.

Figure 1 – α' martensitic microstructure of the Ti-6Al-4V processed by LBM

XEC Determination Method

The method to determine the XEC is based on *in-situ* diffraction measurements during a tensile test within the elastic domain of the given specimen [8]. Three samples were directly fabricated at the desired shape and dimensions by LBM using M2 Cusing machine. The powder used was reused Ti-6Al-4V powder with an average diameter of 35 μm. Samples have been manufactured vertically using standard melting parameters provided by machine manufacturer (layer thickness: 60 μm). The sample thickness is 2 mm, the gauge length is 2 mm wide and 10 mm in length. For a better diffraction signal all the sample surfaces were grinded using SiC grade 600 paper, as a result thickness and width of the gauge length decrease and were measured for each sample: 1.86 x 1.87 mm^2 for sample 1, 1.90 x 1.86 mm^2 for sample 2 and 1.84 x 1.85 mm^2 for sample 3. The tensile tests were performed using a load frame electromechanical driven system that can be adapted to beamline ID11 at the ESRF. A maximal loading of 3 kN is allowed with one mobile jaw. The load was changed step by step (500 N), first increased from 0 up to 2000 N, then decreased back to 0 N. Each step corresponds to a holding time of 25 seconds during which the diffraction rings are recorded. The maximum load (2000 N) corresponds to a tensile stress of 575 MPa for sample 1, 566 MPa for sample 2 and 588 MPa for sample 3, which is well below the yield stress of the alloy in the raw fabrication state estimated to be 1100 MPa. Diffraction data were recorded during both loading and unloading steps showing similar results, further calculations are obtained from the analysis of unloading data. These measurements were performed at the ESRF, on the beamline ID11 with a wavelength of 0.0140891 nm (E=88 keV), a beam size of 300 x 400 μm^2 and diffraction patterns were acquired on a 2D detector (FreLon camera). Such 2D diffraction pattern with indication of the δ and θ angles are represented in Fig. 2a. δ is the angle along the ring, δ=90° corresponding to the tensile axe. The classical ψ angle is simply deduced from δ using $\psi = \delta - \pi/2$ [11]. The θ angle is the Bragg's Angle. Diffraction patterns were integrated from 0° to 90° every 10° in ψ angles, within a sectorial window of $^{+/-}$ 5°. The four quarters were added for a better signal to noise ratio and get an average value for each ψ angle. The resulting diffraction patterns is presented on Fig.2b.

Residual Stresses 2018 – ECRS-10   Materials Research Forum LLC
Materials Research Proceedings 6 (2018) 289-294   doi: http://dx.doi.org/10.21741/9781945291890-46

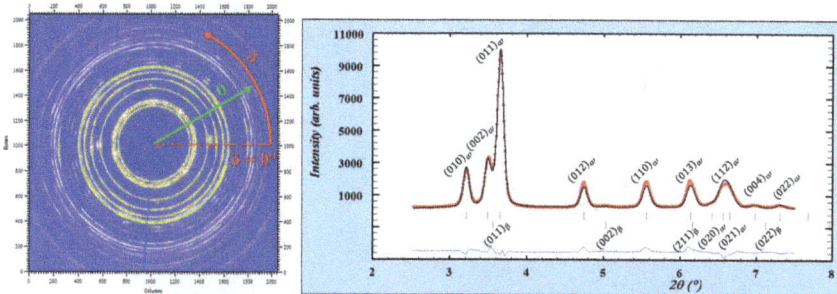

Figure 2 – Example of a diffraction pattern a) as recorded on the 2D detector with the angles δ and θ, and b) integrated over ψ sector.

From these diffraction patterns the lattice spacing d_{hkl} for different orientations were obtained using Bragg's law. The corresponding strain values were calculated using Eq. 2 were $d_{0,hkl}$ is the reference spacing taken from diffraction patterns at 0 N load.

$$\varepsilon_{hkl} \approx \frac{d_{hkl} - d_{0,hkl}}{d_{0,hkl}}. \tag{2}$$

This deformation varies with the angle ψ according to Eq. 3, with σ_{1R} and σ_{2R} being the RS in the tensile and transverse directions respectively and σ_A the applied stress [12].

$$\varepsilon_{\psi,hkl} = \tfrac{1}{2}s_2^{\{hkl\}}(\sigma_A + \sigma_{1R})\sin^2\psi + s_1^{\{hkl\}}(\sigma_A + \sigma_{1R} + \sigma_{2R}). \tag{3}$$

Because the RS state does not change during the loading and due to the high energy of the X-ray beam and the transmission geometry, the whole beam path was analyzed; residual stresses can be neglected as the low load state (0 N) is taken as a reference in order to determine the XEC, resulting in:

$$\varepsilon_{\psi,hkl} = \tfrac{1}{2}s_2^{\{hkl\}}\sigma_A \sin^2\psi + s_1^{\{hkl\}}\sigma_A. \tag{4}$$

Note that in the sections below that concern results on hexagonal phase the lattice plane are designated using the four indices $(hki\ell)$ - where the redundant index i is defined by $h + k + i = 0$.

Results

The procedure used for the determination of XEC values is first described for the lattice plane $(10\bar{1}3)$ on sample 1. The deformation is first plotted against $\sin^2\psi$ for the different stress levels. According to Eq. 4, the deformation may obey a linear relation to the $\sin^2\psi$, the fitting of the experimental points in Fig. 3 shows that this relation is roughly verified. In this plot, one can see that the linear fit corresponding to different values of stress level intersect each other on an invariant point. Indeed, if one sets a zero deformation, Eq. 4 can be simplified and gives:

$$\varepsilon = 0 = \tfrac{1}{2}S_2\sigma_A\sin^2\psi + S_1\sigma_A \Rightarrow \tfrac{1}{2}S_2\sin^2\psi = -S_1 \tag{5}$$

This unique intersection point for several fitting lines is taken as an indication of coherent experimental results. Similar results were obtained, as for Figs. 3 and 4, for all the (hkil) lattice planes and specimens analyzed. Using Fig. 3 and Eq. 4, XEC can be accessed by plotting the slope of fitting lines (Fig. 4a) and their intercept (Fig. 4b) versus the applied stress. The resulting data in the graph of Fig. 4 a and b shows a clear linear dependence against the applied stress. According to Eq. 4 the slope of these lines corresponds respectively to S_1 and $^1/_2 S_2$.

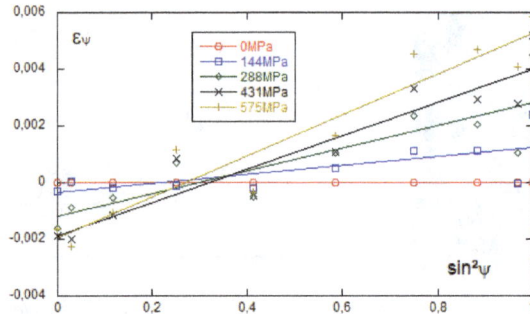

Figure 3 – Deformation ε_ψ versus sin $^2\psi$ for $(10\bar{1}3)$ for different stress values in sample 1

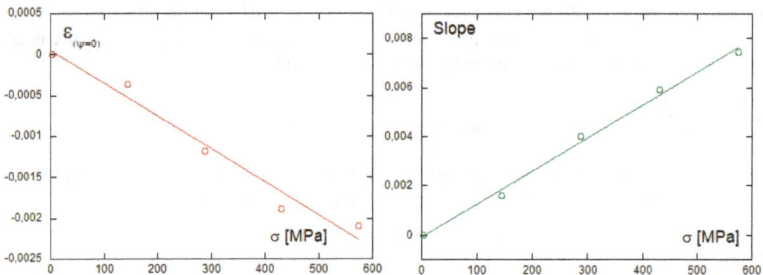

Figure 4 – Line fitting data from Fig. 3: a) Intercept and b) slope versus the applied stress.

Table 1 – XEC in 10^{-6} $(MPa)^{-1}$ experimentally determined for the different specimens.

Sample	$(11\bar{2}0)$		$(10\bar{1}1)$		$(10\bar{1}2)$		$(10\bar{1}3)$		(0002)	
	S_1	$^1/_2S_2$	S_1	$^1/_2S_2$	S_1	$^1/_2S_2$	S_1	$^1/_2S_2$	S_1	$^1/_2S_2$
1	-5.00	16.30	-4.69	15.40	-4.26	14.10	-4.00	13.40	-3.64	12.30
2	-3.82	11.70	-3.71	11.40	-3.55	11.00	-3.45	10.80	-3.32	10.50
3	-2.98	12.40	-2.55	11.70	-1.94	10.60	-1.58	10.00	-1.05	9.14

The same procedure was applied for five different (hkil) planes in the three studied samples, giving rise to the whole set of XEC results showed in Table 1. The values presented in Table 1 shows that XEC are quite different from one specimen to the other. A possible reason for this scattering could be the orientation texture of the samples. Textured materials can no longer be described accurately by XEC, but described by X-Ray Elastic Factors (XEF) F_{ij} [8]. A way to get an idea of the texture is to plot the XEF component F_{11}, which is the slope of the applied stress against the deformation ε_ψ for each value of $\sin^2\psi$. Such a plot is showed in Fig. 5 for the plane $(10\bar{1}3)$ in sample 1. In such a plot, a non-textured material would result in a linear fit of F_{11} as indicated by the dashed line in Fig. 5, whereas the oscillations observed in this plot (full line) may reveals the existence of orientation texture. However, the limited number of stress levels (5 steps) used to draw this plot may contribute to scattering seen in Fig. 5. Other possible origin for the variation of XEC on different samples may arise from a more general variation in

Residual Stresses 2018 – ECRS-10
Materials Research Proceedings 6 (2018) 289-294

Materials Research Forum LLC
doi: http://dx.doi.org/10.21741/9781945291890-46

microstructure or different porosity level due to different position of the specimens on the elaboration plates.

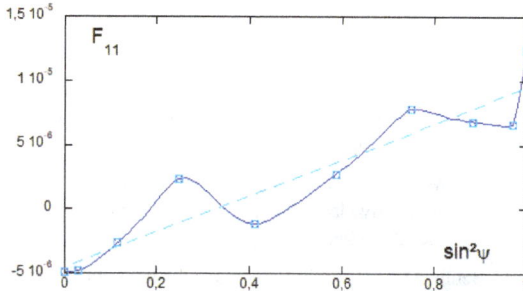

Figure 5 – X-Ray Elastic Factors F_{11} for the loading of sample 1

Nevertheless, in order to pursue the analysis of the experimentally determined XEC values, the mean values of S_1 and $^1/_2 S_2$ were calculated for each lattice plane. Those values are shown in Table 2 together with the XEC values available in the literature for the α phase of pure titanium [9] and α phase from wrought Ti-6Al-4V [10].

Table 2 – XEC experimentally determined and from literature in 10^{-6} $(MPa)^{-1}$

	$(11\bar{2}0)$		$(10\bar{1}1)$		$(10\bar{1}2)$)		$(10\bar{1}3)$		(0002)	
	S_1	$^1/_2S_2$	S_1	$^1/_2S_2$	S_1	$^1/_2S_2$	S_1	$^1/_2S_2$	S_1	$^1/_2S_2$
α' This study	-3.93	13.47	-3.65	12.83	-3.25	11.90	-3.01	11.40	-2.67	10.65
α pure Ti [9]	-2.98	12.03	-2.90	11.80	-2.74	11.28	-2.56	10.83	-2.34	10.16
α W Ti64 [10]			-3.26	13.38						

First, a clear variation of the XEC values with the lattice plane can be noticed, evidencing the elastic anisotropy of the hexagonal structure, with (absolute) values of XEC decreasing when changing from prismatic to basal plane. This tendency is visible for together the α' martensitic phase (this study) and the pure Ti α phase, however the anisotropy seems to be stronger for the previous one. Secondly, large differences between α pure titanium phase and α' martensitic phase of the Ti-6Al-4V in means of XEC are observed. The difference between both sets of values is stronger close to the prismatic lattice plane and can reach 32% for $S_1^{11\bar{2}0}$ and 12% for $\frac{1}{2} S_2^{11\bar{2}0}$. This difference reduces to 14% for S_1^{0002} and 5% for $\frac{1}{2} S_2^{0002}$ for basal plane. This difference in XEC values seems also reduced when the present results are compared to the only value that could be found for the α phase from wrought Ti-6Al-4V (W Ti64) for the $(10\bar{1}1)$ plane. Work is in progress to get a better understanding about the variation of elastic properties of the α hexagonal Ti phase depending on processing method and chemical composition [13].

Conclusion

In this study, the XEC were experimentally determined from the particular α' martensitic phase of the Ti-6Al-4V elaborated through AM process, using a methodology based on XRD and 2D detector. Results obtained from the three different specimens shown a quite large scattering that could be related to microstructural variation of specimens. Measurement of XEC for 5 lattice planes evidenced the orientation variation of S_1 and $^1/_2 S_2$ from prismatic planes to basal planes, in good agreement with the anisotropy of the α phase seen in previous study. Results shown that there are differences between XEC determined for the α' martensitic phase of the Ti-6Al-4V and those of the α pure titanium phase usually used for RS determination. This may lead to errors in RS values determined by $sin^2\psi$ method that could be higher than 10% for the $(11\bar{2}0)$ plane.

Acknowledgement
The author would like to acknowledge M. Dehmas and M. François for their assistance and fruitful discussions.

References

[1] B. Vrancken, V. Cain, R. Knutsen, J. Van Humbeeck, Residual stress via the contour method in compact tension specimens produced via selective laser melting, Scripta Mater. 87 (2014) 29-32. https://doi.org/10.1016/j.scriptamat.2014.05.016

[2] J. Yang, H. Yu, J. Yin, M. Gao, Z. Wang, X. Zeng, Formation and control of martensite in Ti-6Al-4V alloy produced by selective laser melting, Materials and Design 108 (2016) 308-318. https://doi.org/10.1016/j.matdes.2016.06.117

[3] A. Vasinonta, J.L. Beuth, M. Griffith, Process maps for predicting residual stress and melt pool size in the laser-based fabrication of thin-walled structures, J. Manuf. Sci. Eng 129 (2006) 101-109. https://doi.org/10.1115/1.2335852

[4] G. Vastola, G. Zhang, Q.X. Pei, Y.-W. Zhang, Controlling of residual stress in additive manufacturing of Ti6Al4V by finite element modeling, Add. Manufacturing 12 (2016) 231-239. https://doi.org/10.1016/j.addma.2016.05.010

[5] Y. Liu, Y. Yang, D. Wang, A study on the residual stress during selective laser melting of metallic powder, Int. J. Adv. Manuf. Tech. 87 (2016) 647-656. https://doi.org/10.1007/s00170-016-8466-y

[6] J. Grum, R. Sturm, A new experimental technique for measuring strain and residual stresses during a laser remelting process, J. of Mat. Processing Tech. 147 (2016) 351-358. https://doi.org/10.1016/j.jmatprotec.2004.01.007

[7] P. Mercelis, J.-P. Kruth, Residual stresses in selective laser sintering and selective laser melting, Prototyping Journal, 12 (2006) 254-265. https://doi.org/10.1108/13552540610707013

[8] V. Hauk, Structural and Residual stress analysis by nondestructive methods, Elsevier, Amsterdam, 1997, Chapter 13

[9] B. Eigenmann, E. Macherauch, Rontgengraphische Untersuchung von Spannungszustanden in Werkstoffen, Teil III, Material-wissenschaft und werkstofftechnik, Bd.27, (1995) 426-437. https://doi.org/10.1002/mawe.19960270907

[10] G. Bruno, B. Dunn, The precise measurement of Ti6Al4V microscopic elastic constants by means of neutron diffraction, Meas. Schi. Technol. 8 (1997) 1244-1249. https://doi.org/10.1088/0957-0233/8/11/006

[11] L. Vautrot, G. Geandier, M. Mourot, M. Dehmas, E. Aeby-Gautier, B. Denand, S. Denis, Internal stresses in Metal Matrix Composites in relation with matrix phase transformations, Advanced Materials Research, 996 (2014) 944-950. https://doi.org/10.4028/www.scientific.net/AMR.996.944

[12] V. Ji, Contribution à l'analyse par diffraction des rayons X de l'état microstructural et mécanique des matériaux hétérogène, Habilitation à Diriger des Recherche, Université des Sciences et Technologies de Lille (2003)

[13] N. Dumontet, D. Connetable, B. Malard, B. Viguier, Elastic properties of the α' martensitic phase of the Ti-6Al-4V, to be published

Keyword Index

Author Index

About the Editor

Marc Seefeldt is Associate Professor at KU Leuven and works on defects in crystalline solids, polycrystal plasticity, textures, multiscale modelling and residual stress analysis. Since 2010, he is member of the Scientific Committee of ECRS. The hosting Department of Materials Engineering of KU Leuven has a strong research record in stress analysis on textured materials, on microelectronic components, as well as in Additive Manufacturing.

www.ingramcontent.com/pod-product-compliance
Lightning Source LLC
Chambersburg PA
CBHW071330210326
41597CB00015B/1395